ABOUT THE AUTHOR

Prof. T. Pullaiah obtained his M.Sc. (1973) and Ph.D. degrees in Botany from Andhra University. He was a Post Doctoral Fellow at Moscow University, Russia, during 1976-78. He traveled widely in Europe and visited Universities and Botanical Gardens in about 16 countries. Professor Pullaiah joined Sri Krishnadevaraya University as Lecturer in 1979 and became Professor in 1993. He held several positions in the University, which include Dean, Faculty of Life Sciences, Head of the Department of Botany, Chairman, BOS in Botany, Head of the Department of Sericulture, Co-ordinator and Chairman, BOS in Biotechnology, Principal, S.K.University College and Warden S.K.U.Hostels. Prof. Pullaiah has published 44 books, 245 research papers and 35 popular articles. He is Principal Investigator of 20 major Research Projects totaling more than a Crore of Rupees funded by DBT, DST, CSIR, UGC, BSI, WWF, GCC, BRNS etc. Under his guidance 50 students obtained their Ph.D. degrees and 34 students their M.Phil. degrees. He is recipient of Best Teacher Award from Government of Andhra Pradesh, Prof. P. Maheshwari Gold Medal of Indian Botanical Society, Prof. Y.D.Tyagi Gold medal of Indian Association for Angiosperm Taxonomy and Scientist of the Year Award from National Environmental Science Academy. He was Vice President of Indian Botanical Society and presently Councilor of the same. He was a member of Species Survival Commission of IUCN.

Biodiversity in India

Volume 5

Biodiversity in India
Volume 5

— *Editor* —
T. Pullaiah
Department of Botany
Sri Krishnadevaraya University
Anantapur – 515 003
Andhra Pradesh

2011
REGENCY PUBLICATIONS
Delhi - 110 002

ISBN 9788170359715

Published by : **REGENCY PUBLICATIONS**
(*Division of Daya Publishing House*)
4760-61/23, Ansari Road, Darya Ganj,
New Delhi - 110 002
Phone: 23245578, 23244987
Fax: (011) 23260116
e-mail : info@regency-books.com
website : www.dayabooks.com
website : www.regencybooks.com

Laser Typesetting : **Classic Computer Services**
Delhi - 110 035

Printed at : **Chawla Offset Printers**
Delhi - 110 052

PRINTED IN INDIA

Preface

The overwhelming response from readers to the volumes 1-4 of Biodiversity in India prompted us to continue the series. The present volume 5 includes 10 articles on millipedes, Ethnobotany, medicinal plants, Eragrostidae and cultivated plants in Andhra Pradesh. We would like to continue the series in future.

Prof. T. Pullaiah

Contents

Chapter 1

Diversity, Restoration and Conservation of Millipedes

K.R. Sridhar* and K.M. Ashwini

Department of Biosciences, Mangalore University, Mangalagangotri, Mangalore – 574 199, Karnataka, India

ABSTRACT

Diplopoda (millipedes) is one of the largest classes of animal kingdom belonging to the Devonian period constitutes the third largest group among terrestrial arthropods following Insecta and Arachnida. They are widely distributed and show a wide range of life styles. Global estimate of described millipedes is about 12,000 species (out of a total estimate 80,000 species) distributed in all the continents except for Antarctica. Five ecomorphotypes have been distinguished based on their body structure (bark-dwellers, pill-millipedes, rammers, litter splitters and borers). Millipedes are known to associate with a wide range of fauna and microbes, thrive on a variety of organic matter of low to high nutrient value, disintegrate organic matter, mix faecal pellets with microbes and transform into humic substances. They are also known for mycophagy and coprophagy. The rate of organic matter ingestion varies from 5 to 10 per cent of body mass within 24 hr. In forests, millipedes are known to ingest 20-100 per cent plant detritus and return 60-90 per cent organic matter in the form of faecal pellets. They are conservative and very sensitive to water deficit and fail to overcome the limitation of even a single edaphic factor particularly soil texture and litter thickness. The climate desiccation and human interference are the major threat lead to millipede extinction. Limited mobility of millipedes might

* Corresponding Author: E-mail: sirikr@yahoo.com

have resulted in high degree of speciation and endemism. Increased population of saprophagous millipedes in managed agricultural or forest ecosystem indicates the restoration of soil health. Special attention is required for conservation of millipedes in native forests and forest plantations. In view of importance of millipedes in soil fertility and organic farming, the present review mainly focuses on their diversity, interaction with fauna and microbes, food and feeding habits, pattern of organic matter processing, endemism, restoration and conservation.

Keywords: *Millipedes, Richness and diversity, Food and feeding, Organic matter processing, Faecal pellets, Endemism, Human interference, Restoration and conservation.*

Introduction

Integrated agriculture and agroforestry aims at optimum utilization of natural resources for farming to achieve sustainability with maximum benefit. In tropical agroforestry, Vohland and Schroth (1999) advocated suitable tree plantation to maintain adequate amount of litter layer to encourage the activities of diverse and abundant litter fauna. Considering sensitive environmental indicators in such practices is a positive step towards self-assessment. Although there is a growing concern of environmental benefits and disadvantages resulting from agriculture and industries, little attention has been focused on the assessment of environmental impact of such changes (Bailey *et al.*, 1999). Forest ecosystems have been considered as mosaic sylvogenetic phases (Oldeman, 1990), the humus profile development in forest floor is the result of functional connections between vegetation, soil microbes and saprophagous fauna. Study of a few centimetres of topsoil reveals the dynamic process of forest ecosystem (Bernier and Ponge, 1994). Decomposition of litter on the forest floor is strongly dependent on humus type (*e.g.*, in moder and mor humus, the rate of litter disappearance is slower than mull humus) (van der Drift, 1963). Comparison of humus profiles between natural forests, degraded forests and plantations reveals the influence of human interference on the balance of natural ecosystem. The world annual production of renewable lignocellulosic residues amounts to approximately $20\text{-}50\times10^9$ tonnes, of which 4×10^9 tonnes are available for use (Kelley and Peterson, 1997). The major technical limitation of digestibility of such residues is the conversion of lignin to cellulose. One of the potential outlets to transform lignin (wide C/N ratio to narrow C/N ratio) is composting. Composting is possible only by the action of a consortium of soil fauna and microorganisms, where millipedes constitute a major segment.

Majority of the earth's biological diversity is constituted by the Arthropoda (Golovatch *et al.*, 1995; Stock, 1993; Wilson, 1992). Their vital role in ecosystem services includes plant litter decomposition and humus formation (Giller, 1996, Plowman, 1990). Important role of arthropod assemblages in soil food webs qualifies them as indicators to assess the impacts of forest management practices (Kremen *et al.*, 1993). Millipedes being highly sensitive for a single factor (*e.g.*, litter thickness), their abundance, richness and diversity would precisely assess the positive or negative impacts of agricultural practices. Species abundance, dynamics, diversity and biomass of millipedes are reliable indices to assess the environmental perturbations. The current review aims at consolidating information on important features of millipedes such as habitats, richness, diversity, association with fauna and microbes, food and feeding habits, processing of organic matter and endemism. In view of importance of millipedes in soil structure and function, impacts of human interference, restoration and conservation measures are also discussed.

Millipedes

'Millipede' is a misnomer and certainly not thousand-legged, they have less than 50 pairs or a maximum of 375 pairs (Enghoff, 1990), with two pairs of legs per body segment and hence classified as Diplopoda. *Kampecaris tuberculata* has been considered as one of the possible oldest millipedes in Silurian Old Red Sandstone (Cloudsley-Thompson, 1988). Wide range of millipedes was recovered from Carboniferous period including pill-millipedes (order, Amynilspedida) (Hannibal and Feldmann, 1981). Decline in fossil records was noted after Carboniferous period. Late Paleozoic fossils resemble the recent spirobolids (Kraus, 1974). In late Cretaceous, many spirobolids have been described from Mangolia (Dzik, 1975). In Oligocene, substantial amber preserved millipedes were collected (Shear, 1981). Diplopoda is a largest class among animal kingdom constituting the third largest group in terrestrial arthropods following Insecta and Arachnida (Golovatch *et al.*, 1995). Head of millipedes are equipped with a pair of antennae, which are often tapped on the ground while walking. Some millipedes are blind, while some possess a number of rudimentary eyes on either side of their head. Double leg pairs exert forward thrust, calcified head and leverage of labrum drives millipedes between fibres of rotten wood or packed soil particles which is the main advantage than other terrestrial invertebrates. Millipedes with weaker mouthparts tend to confine their activities and feeding on highly moist and partly rotten leaf and woody debris. Cave dwelling millipedes have modified mouthparts to remove organic clay from lime stone particles. Millipedes have long lifespan (1-11 years), show a wide range of life styles and reproductive strategies (Dangerfield and Telford, 1991; Carrel, 1990). Such extended longevity has been linked to the availability and consumption of poor quality food (Blower, 1985). The body size of millipedes ranges from 0.2-30 cm length and 0.1-4 cm width (Hopkin and Read, 1992). Colonization of higher altitudes is generally accompanied by increased body size (Enghoff, 1993). Most studies on diplopods are focussed on temperate than tropical species (Cloudsley-Thompson, 1988). Tropical millipedes have evolved large body size compared to their temperate counterparts (*e.g.* Spirostreptidae, 40 g vs. Julidae, 2 g per animal) (Dangerfield and Telford, 1989; Lawrence, 1984). The large size confers several advantages including increased tolerance to moisture stress (Baker, 1980) and increased ability to burrow (Manton, 1977). Body size affects energy acquisition and utilization and ultimately influences the rates of organic matter decomposition (Peterson and Luxton, 1982). The global estimates of described millipede nominal species are up to 12,000 (out of 80,000) assigned to 2947 genera and 145 families distributed on all continents except for Antarctica (Bueno-Villegas *et al.*, 2003; Golovatch *et al.*, 1995; Heywood and Watson, 1995; Hoffman 1980; Hoffman *et al.*, 2002; Sierwald and Bond, 2007; Shelly, 2003; WCMC, 1992). The genus *Arthrosphaera* are endemic tropical inhabitants, abundant in the upper soil horizons of forests in moist than in dry regions. They are confined to widely separated lands on tropical biomes of Indo-Australia, South Africa and Madagascar (Pocock, 1892). The individuals belonging to the order Sphaerotheriidae once inhabited the countries between the two areas from Indo-Australia to Madagascar and South Africa, but later wiped off in the intermediate regions due to unsuitable climatic conditions.

Based on the structures of millipedes, five morphotypes have been distinguished (Kime and Golovatch, 2000): 1) Polyxenidae (bristly millipedes or bark-dwellers); 2) Glomeridae (pill-millipedes or rollers); 3) Julidae (bulldozers or rammers); 4) Polydesmidae (wedge or litter-splitters) and 5) Platydesmidae (borers). Hopkin and Read (1992) grouped millipedes in to five ecomorphological types by: 1) Bulldozers or rammers (Julidae, Spirobolidae and Spirostreptidae) are long cylindrical forms, use their broad head as a ram which facilitate to burrow the soil; 2) Wedge types (Polydesmidae) are short forms with laterally expanded segments have keels with tapered anterior part facilitates burrowing; 3) Borers (chordeumatid and colobognath millipedes, *e.g. Polyzonium*) have free sternites and pleurites, strongly tapered anterior portion facilitates forward movement and help to widen the crevice; 4) Rollers (Glomeridae and Sphaerotheriidae: *e.g. Glomeris* and *Arthrosphaera* spp.) roll into sphere as a defensive mechanism, which reduces the water loss and minimize the surface area exposed to air; 5) Bark dwellers (*e.g., Polyxenus lagurus*) are small and do not possess burrowing ability. The flattened nature helps to live under bark.

Habitats

Based on their habitat preference, millipedes have been classified mainly into stratobionts and pedobionts (or geobionts). Three main habitats have been distinguished in woodlands: 1) floor and aerial parts of vegetation; 2) plant litter on floor and soil layers and 3) beneath the bark and rotten wood, tree stumps and fallen logs (Attems, 1936). Forest floor with large amount of decaying litter, stones, rotting stumps and fallen logs support a variety of millipedes (Blower, 1955). The forest floor or the soil strata shows a horizontal distribution of millipedes depending on the variety and quantity of accumulated litter. Polydesmids of Nigeria (*Habrodesmus dubosequi, Xanthodesmus physkon* and *Xanthodesmus* sp.) are surface dwellers (Lewis, 1971a). *Ophyiulus pilosus* inhabits deeper litter layers, while *Julus scandinavius* prefers upper layers of litter. Flatback millipedes, *Polydesmus* and *Polymicrodon* are the litter dwellers, Julids mainly prefer deep layers of soil, glomerids are intermediaries in habitat preference. Some species of millipedes are arboreal as they climb trees (Hopkins and Read, 1992) (*e.g. Tachypodoiulus niger* and *Nemasoma varicorne*) (Hopkins and Read, 1992). Enghoff (1982a) predicted that shortened accessory claw of *Cylindroiulus lundbladi* might be an adaptation for tree climbing. Millipedes are also known from the canopy of neotropical montane forest of Costa Rica (Nadkarni and Longino, 1990). Giant pill-millipedes (Sphaerotheriidae) are commonly associated with trees in Africa (Haacker and Fuchus, 1972). *Polyxenus lagurus* and *Proteroiulus fuscus* are most frequent in birds' nests (Tajvoský *et al.*, 2001) and eighteen species have been recovered from seven types of bird nests in Slovakia. A highest of 10 species of millipedes was associated with nests of blackbird (*Turdus merula*) and mallard (*Anas platyrhynchos*). Trees of *Acacia karroo* and broad-leaved trees of KwaZulu-Natal,

Richards Bay, South Africa support millipedes (*Centrobolus fulgidus* and *C. richardii*) on their canopy (Greyling *et al.*, 2001). *Polydesmus denticulatus* is an underwater millipede known to survive winter flooding by submersion (Hopkin and Read, 1992). In Amazon floods, sub-adults of tropical millipedes reach a depth of 15 m each year (Adis, 1986), during inundation period they feed on algae. Millipedes are also cave dwellers and they were extensively studied in USA (Shear, 1969, 1984) and France (Geoffroy, 1984). Some millipedes are common in deserts (Hopkin and Read, 1992; Crawford, 1976; Crawford *et al.*, 1987), *Orthoporus ornatus* (New Mexican and Texas desert millipede) remains on surface up to a temperature 35°C, hides under stones and refuge on further elevation of temperature (Wooten and Crawford, 1975). Quiescent behaviour of millipedes has been linked to low rainfall and high temperatures in Julidae (Bailey and Kovaliski, 1993).

Richness and Diversity

A wide range of landuse, agricultural, forestry and wild life conservation practices influences the richness and diversity of soil and litter fauna (Curry, 1944; Didham *et al.*, 1996; Edwards and Bohlen, 1995; Wardle, 1995). Nakamura *et al.* (2003) studied the millipede population in five rainforests, five pasture and 10 revegetated forests of Mary River catchment of subtropical Australia. The richness was highest in rainforest locations and it was significantly correlated with litter depth, tree spacing and foliage/ litter index. Julid millipede diversity was rich in heterogeneous natural woodland than homogeneous pine and eucalypt plantations in Zimbabwe (Dangerfield and Telford, 1992). Dangerfield and Telford (1992) are of the opinion that species composition, richness and relative abundance of millipedes will influence the decomposition and nutrient dynamics of Savanna ecosystem. Millipedes were abundant in number as well as biomass in closed canopy woodlands than in open forest areas (Dangerfield, 1990). Mixed tree sites showed highest diplopods than eucalypt forests. Survey of 27 sites of Mediterranean landscape revealed that the occurrence of millipede communities was dependent on vegetation structure (mainly the degree of openness of the sites) (David *et al.*, 1999). In Miombo forests of South Africa, millipedes account up to 36 per cent of the total density of fauna and up to 75.5 per cent of the biomass (Dangerfield, 1990). Substrate palatability is known to influence the distribution of temperate millipede, *Julus scandinavius* (Kheirallah, 1979). Members of Julidae occurred in open grassland, while members of Glomeridae and Porcellionidae were found in the open sites with high oak cover. Caves of semi-arid West Australia supported considerable millipede population (Humphreys, 1989; Enghoff, 1982b). Marked increase in cave-dwelling millipede population was seen on addition of organic matter and water in Western Australia (Humphreys, 1989). Table 1.1 reveals the population density of millipedes in temperate, tropical and Mediterranean habitats. In temperate habitats, millipede population is up to 3,000/ m^2, while in tropics it increased up to 8,000/m^2.

Association with Fauna and Microbes

Diplopods are known to associate with a wide variety of animals. Millipedes were recovered in rat (Humphreys, 1991) and ant nests (Loomis, 1972) in temperate and tropical regions. Several species of millipedes are known to associate with ants, army ants (Loomis, 1972; Cloudseley-Thompson, 1958; Hölldobler and Wilson, 1990; Rettenmeyer, 1962) and termite nests (Hopkin and Read, 1992). Arboreal ant nests of a maritime forest in Japan consisted of species of Ployxenidae (mean, 2.4 millipedes/nest) (Ishii and Yamaoka, 1982). In Britain, *Blaniulus guttulatus, Proterioiulus fuscus* and *Polyxenus lagurus* are frequently associated with ants and termites (Hopkin and Read, 1992). Ants protect millipedes from predators, while millipedes are known to clean up detritus of ant nests.

Table 1.1: Population Density of Millipedes from Various Parts of the World

Millipede	*Location*	*Density (number/m²)*	*Reference*
Cylindroiulus punctatus	Temperate	80-177	van der Drift (1951)
Glomeris marginata	Temperate	2.3-7.5	Thiele (1956)
Narceus annularis	Temperate	18.4-70	Blower (1956)
Parafontria laminata	Temperate	100-1000	Niijima (1998)
Parafontaria tonominea	Temperate	25.6-72	Kaneko (1999)
Millipedes	Temperate	100	Edwards (1967)
Millipedes	Temperate	160	Striganova & Rachmanov (1972)
Millipedes	Temperate	230	Saulnier & Binche, (1986)
Millipedes	Temperate	2,000-3,000	Tracz (1987)
Millipedes	Temperate	0-40	Erwin & Andreas (1997)
Jonespeltis splendidus	Tropical	8000	Bano & Krishnamoorthy (1985)
Seychelleptus seychellarum	Tropical	4.4	Lawrence & Samways (2003)
Millipedes	Tropical	8,000	Aouti (1978)
Millipedes	Tropical	20-40	Dangerfield (1990)
Millipedes	Tropical	136	Loranger (2001)
Millipedes	Tropical	15.3-56	Warren & Zou (2002)
Millipedes	Mediterranean	114	Iatrou & Stamou (1989)

Stomodaeal region of gut of *Jonespeltis splendidus* (Polydesmidae) are rich in bacterial population, while proctodaeal region is rich in fungal population (Bano *et al.*, 1976). Cocci and spore forming bacteria were totally absent in proctodaeal region (Bano *et al.*, 1976). From hindgut of *Glomeris* sp. six bacteria, six actinomycetes and two fungal strains were isolated (Gebhardt *et al.*, 2002). *Pseudomonas stutzeri* and *Pseudomonas putida* survived the passage through the gut of *Pachyiulus flavipes* and increased in fresh excrements (Byzov *et al.*, 1996). Bacteria recovered from millipede gut (*Ommatoiulus sabulosus*) included species of *Klebsiella, Bacillus* and *Corynebacterium*. Actinomycetes (*Micromonospora* sp.) are known to accumulate in the hindgut and likely to participate in the breakdown of chitin (Zenova *et al.*, 1996). Millipedes are known to transport *Streptomyces* sp. adhered to the tarsal region of their legs (Szabo *et al.*, 1983). Faecal pellets consist of dense populations of *Micromonospora*.

Arthropods are known to establish intimate associations with several fungi. Seventy six species of fungi have been isolated by Taylor (1982) from the gut of *Orthoporus ornatus*. Such associations might be for food (mycophagy) or for shelter or for sites of mating and oviposition. Some fungal species have symbiotic associations with insects for their growth and dispersal. Besides nutritional advantages (*e.g.*, carbohydrates, fungal sterols, urea and ammonia), acquired fungal enzymes serve in digestion in insects. *Triainomyces hollowayanus* (Laboulbeniales) has been isolated from pill-millipede, *Procyliosoma tuberculatum* (Sphaerotheriidae) in New Zealand (Rossi and Weir, 1998). Some species of Laboulbeniales are associated with specific millipedes. So far, three genera of Laboulbeniales are known to infect millipedes (*Diplopodomyces callipodos*, *Callipus foetidissimus* sub sp. *sorrentinus* and *Callipus foetidissimus* sub sp. *gallicus* and *Callipus hamulger*) from Italy, France and Sardinia. Similarly, *Troglomyces manfrediae* were recorded from Italy on *Trogloiulus mirus* (Julidae) and four species of *Rickia* were found exclusively on millipedes. *Rickia dendroiuli* was on *Dendroiulus latzeli* (Julidae) from Italy; *Rickia pachyiuli* on *Pachyiulus hungaricus* (Julidae) from Rumania; *Rickia siddhartha* on *Ktenostreptus lankaensis* (Harpagophoraceae) from Sri Lanka; *Rickia uncigeri* on *Unciger foetidus* and *Unciger transsilvanicus* from Germany and Poland. Ammonia and urea rich castings of *Jonespeltis splendidus* possess more humus than soil (Bano *et al.*, 1976). Fungal isolates of gut of *Jonespeltis splendidus* humified the soil *in vitro*, while bacteria failed (Bano *et al.*, 1976). Fungi such as *Cladosporium*, *Cylindrocephalum*, *Diplocladium*, *Fusidium*, *Monosporium*, *Mucor*, *Phoma* and *Torula* were isolated from the guts of *Jonespeltis splendidus*. *Cylindrocephalum* and *Torula* were confined to proctodaeal region, while *Diplocladium*, *Monosporium* and *Mucor* to stomodaeal region. Although *Fusarium*, *Helminthosporium* and *Trichoderma* were seen in soil, they were absent in the millipede guts. Faecal pellets of millipedes (*Phyllogonostreptus nigrolabiatus*) contained Arbuscular Mycorrhizal (AM) fungal propagules, they were viable up to four days of storage (Harinikumar and Bagyaraj, 1994).

Food and Feeding Habits

Organic Matter

Millipedes are known to feed on a variety of organic matter, which includes low to highly nutritious substances. Such dietary specialization is often considered as an impact of competitor avoidance, thus enabling biotic interactions to influence the assemblage of different species (Hannibal and Feldmann, 1981). Although millipedes consume plant detritus (Lawrence, 1984; Wooten and Crawford, 1975), some millipedes feed on soil (*Habrodesmus dubosequi* and *Xanthodesmus* spp) (Lewis, 1971a, 1971b), algae (Wooten and Crawford, 1975; Lewis, 1971a; Bocock and Heath, 1966; Saito, 1968) and bark (*Orthoporus ornatus*) (Wooten and Crawford, 1975). First instar larvae of *Luminodesmus sequoiae* consume humus (Davenport *et al.*, 1952), while *Apheloria coriacea* feeds on woody peat (Eaton, 1943). The range of food consumed by tropical Spirostreptidae is wider than that recorded for temperate millipedes (Dangerfield, 1995). *Blaniulus guttulatus* has been known to eat potato tubers in gardens (Kuhnelt, 1976; Lehtinen and Terhivuo, 1996). Millipedes are also known to consume highly nutritious stuff such as fallen fruits (Dangerfield and Telford, 1991,

1993), seeds (Dangerfield *et al.*, 1992; Dangerfield and Kaunda, 1994) and fungi (Dangerfield, 1993a; Dangerfield and Telford, 1991; Dangerfield and Kaunda, 1994). Millipedes are also known to feed on dead bodies of snails, insects, earthworms and vertebrates (Dangerfield, 1993; Dangerfield and Telford, 1989; Cole, 1946). A few South-European forms are carnivorous, feeding earthworms, insects and individuals of their own species (Kuhnelt, 1976; Gromysz-Kalkowska *et al.*, 1994; Edwards, 1974). Some members of the order Callipodida eat food of animal origin (Hoffman and Payne, 1969). Although coprophagy is rare phenomenon in millipedes, some are known feed on their own faecal pellets, some consume mammalian faecal matter (Dangerfield and Telford, 1991; Kuhnelt, 1976; Dangerfield *et al.*, 1992). Sibly (1981) is of the opinion that the ingestion of mineral soil helps millipedes in rapid turnover of low quality diet and facilitate high nutritional benefit. Kime and Wauthy (1984) interpreted that mineral soil provides nutrients, particularly salts (*e.g.*, Ca^{++}) for the production of exoskeleton (such salts are not available in other food sources).

Millipedes are known to ingest more than 5-25 per cent of annual litter (Bocock, 1963; Blower, 1970a, 1970b; van der Drift, 1975). Palatability of decaying leaves to millipedes was related to the nitrogen, carbohydrate and moisture (Sakwa, 1974). Neither phenolics nor water-soluble content of leaf litter belonging to 25 plant species was related to palatability of millipedes (Neuhauser and Hartenstein, 1978). But leaf palatability to *Iulus* and *Cylindroiulus* depends on the polyphenol levels of beech leaf litter (van der Drift, 1951). Preference of decomposed leaves of *Artocarpus, Macaranga* and *Terminalia* was related to low polyphenols and high sugar content. Feeding activity of *Diploiulus caeruleocinctus* correlated with calcium in leaf litter (van der Drift, 1951). *Jonespeltis splendidus* consumes leaf litter of advanced stage of decay over partly decayed or fresh (Devi and Prabhoo, 1990). The extent of microbial colonization influences the palatability of litter in *Glomeris marginata* (Anderson and Bignell, 1982). Bacteria and fungi of leaf litter became valuable source of nutrition to *G. marginata* (Bignell, 1984). Bacterial load increased, while fungal load decreased from food to faecal material in *G. marginata* (Anderson and Bignell, 1980). Tajovský (1992) demonstrated doubled microbial activity between leaf litter and faeces in *G. marginata*. It has been suspected that microbes killed in the midgut might be digested in further gut passage (Byzov *et al.*, 1998). Nicholson *et al.* (1966) indicated that a shift in microbial population from fungi to bacteria was a consequence of leaf litter passage through the guts of millipedes.

Mycophagy

About 20-75 per cent of soil fauna belong to fungivore trophic category (McGonigle, 1995). Ponge (1999b) revealed that fungal odour attracts different soil animals towards the detritus. Are fungi in food or in soil meet the nutritional requirements of millipedes? Fungal hyphae are important source of food of millipedes (Cromack *et al.*, 1977a, 1977b). Studies on feeding habits of millipedes belonging to Polymixidae, Siphonotidae and Siphonophoridae showed mycophagy (Vohland and Schroth, 1999). Taylor (1982) isolated fungi from the gut of desert millipede, *Orthoporus ornatus* and deduced that it selectively feeds on *Bispora* sp., *Cunninghamella bertolothea, Cunninghamella echinulata* and *Aspergillus fumigatus*. *Pachyilulus flavipes* is reported to selectively digest dematiaceous (pigmented) fungi (*Cladosporium cladosporoides*, *Alternaria* sp. and *Acremonium* sp.) than hyaline (colourless) forms (*Mucor* sp., *Trichoderma* sp. and *Penicillium* sp.) (Byzov *et al.*, 1998). When dematiaceous and hyaline fungi were offered to three millipede species simultaneously, they preferred dematiaceous to hyaline (Maraun *et al.*, 2003). However, millipedes consumed hyaline fungi when offered separately indicating the broad preference of mycophagy.

In Northern Missouri about 45 per cent of the fruit bodies of the fungus *Epichloe typhina* were damaged by the millipede *Aniulus bollmani* (Parajulidae) (Bultman and Mathews, 1996). Over 50 per cent of millipedes fed ascospore-releasing structures of fungi. Based on feeding ^{14}C-labelled bacterium and fungus along with litter to *Glomeris marginata* and analysis of its faecal pellets and residual gut content, it has been ascertained that microbes colonizing leaf litter were valuable nutritional source (Bignell, 1989). Gut passage of beech leaf litter by *G. marginata* decreased the fungal biomass (leaf litter, 445.6 ìg/g; faecal pellets 131.2 ìg/g) showing partial digestion in the gut (Maraun and Scheu, 1996; Anderson *et al.*, 1983). Maraun and Scheu (1996) measured ergosterol (lipid, specific to filamentous fungi) of faecal pellets of *G. marginata* and concluded that fungi were more intensively digested than bacteria in gut passage. It has been suspected that *Orthoporus ornatus* acquires enzymes from fungi that degrade the polysaccharides in the gut (Dangerfield *et al.*, 1992). British millipede population is known to harbour the fungus, *Enterobryus* (Eccrinales) in the hindgut (Moss and Taylor, 1996). The sporangiospores pass through the gut and such gut passage facilitates spore germination. Polydesmids are known to consume highly moist fungal colonized leaves (Kuhnelt, 1976). Calcium incorporation to millipedes takes place through fungi as calcium oxalate, such event signifies the importance of mycophagy in millipedes.

Coprophagy

Coprophagy is ingestion of faeces of the same or trophically similar species (McBrayer, 1973). Coprophagy is known to be essential for the survival of millipedes *Apheloria montna* (McBrayer, 1973). Millipedes access to faeces as well as leaf litter increased their body mass by about 16 per cent in 30 days compared to <1 per cent in millipedes deprived of access to faeces. Level of calcium increased several folds between leaf litter and faeces (49.9 vs. 226.6 mg/g ash) (Aerts, 1997). Millipedes deprived of faeces showed higher Ca^{++} than those accessed to faeces (462.4 vs. 419 mg/g ash), while Ca^{++} assimilation (8.3 vs. 9.8 per cent) and uptake (6.25 vs. 21.15 mg/g/day) differed drastically between animals access to litter alone and litter with faeces. The trend is also similar for Na^{+} and K^{+}.

Processing of Organic Matter

Breakdown of Plant Litter

Climate and litter chemistry are the most important factors, which regulate the rates of decomposition in global to continental scale (Aerts, 1997; Meentenmeyer, 1978). Saprophagous soil invertebrates such as millipedes, woodlice and larvae of Diptera in temperate regions are known to consume 20-100 per cent plant litter input per annum (Tajovský *et al.*, 1992). Among arthropods, millipedes are the major detritus feeding macroarthropods contribute most to breakdown of organic matter (Edwards *et al.*, 1968). They play an important role in mechanical disintegration of organic matter, mixing it with microbes and decomposition of starch and cellulose in their digestive tract (Edwards *et al.*, 1968; Bano, 1979). Leaf breakdown, stimulation of microbial activity and deposition of faecal pellets by millipedes have important ecological impacts (Hanlon, 1981). Millipedes contribute for the breakdown of organic matter in several ways: disintegrate plant tissue which easily degradable by microorganisms; selectively decompose plant residues; transform plant residue to humic substances; increase the surface area which facilitate the microbial activity; aggregate organic matter with soil minerals; mix organic matter in upper layers of soil (Edwards *et al.*, 1970). The process of plant detritus decomposition has been interpreted in terms of sequence of steps mediated by microbes and invertebrates (Anderson and Ineson, 1984). Millipedes fragment plant litter, generate faeces and provide more surface area for further microbial attack (Hopkin and Read, 1992). Bacterial population increased between 10-100 fold in faeces on litter after passage through gut of *Glomeris marginata* (Anderson and Bignell, 1980). It is reported that the bacterial proliferation takes place in the hindgut and faeces. The viable count of bacteria was higher in the gut and faeces than in leaf litter. Feeding activity of millipedes stimulates microorganisms to substantial breakdown of organic matter (van der Drift, 1951; Anderson and Bignell, 1980).

Rate of consumption of conifer litter by millipedes ranged between 10-20 per cent (about 36 per cent annual litter fall) in British Columbia (Cárcamo *et al.*, 2000; 2001). Warren and Zou (Warren and Zou, 2002) quantified the impact of three plantation types on the abundance and biomass of millipedes and earthworms in Puerto Rico. Abundance and biomass of millipedes and earthworms are higher in plantations of *Leucaena leucocephala* than in plantations of *Casuarina equisetifolia* and *Eucalyptus robusta*. Two millipedes, *Leptogoniulus nearsii* and *Trigoniulus lumbricinus* were most abundant in plantations of *Leucaena*. Millipedes in *Leucaena* plantations increased decomposition and mobilization of nutrients. Millipede biomass was greater in top 10 cm soil than in 10-20 cm depth and it was significantly correlated to nitrogen concentration and C/N ratio in litter. Millipedes belonging to Spirostreptidae breakdown different plant residues at different rates at 10 weeks duration (*Dactyladenia* > *Oryza* > *Zea* > *Leucaena* > *Glyricidia*) (Tian *et al.*, 1995). In temperate region, millipedes are known to ingest 5-10 per cent of annual litter (van der Drift, 1975). It has been estimated that in temperate woodlands the mean density of millipedes is 100/m^2, mean daily consumption of litter is about 4 mg/animal (Edwards, 1967). A population of 100/m^2 would transform 400 mg litter per day or up to 146 g per annum. Striganova (1971) estimated that millipedes consume 4,800 kg/ha/yr. Bocock (1963) estimated that *Glomeris* consume 1.7-10 per cent of annual litter. In spite of importance of soil fauna in litter decomposition (except for termites and earthworms), tropical ecosystems are relatively less studied than temperate ecosystems (Lavelle, 1983, 1988). Under humid tropical conditions millipedes on an average accounted for about 28.4 per cent of plant residues (Striganova, 1971).

Millipedes are capable of digesting lipids, proteins and simple carbohydrates. *Oxidus gracilis* were unable to digest lignin, while *Glomeris* were unable to digest cellulose (Neuhauser and Hartenstein, 1978; Jocteur Monrozier and Robin, 1988). But, the gut extract of *Polydesmus angustus* hydrolysed celluloses, hemicelluloses and pectin possibly through the enzymes of microbial origin. According to Neuhauser and Hartenstein (1978) millipedes degrade substantial amount of cinnamic acid and vanillin. Breakdown of cellulose and hemicellulose takes place in millipede gut by rich aerobic microbes (Taylor and Crawford, 1982). However, xylanase, α-amylase and cellulase enzymes were detected in the gut walls of *Glomeris marginata* after bacteriostatic treatment indicating such enzymes are the product of millipedes (Urbasek and Tajovský, 1991). In *G. marginata*, optimum temperature for carbon mineralization and consumption was 15-23°C, which enhanced the litter decomposition limit rather than its decomposition rate; cellulose loss was independent of temperature from 4-23°C and it was preferentially decomposed by microorganisms in faeces rather than in *Pinus* needles; cellulose decomposition in faeces suggests the occurrence of cellulolytic microorganisms in the gut (Coûteaux *et al.*, 2002). These results show that the structural components of plant litter are degraded in millipede-harboured microbes. Organic matter of soil declined to 50 per cent during its passage through the gut of *Jonespeltis splendidus* (Bano *et al.*, 1976). Mango (*Mangifera* sp.) leaf litter attracted *J. splendidus* and it was humified to a greater extent than coffee and *Cassia* leaf litter in Southern India (Bano, 1979). Anderson *et al.* (1985) showed that feeding activity of *G. marginata* significantly increased nitrogen mineralization. Three species of glomerids fed on leaf litters: chestnut, beech and holm-oak (C/N ratio, 28:39.1) produced faecal pellets with considerably narrow C/N ratio (20.25:27.95) (Marcuzzi, 1970). Millipedes are also important in storing NO_3^- and PO_4^{3-} in their biomass and NO_3^- in their faecal pellets that enrich the litter layers (Teuben and Verhoef, 1992). The total nitrogen in faecal pellets was 0.54-1.2 times higher than the food, so also the total phosphorus (Dangerfield and Telford, 1989). Webb (1977) found greater amount of phosphorus in faeces than in litter. Castings of *J. splendidus* were rich in urea and ammonia (Bano, 1976).

Millipedes and Earthworms

Enhanced nutrient release due to the impact of millipedes and soil fauna has been documented by Scheu and Wolters (1991). Millipedes along with earthworms further enhance the breakdown of plant litter. Nitrogen labelling (^{15}N) studies indicated the increased rate of decomposition of rape litter by the impact of earthworms and millipedes (Martens *et al.*, 2001). Bonkowski *et al.* (1998) studied the pattern of leaf litter decomposition in microcosms with and without earthworms (*Octolasion lacteum*) and millipedes (*Glomeris marginata*). Presence of millipedes with earthworms significantly increased litter decomposition. Leaf litter and faecal pellets of millipedes remained as such on soil surface without earthworms, while in the presence of earthworms they were mixed with soil. Faecal pellets of millipedes are highly attractive food source for endogeic earthworms (Bonkowski *et al.*, 1998). Faecal pellets of *G. marginata* as nutrient source for earthworms resulted in increase of biomass of

O. lacteum on beech leaf litter (Bonkowski *et al.*, 1998). Bonkowski *et al.* (1998) consider *O. lacteum* as a secondary decomposer, which feeds on organic matter fragmented by saprophagous macroinvertebrates. Occurrence of *G. marginata* along with *O. lacteum* significantly increased the beech litter breakdown and ultimately improved the soil fertility. Earthworms: *Lumbricus castaneus* and *O. lacteum* incorporated faecal pellets of *G. marginata* into soil and increased carbon mineralization of pellets within 14 days (Scheu and Wolters, 1991). The soil processed by earthworms and millipedes consists of two-thirds more bacteria than unprocessed soil (Kollmansperger, 1956).

Food Ingestion and Assimilation

Lumen of millipedes acts as natural fermenter for microorganisms with sufficient moisture, buffered with constant pH, constant redox potential and regular mixing (Bignell, 1984). Millipedes are known to process food in two stages: 1) rapid ingestion by mechanical breakdown using mandibles and nutrient absorption through microvilli in the midgut; 2) the nutrient release by digestive enzymes of salivary glands and midgut epithelial cells. Additional enzymes released from the microbes mixed with food in the lumen (Hopkin and Read, 1992). Köhler *et al.* (1991) demonstrated that the assimilation is greater in millipedes with mouthparts, which grind food into smallest particles. The ability of digestion reduced in millipedes on incorporation of antibiotics, demonstrated the importance of gut microbes in digestion and assimilation (Taylor, 1982). In *Glomeris marginata* a 100-fold increase in bacteria from food to faeces was seen (Anderson and Ineson, 1983). The rates of food ingestion in millipedes range from 5-10 per cent of body mass in 24 hr (Bocock, 1963; Hopkin *et al.*, 1985; Reichle, 1968). It is estimated that millipedes eat leaf litter five times of their mass at maturity (Blower, 1974). The *G. marginata* consumes leaf litter about 10 times of its mass per year (van der Drift, 1975) and assimilated 18 per cent of beech litter (Bignell, 1989). Smaller millipedes have high assimilation rates than larger ones (Crawford *et al.*, 1987). Although large species have high ingestion rates, it will not result in high assimilation efficiency (Dangerfield and Milner, 1993; Striganova and Prishutova, 1990), indicating the efficient digestive tactics of millipedes to exploit poor quality food (Sibly, 1981). A maximum of 30 per cent assimilation on dry mass basis was reported in millipedes. Microbes colonizing the leaf litter are valuable nutritional source to millipedes and assimilate with high efficiency. Labelled (^{14}C) microbial (bacteria and fungi) and plant fibre ingestion by *G. marginata* was studied by Bignell (1989). Uptake of microbial tissue (bacteria, 72.2-95.9 per cent ; fungi, 69.2-73.5 per cent) into the millipede after feeding was found to be over three times (18.3 per cent) that of plant fibre assimilation (Bignell, 1989). Food consumed for one-third of the lifetime by *Ophyiulus pilosus* will be spent mainly for its development from egg to adult (Blower, 1974). Table 1.2 deals with consumption rate of litter by different millipedes. The average litter consumption ranged between 1 and 157 mg/animal/day. Consumption rate of litter was highest (mean, 157 mg/animal/day) in *Seychelleptus seychellarum*, endemic millipede of Seychelles Islands (Lawrence and Samways, 2003). Efficiency of litter assimilation by some millipedes is given in Table 1.3.

Table 1.2: Rate of Litter Consumption by Millipedes

Millipede	*Consumption (mg/animal/day)*	*Reference*
Alloporus uncinatus	7	Dangerfield (1990)
	75	Blower (1974)
Amblyiulus continentalis	9.6	Striganova & Rachmanov (1972)
Calostreputs sp.	17	Blower (1974)
Chromatojulus projuctus	2.1	Gere (1956)
	5	Pobozsny (1985)
	8	Pobozsny (1986)
Cylindroiulus luridus	7	Pobozsny (1986)
	9.4	Pobozsny (1997)
Glomeris balcanica	11	Iatrou & Stamou (1989)
Glomeris hexasticha	4	Gere (1956)
Glomeris marginata	23-30	Dangerfield (1993b)
Leptoiulus proximus	6	Pobozsny (1986)
Odontopyge sp. 2	34	Blower (1974)
Odontopyge sp. 3	30	Blower (1974)
Orthomorpha gracilis	12	Kheirallah (1979)
Orthoporus ornatus	65	Wooten & Crawford (1975)
Paratophilus sp.	32	Dangerfield & Milner (1996)
Protracheoniscus politus	1.02	Gere (1956)
Rossiulus kessleri	23.3-78.4	Dangerfield & Milner (1993)
Sarmatiulus kessleri	63	Striganova (1972)
Schizophyllum caspium	25	Striganova & Rachmanov (1972)
Seychelleptus seychellarum	157	Lawrence & Samways (2003)
Unciger foetidus	3.3	Gere (1956)

Table 1.3: Rate of Assimilation of Litter by Millipedes

Millipede	*Assimilation Efficiency (mg/animal/day)*	*Reference*
Alloporus uncinatus	18	Blower (1974)
Calostreputs sp.	25.7	Blower (1974)
Odontopyge sp. 2	7	Blower (1974)
Odontopyge sp. 3	13.7	Blower (1974)
Paratophilus sp.	16.4	Blower (1974)
	Assimilation Efficiency (%)	
Amblyiulus continentalis	83.3-83.4	Striganova & Rachmanov (1972)
Glomeris marginata	35.8	Anderson & Bignell (1982)
Rossiulus kessleri	18.1-73.3	Dangerfield & Milner (1993)
Schizophyllum caspium	32.7-36	Striganova & Rachmanov (1972)

Faecal Castings

Saprophagous fauna (millipedes, woodlice and larvae of Diptera) feed 20-100 per cent plant litter on the forest floor and return the organic matter (60-90 per cent) in the form of faecal pellets (Tajovský, 1992). Temperate millipedes generate faeces 0.6-2.1 per cent of total annual litter Dangerfield and Milner, 1996; Pobozsny, 1985, 1986). In Zimbabwe, 18-28 per cent of annual litter converted into millipede faeces and faeces itself amounts to 12 per cent of litter standing crop (Dangerfield, 1993b). Faecal pellet production by millipedes will be higher under natural conditions than in laboratory conditions (Dangerfield, 1994). Rate of faecal pellet production by a few millipedes has been documented in Table 1.4. The South African millipede, *Paratophilus* sp. produced maximum faecal mass (126.1 mg/animal/day) followed by endemic millipede of Seychelles, *Seychelleptus seychellarum* (111 mg/animal/day). In habitats with high rates of litter decomposition, faecal pellets of millipedes store more nutrients (*e.g.* Cousine Island, Seychelles) (Lawrence and Samways, 2003) and such storage retard rapid nutrient loss through leaching (Dangerfield, 1990). In faecal pellets, particle size of organic matter decreases, which results in increased capacity of moisture retention (McBeyer, 1973) and increased microbial growth (McGonigle, 1995; Hanlon, 1981; Witkamp and Crossley, 1966).

Table 1.4: Extent of Faecal Pellets Production by Millipedes

Millipede	*Faecal Pellets (mg/animal/day)*	*Reference*
Alloporus uncinatus	61.5	Blower (1974)
Calostreputs sp.	12.6	Blower (1974)
Chromatojulus projuctus	2	Gere (1956)
Cylindroiulus luridus	8.1	Pobozsny (1997)
Glomeris hexasticha	4.5	Pobozsny (1985)
Orthoporus ornatus	46.7	Wooten & Crawford (1975)
Paratophilus sp.	126.1	Dangerfield & Milner (1996)
Protracheoniscus politus	0.97	Gere (1956)
Seychelleptus seychellarum	111	Lawrence & Samways (2003)
Unciger foetidus	2.2	Pobozsny (1997)
Millipedes	8-85	Dangerfield & Milner (1996)

Dangerfield (Dangerfield, 1995) tested the impact of starvation and mechanical grinding of leaf litter on rate of assimilation and faecal pellet production in three species of spirosteptid millipedes (*Chaleponcus digitatus, Chaleponcus limbatus* and *Zinophora* sp.) common in Savannas of Southeastern Botswana. In the presence of food, millipedes produced 6-20 faecal pellets/day/animal, pellet production stopped within 48 hr of starvation, but retained some food in the gut. Whole leaves, medium and fine ground leaf litter did not change the rate of faecal pellet production, but the mass of pellets increased with a decrease in food particle size. When ashed soil, leaf litter of *Cupressus* sp. and *Cupressus* leaf litter plus ashed soil (1:1 v/v) offered to

Alloporus uncinatus resulted in stable pellet production in animals fed with leaf litter plus ashed soil (Mwabvu, 1966b). Faeces consisting of pulverized litter offer greater surface area for leaching and microbial activity. The faeces of millipedes constitute the undigested dead plant tissues, mineral particles and microflora with higher water holding capacity and surface to volume ratio than raw leaf litter (Tajovský *et al.*, 1992). The surface to volume ratios of faecal pellets was 1.7:1 in *Narceus* sp., 3.2:1 in *Glomeris* (Nicholson *et al.*, 1966, Webb, 1977). *Glomeris* faeces consisted of 30 times more bacteria than litter and humification occurs due to bacterial action on the defecated faeces of *Glomeris marginata* (ven der Drift (1951). Faecal pellets of *G. marginata* consisted of leaf fragments (0.001-0.1 mm diameter), pollen, algae, bacteria and fungal mycelia (Nicholson *et al.*, 1966). Millipede faecal pellets produced in the field consisted of about 11 types of materials (plant fragments, mineral particles, free amorphous organic matter, bound amorphous organic matter to minerals, coal, pollen, plant epidermal cells, fungal hyphae, bacteria, earthworm cells) in semievergreen woodlands of Guadeloupe, French West Indies (Loranger *et al.*, 2003).

Tajovský *et al.* (1992) studied the pattern of decomposition of faecal pellets of *Glomeris hexasticha* in forest soil. Microbial succession was seen during decomposition, carbon and nitrogen of faecal pellets declined up to 44-50 per cent along with total mass and the C/N ratio declined from 23.7 to 20.9. Heterotrophic bacteria utilized easily degradable predominant organic substrates at first stage, followed by domination of actinomycetes and oligotrophic bacteria, which degrade more complex substrates in second stage. On faecal pellets, nematodes increased followed by nematophagous fungi. Decomposition leaf litter (*Cupressus* sp.) by millipedes (*Alloporus uncinatus*) when offered alone and leaf litter plus faecal pellets in the laboratory and field conditions revealed that the mass loss was less in field pellets than in laboratory pellets and predicted due to 52 per cent mass of field pellets consisted of mineral soil (Mwabvu, 1996a). Webb (1977) studied the mass loss, decomposition rate, nitrogen fixation and microbial production of faecal pellets of *Narceus annularis*. Changes in physical and chemical features and microbial succession of faecal pellets of *Glomeris marginata* under field conditions were studied by Nicholson *et al.* (1966). Several microorganisms have been isolated from the excrement of millipede *Pachyiulus flavipes* (Byzov *et al.*, 1998). Yeasts and fungi were considerably declined in the gut passage, which might be due to the action of midgut fluid (Byzov *et al.*, 1998). Faecal bacterial community of African millipede belonging to Spirostreptidae was studied based on amplified rDNA-restriction analysis (ARDRA) and found the dominance of actinomycetes. Most of the bacteria in millipede faeces belonged to the genera *Rhodococcus*, *Leifsonia* and *Cellulomonas* (Oravecz *et al.*, 2002). Faecal pellets increased the phosphorus and nitrogen availability to microorganisms for further degradation.

Nutrient immobilization is reduced by the millipede abundance due to increase in nitrogen concentration and decrease in C/N ratio between millipede food and faeces (Edwards, 1974; Anderson *et al.*, 1983). Field pellets of millipede *Alloporus uncinatus* consisted of 52 per cent mineral soil and 48 per cent organic matter (Mwabvu, 1996a). Millipedes retard nutrient loss from the persistent faecal pellet, which is beneficial for crop production (Dangerfield, 1990). In calcareous soils faecal pellets of

Glomeris marginata were stable, such stability was due to Ca^{++} (Kubikov and Rusek, 1976). Field pellets were more stable than the laboratory generated pellets (Mwabvi. 1966b). Larger pellets with mineral soil appeared more friable than organic rich pellets (Dangerfield, 1993b) and large pellets were less likely to persist in the environment and would potentially release nutrients soon. Fungal mycelial mat developed on the surface of faecal pellet (Tajovský *et al.*, 1992) might be responsible for its further stability and slowdown the release of nutrients.

Endemism

Limited mobility of millipedes resulted in high degree of speciation and evolution of a large number of endemic species (Hopkin and Read, 1992). Unlike other larger animal groups, which have a wide ecological range and morphologically well-diverged habitats, millipedes inhabit a narrow adaptive zone and are relatively morphologically uniform. Most endemics have a narrow distributional range, often exhibiting single-site endemism or scattered or disjunct pattern of distribution. Coastal forests of Eastern Africa are known for endemism of plants and animals including millipedes (Burgess, 1998; Hoffman, 1993). Habitat preference of two millipedes, *Centrobolus richardii* and *C. fulgidus* of coastal dune forests of South Africa was influenced by season (Greyling *et al.*, 2001). The climate desiccation by human interference is the major threat for these endemic species. Fragmentation by roads acts as ecological edges. The composition of millipedes differed between the edge and core assemblages. Fragmentation of Northern KwaZulu-Natal coastal dune resulted due to post-mining activity (Weiermans and van Aarde, 2003).

Giant millipede, *Seychelleptus seychellarum* (Sprostreptidae) (length, up to 24 cm) is endemic to islands Seychelles (Lawrence and Samways, 2003). Low quality diet result in high ingestion and high throughput rates in these millipedes (Dangerfield and Milner, 1996). Hence, the high density and large body size of giant millipedes of Seychelles have a major impact on litter consumption and production of faecal material (Lawrence and Samways, 2003). Abundance of *S. seychellarum* was 4.44/m^2 (~1,198,800 individuals in whole Cousine Island) (females, 66.44 per cent ; males, 13.15 per cent ; immatures, 20.41 per cent). They are nocturnal and females feed more than males (Lawrence and Samways, 2003). Average litter consumption rate is 157 mg/day, it amounts to 0.28 and 3.37 per cent of body mass and generate mean faecal pellets of 111 mg/day. It has been estimated that *S. seychellarum* consume 4.55 per cent litter standing crop and 17.19 per cent of daily litter fall in 24 hr. In entire Cousine Island, *S. seychellarum* ingest 73.57 kg litter/day and became an important 'key stone species' of Cousine Island deserves conservation. Canary Islands of North Africa is known for rich endemic flora and fauna. Millipede belonging to the genus, *Dolichoiulus* (Julidae) is endemic to Canary Islands (Enghoff and Báez, 1993). Out of 46 species of *Dolichoiulus*, most species live in one island. Cave millipedes have been extensively studied in USA (Shear, 1969, 1984) and France (Geoffroy, 1984) and found high endemism due to isolated habitat. Peninsular India inhabits about 44 species of *Arthrosphaera* (Pocock, 1892; Attems, 1936; Achar, 1986, 1980; Chowdaiah, 1969; Pocock, 1899; Ashwini, 2003) (Table 1.5). They are inhabitants of regions with adequate rainfall in forests of Western Ghats and Eastern Ghats (Maharashtra, Karnataka,

Kerala, Tamil Nadu and Andhra Pradesh) of India and Sri Lanka. Some species are high altitude inhabitants. *Arthrosphaera magna* is endemic to forest and coastal locations of Southern India (Maharashtra, Karnataka and Tamil Nadu) (Attems, 1936; Sakwa, 1974; Achar, 1980; Ashwini, 2003). Some representative species of millipedes *Arthrosphaera* seen in Western Ghat forests have been given in Figures 1.1 and 1.2.

Table 1.5: Checklist of *Arthrosphaera* Species Reported from India and Sri Lanka (?–Not defined)

Millipede	*Distribution*	*Altitude (m)*	*Reference*
Arthrosphaera atrisparsa Butler	Bombay, India	?	Pocock (1899)
Arthrosphaera aurocincta Pocock	East India	?	Pocock (1899)
Arthrosphaera bicolor Pocock	Salem and Shevaroy Hills, India	?	Pocock (1899)
	Southern India	?	Chowdaiah & Kanaka (1974)
Arthrosphaera brandti Humbert	Sri Lanka	?	Pocock (1899)
	Tamil Nadu, India	?	Attems (1936)
Arthrosphaera carinata Attems	Karnataka, India	1370-1880	Attems (1936)
Arthrosphaera corrugata Butler	Sri Lanka	?	Pocock (1899)
Arthrosphaera craspedota Attems	Kerala, India	520-980	Attems (1936)
	Southern India	?	Chowdaiah & Kanaka (1974)
Arthrosphaera dalyi Pocock	Palani Hills, India	1830	Pocock (1899)
	Southern India	?	Chowdaiah & Kanaka (1974)
	Kerala, India	?	Janardanan & Ramachandran (1983)
	Algarkovil Hills, India	?	Achar (1986)
Arthrosphaera davisoni Pocock	Annamalai Hills, India	1430	Pocock (1899)
	Khandla hills, India	?	Achar (1986)
Arthrosphaera dentigera Verh.	Sri Lanka	?	Attems (1936)
Arthrosphaera disticta Pocock	Shevaroy Hills, India	?	Pocock (1899)
	Southern India	?	Chowdaiah & Kanaka (1974)
	Kerala, India	?	Janardanan & Ramachandran (1983)
	Alagarkovil Hills, India	?	Achar (1986)
Arthrosphaera fumosa Pocock	Coimbatore, India	1430	Pocock (1899)
Arthrosphaera gracilis Attems	Kerala, India	?	Attems (1936)
	Southern India	?	Chowdaiah & Kanaka (1974)
Arthrosphaera hendersoni Pocock	Kodaikanal, India	2130	Pocock (1899)
	Southern India	?	Chowdaiah & Kanaka (1974)

Contd...

Table 1.5–Contd...

Millipede	*Distribution*	*Altitude (m)*	*Reference*
Arthrosphaera heterosticta Newp.	Madras, India	?	Pocock (1899)
Arthrosphaera inermis Humbert	Madras, India	?	Pocock (1899)
	Sri Lanka	?	Attems (1936)
	Tamil Nadu and Kerala, India	?	Attems (1936)
Arthrosphaera leopardina Butler	Sri Lanka	?	Pocock (1899)
Arthrosphaera lutescens Butler	India	?	Pocock (1899)
	Tamil Nadu and Kerala, India	?	Attems (1936)
	Karnataka, India	?	Achar (1986)
Arthrosphaera magna Attems	Karnataka, India	?	Achar (1986)
	Maharashtra, Tamil Nadu and Andhra Pradesh, India	?	Attems (1936)
	Karnataka, India	156-200	Ashwini (2003)
	Karnataka, India	91	Ashwini (2003)
Arthrosphaera marginella Silv.	India	?	Attems (1936)
Arthrosphaera marmorata Butler	India	?	Pocock (1899)
Arthrosphaera nitida Pocock	Palani Hills, India	2130	Pocock (1899)
	Kerala, India	?	Janardanan & Ramachandran (1983)
	Kerala, India	?	Achar (1986)
Arthrosphaera noticeps Butler	Sri Lanka	?	Pocock (1899)
Arthrosphaera rugosa Verh.	Sri Lanka	?	Attems (1936)
Arthrosphaera pilifera Butler	Sri Lanka	?	Pocock (1899)
Arthrosphaera pygostolis Attems	Tamil Nadu, India	?	Attems (1936)
Arthrosphaera rugulosa Butler	Sri Lanka	?	Pocock (1899)
Arthrosphaera scholastica Attems	Tamil Nadu, India	2130	Attems (1936)
Arthrosphaera severa Attems	Tamil Nadu, India	?	Attems (1936)
Arthrosphaera thurstoni Pocock	Nilgiri Hills and Shevaroy Hills, India	305	Pocock (1899)
Arthrosphaera transitiva Attems	Tamil Nadu, India	?	Attems (1936)
Arthrosphaera versicolor White	Sri Lanka	?	Pocock (1899)
Arthrosphaera wroughtoni Pocock	Maharashtra, India	?	Pocock (1899)
Arthrosphaera zebraica Butler	Bombay, India	?	Pocock (1899)
	Karnataka, India	?	Chowdaiah (1966)
Arthrosphaera sp.	India	?	Chowdaiah (1962)
Arthrosphaera sp.	Karnataka, India	?	Chowdaiah (1966)
Arthrosphaera sp.	Karnataka, India	?	Chowdaiah (1969)

Contd...

Table 1.5–Contd...

Millipede	*Distribution*	*Altitude (m)*	*Reference*
Arthrosphaera sp.	Kerala, India	?	Janardanan & Ramachandran (1983)
Arthrosphaera sp.	Karnataka, India	?	Achar (1986)
Arthrosphaera sp.	Karnataka, India	?	Achar (1986)
Arthrosphaera sp. 1	Karnataka, India	863	Ashwini (2003)
Arthrosphaera sp. 2	Karnataka, India	167-909	Ashwini (2003)
Arthrosphaera sp. 3	Karnataka, India	182-870	Ashwini (2003)
Arthrosphaera sp. 4	Karnataka, India	450	Ashwini (2003)

Human Interference

In woodlands, millipedes were abundant, while declined drastically in disturbed lands and agricultural pasture (Shear, 1981). Use of naphthalene to exclude soil

Figure 1.1: Pill Millipedes Found in Western Ghat Forests
***Arthrosphaera* sp. 3 (A), *Arthrosphaera* sp. 4 (B)**

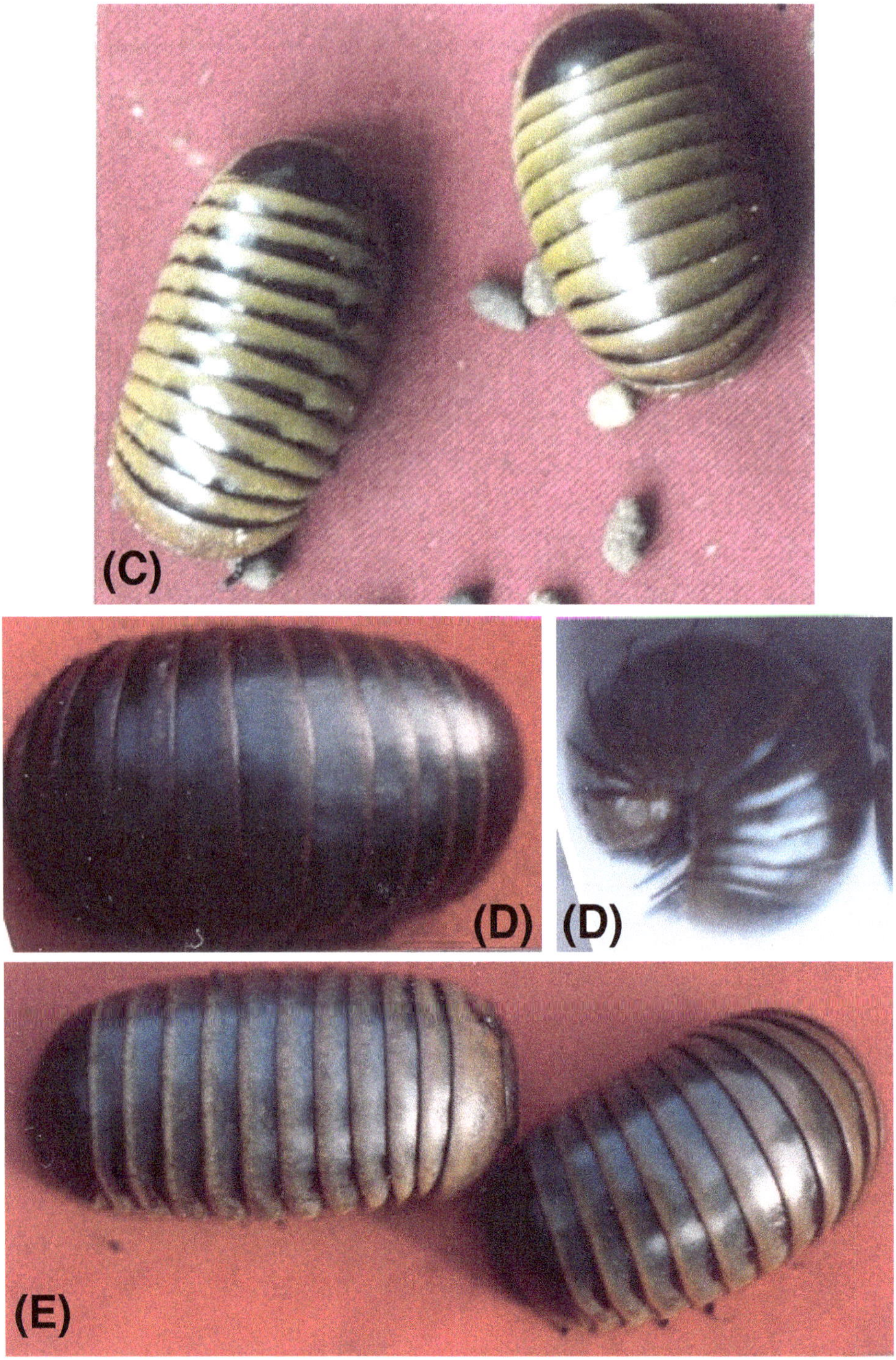

Figure 1.2: Pill Millipedes Found in Western Ghat Forests
Arthrosphaera magna **(C),** ***Arthrosphaera*** **sp. 1 (D) and** ***Arthrosphaera*** **sp. 2 (E)**

fauna in plots with *Quercus robur* retained leaf mass up to 91 per cent, while control plots with 45 per cent (Kurcheva, 1960) is a clear proof to show how important the soil fauna in decomposition process. Experiments with litter in fine-mesh bags to exclude soil invertebrates (as happens due to pollution) resulted in greatly reduced rate of decomposition of plant litter (Dangerfield and Milner, 1996; Anderson, 1988; Swift *et al.*, 1979). Although millipedes are susceptible to pollutants, studies have not been directed to understand such impacts on millipedes' role in litter decomposition. Decline in millipede fauna was seen in heavy metal polluted woodlands than unpolluted ones (Hopkin *et al.*, 1985). Highly contaminated soils with heavy metals (Cd, Zn and Pb) in France resulted in elimination of millipedes (Grelle *et al.*, 2000). Concentration of lead was highest in millipedes (*Julus scandinavius*) in spruce stand in Southern Germany, which was polluted with lead emission from car exhausts (Roth, 1993). Detritus feeding invertebrates had higher lead levels than zoophagous groups. Millipedes require high calcium for the development of exoskeleton, as lead follows biochemical pathways similar to that of calcium, its accumulation is high in millipedes. The heavy metal contaminated grasslands of Saxony-Anhalt, Central Germany, consisted of millipede, *Craspedosoma rawlinsii* (Voigtländer and Düker, 2002). However, Read and Martin (1990) did not find correlation between heavy metal vs. number of individuals and species of millipedes.

Saulnier and Binche (1986) investigated the impact of fire on millipede fauna at a cork-oak forest in the Pyrenees, France. Abundance of millipedes was 230/m^2 (biomass, 625g/m^2) in control plot. The number declined five times, while the biomass declined 10 times in burnt plot. The large decomposers, mainly millipedes inhabiting the litter layer were most affected by fire. Habitat fragmentation affects millipedes extensively as many millipedes are endemic. In addition, fragmentation restricts the millipedes' activities and their food availability or choice. Ten species of millipedes belonging to six families in Wog Wog, Australia responded to the habitat fragmentation in relation to uncleared forest in different ways (Baker, 1998). Some were susceptible for change over from control (pine plantation) to eucalypt plantations. Cultivation of mixed plantations would facilitate to maintain abundance, richness and diversity of millipedes. If the system is disturbed, the whole community has to undergo change for readjustment to new equilibrium state, which results in reduced or increased activities (Anderson, 1987).

Restoration and Conservation

Climate and altitude have considerable effects on life cycle of millipedes. The millipedes occupied 150-1900 m altitude in pine forest or in humid grasslands in temperate region. Higher altitude inhabitation was accompanied by increased body size. Larger species of millipedes (*e.g. Dolichoiulus*) eat larger, tougher or harder food items than small species (Enghoff, 1990). Unlike other soil macrofauna (*e.g.* earthworms, woodlice), millipedes are conservative: sensitive to water deficit and fail to overcome the limitation of edaphic factors, particularly soil texture and litter thickness although other ecological conditions are favourable (Kime and Golovatch, 2003). *Centrobolus fulgidus, C. richardii* and *Spinotarsus* sp. increased the concentration of selected elements in soil (C, N, P, K and Mg) in a microcosm study reveals the importance of millipedes in regeneration of vegetation in coastal dunes of South Africa (Smit and van Aarde, 2001). This study was correlated with the directional changes in selected soil elements with age of regeneration of post-mining areas of sand dunes of Richards Bay, South Africa (van Aarde *et al.*, 1996a, 1996b, 1998). The regenerating coastal dune forests of Richards Bay consist of 13 species of millipedes (van Aarde *et al.*, 1996c). Mean species richness and diversity of millipedes increased with age of the rehabilitating stand in coastal sand dune forests (>30 years stands showed maximum millipedes) in Northern KwaZulu-Natal province of South Africa (van Aarde *et al.*, 1996a, 1996b, 1996c). Conversion of native forest to plantations (conifer and eucalypt) in Tasmania led to concern of conservation of native ground-dwelling invertebrates largely depending on rotten logs (Bonham *et al.*, 2002). Alder (*Alnus*) afforestation proved to be more suitable for the colonization of millipede population in Sokolov region of Czech Republic (Tajovský, 2002). Millipedes were less diverse in plantations than in native forests. Several plantation practices *viz.*, windrowing (provides shelter in the form of accumulation of rotting logs), mound ploughing (creates long moist furrows of leaf and twig litter in eucalypt plantations), pruning (generates large amount of rotting wood in pine plantation) and thinning (adds rotting logs with persistent bark and leaves in pine plantations) helped the soil invertebrates to reinvade adjacent plantations from native forests and lowered the risk of extinction from native forests. These examples show the impacts of extensive forest devastations, limits and management practices necessary to retain the soil fauna.

Organic matter turnover by invertebrates facilitates biogeochemical cycle of elements (Krivolutzky and Pokarzhevsky, 1977). Several oxalate decomposers have been distinguished among macroarthropods (Cromack *et al.*, 1977b). Calcium oxalate and oxaloacetic acid decomposed into ionic calcium by microbes in the digestive system that influence the calcium cycle in soil. Edaphic factors, which influence millipedes, include the soil texture, soil water content, temperature, mineral content (mainly Ca^{++} and Mg^{++}), humidity and type of humus. Survey of a wide geographical conditions revealed that most diverse millipede population was confined to calcareous soils of Belgium (Branquart *et al.*, 1995). Pramanik *et al.* (2001) related the rich population of arthropods (including Diplopoda) to the nutrient content, mainly calcium of soil in *Cassia* plantation. Palatability of several leaf litters to millipede, *Diploiulus londinensis caeruleocinctus* was correlated to high calcium content (Neuhauser and Hartenstein, 1978). Leaf litter feeding activity of *Glomeris marginata* in deciduous woodlands resulted in marked mobilization of Ca^{++} and K^{+}, which amplified the nutrient fluxes (Anderson *et al.*, 1983). The most important saprophagous macroarthropod of the beech wood sites on a basalt lime stone gradient in Germany is *G. marginata* (Maraun and Scheu, 1996). Diplopoda represents one of the common trophic levels in food chain distribution of Ca^{++}. To assess the nutrient cycling, it is necessary to focus on the nutrients in food, faeces and animal biomass. Calcium assimilation takes place in millipedes through fungi as calcium oxalate. The calcium concentration of millipedes was 122 mg/g drymass due to highly calcareous exoskeletons indicating the importance of millipedes in calcium cycling in Rocky Mountains of Canada (Carter and Garg, 1976). Diplopods significantly increase the Ca^{++} and Mg^{++} concentrations in their biomass (Teuben and Verhoef,

1992). High calcium content was associated with diplopods (mean, 327.33 mg/g ash-free drymass) with highly calcified exoskeletons (Reichle *et al.*, 1969). High concentration of specific nutrient levels in whole body acts as buffer stock.

Future Outlook

Ponge (1999a) hypothesized that mull, moder and mor humus forms are the result of multiple interactions of soil organisms (fauna, free and symbiotic microbes) and plants, which have coevolved. Ponge (1999b) further indicates that if the above hypothesis is correct, the events such as invasion of foreign organisms or addition of new toxic substances or disturbances to which organisms have not adapted results in destabilizing the ecosystem (*i.e.* negative feedback loop, which is catastrophic). Human interferences should be in limits and should not lead to destabilization of soil ecosystem or tilt the positive feedback loop to negative feedback loop. In forests, humus formation is the result of intricate long-term biological processing through plants, animals and microorganisms. In Scots pine forest, Ponge (1999b) found that the pine needles were initially invaded by fungi, followed by feeding of soil fauna to transform the tissue into faecal mass. The faecal mass forms an important part of the humus, which is the main site of ion exchange and buffering agent for roots. In forests, the tree growth is conditioned by both autotrophic process (photosynthesis) and heterotrophic process (litter decomposition). Hence, the forest ecosystem is a feedback loop involving vegetation, soil animal communities and microbes within and around the humus, altogether forming a mosaic pattern. The complexity of bacterial, fungal and faunal community structure, their interaction and function in organic matter decomposition and nutrient cycling are essential aspects of litter studies, particularly in tropical forests (Ananthakrishnan, 1996). Soil invertebrates enhance the rate of decomposition and microbial activity through a number of mechanisms including enhancement of nitrogenous energy materials and elimination of bacteriostasis and mycostasis (Pramanik *et al.*, 2001). Millipedes are very sensitive to even a single edaphic factor (*e.g.* soil texture, litter thickness) (Kime and Golovatch, 2000). They are unable to overcome such differences although other ecological conditions are favourable. Conservation of millipedes and their activities needs simulation of precise conditions of litter strata on soil.

The role of different trophic levels to nutrient release varies in relation to the density, biomass and longevity of soil fauna. Decomposer fauna acts as catalyst of microbial activity in soil. Decomposition represents an ecological service for the whole ecosystem as 60-90 per cent of terrestrial primary production is decomposed in the soil (Giller, 1996). Although the ecological gains occur at a slow pace, they are significant for soil conservation with an increase in nutrient status, stability and diversity of decomposer community. Physical changes in litter decomposition were governed by chemical composition of litter, range of decomposer organisms and environmental regimes where the decomposition occurs (Ambasht and Srivastava, 1995). Millipedes ingested leaf litter and soil, increased soil respiration, leaching (*e.g.*, Ca^{++}, Mg^{++} and NO_3^-) from the soil and enhanced microbial activity in forest soil (Kaneko, 1999). Mean carbon mineralization rates of faecal pellets of *Glomeris marginata* and earthworm (*Octolasion lacteum*) fed with mechanically fragmented litter showed almost equal mean rates of CO_2-C production (Scheu and Wolters, 1991).

Millipede gut microflora has not been thoroughly characterized. There are many microorganisms living exclusively in the intestine of millipedes and are unable to survive in the soil (Márialigeti *et al.*, 1985). Most likely, millipede gut provides a suitable environment for microbial function that cannot be achieved otherwise. Recent study shows that endosymbiotic bacteria (*Bacillus* sp.) isolated from hind gut of *Glomeris* sp. produced a metabolite known as bacillomycin D, which possesses a strong antifungal activity (Gebhardt *et al.*, 2002). Yeast (*Saccharomyces cerevisiae*) fed at 5 per cent of leaf dry mass to millipede, *Polydesmus angustus* significantly increased the growth rate (David and Celerier, 1996). Adult females became large and attained fertility 4.3 folds higher over controls. Marcus *et al.* (1987) considered millipede, *Triaenostreptus* as an ideal model to maintain in laboratory to understand physiology as it grew up to 30 cm length and provided repeated blood samples. Millipedes are the most ideal individuals to understand the developmental morphology and molecular mechanisms of control with specific reference to segmentation (Hopkin and Read, 1992).

Acknowledgement

Authors are grateful to Mangalore University for grant of permission to carry out studies on pill millipedes of Western Ghats and west coast of Karnataka, which was the stimulus to write this review.

References

Achar, K.P. 1980. The use of Air-drying and Giemsa Differential Banding Techniques in the Cytological Studies of some Indian Diplopoda (Myriapoda). Ph.D. Thesis, Bangalore University, Bangalore, India.

Achar, K.P. 1986. Analysis of male meiosis in seven species of Indian pill-millipede (Diplopoda: Myriapoda). Caryologia 39: 89-101.

Adis, J. 1986. An aquatic millipede from a Central Amazonian inundation forest. Oecologia 68: 347-349.

Aerts, R. 1997. Climate, leaf litter chemistry and leaf litter decomposition in terrestrial ecosystems: a triangular relationship. Oikos 79,: 439-449.

Ambasht, R.S. and Srivastava, K. 1995. Tropical litter decomposition: A holistic approach. In: Reddy, M.V. (ed.), Soil Organisms and Litter Decomposition in Tropics Oxford & IBH Publ. Co. Ltd., New Delhi, India, pp. 225-247.

Ananthakrishnan, T.N. 1996. Forest Litter Insect Communities – Biology and Chemical Ecology. Oxford & IBH Publ. Co. Ltd., New Delhi, India.

Anderson, J.M. 1987. Interactions between invertebrates and microorganisms: Noise or necessity for the soil processes? In: Fletcher, M., Gray, T.R.G. and Jones, J.G.(eds.), Ecology of Microbial Communities. Cambridge University Press, Cambridge, pp. 124-145.

Anderson, J.M. 1988. Invertebrate-mediated transport processes in soils. Agriculture, Ecosystems and Environment 24: 5-19.

Anderson, J.M. and Bignell, D.E. 1980. Bacteria in the food, guts and faeces of the millipede *Glomeris marginata* (Villers). Soil Biol. Biochem.. 12: 251-255.

Anderson, J.M. and Bignell, D.E. 1982. Assimilation of ^{14}C-labelled leaf fibre by the Millipede *Glomeris marginata* (Diplopoda, Glomeridae). Pedobiologia 23: 120-125.

Anderson, J.M. and Ineson, P. 1983. Interactions between soil arthropods and microbial populations in carbon, nitrogen and mineral nutrient fluxes from decomposing leaf litter. In: Lee, J. and McNeill, S. (eds.), Nitrogen as an Ecological Factor. Blackwell Scientific, Oxford, pp. 413-432.

Anderson, J.M. and Ineson, P. 1984. Interactions between microorganisms and soil invertebrates in nutrient flux pathways of forest ecosystems. In: Anderson, J.M., Rayner, A.D.M. and Walton, D.W.H. (eds.), Invertebrate-Microbial Interactions. Cambridge University Press, Cambridge, pp. 59-88.

Anderson, J.M., Ineson, P. and Huish, S.A. 1983. Nitrogen and cation mobilization by soil fauna feeding on leaf litter and soil organic matter from deciduous woodlands. Soil Biol. Biochem. 15: 463-467.

Anderson, J.M., Leonard, M.A., Ineson, P., Huish, S., 1985. Faunal biomass: a key component for a general model of nitrogen mineralization. Soil Biol. Biochem. 17: 735-737.

Aouti, A. 1978. Etude comparée des peuplements de Myriapodes Diplopodes d'une forét hygrophile et d'une plantation d'Hevéa en Basse côte d"voire. Annales de 'Université Abidjan, Série E 11, 7-32. (H and R book, 1992).

Ashwini, K.M. 2003. Ecological Studies on Indian Pill Millipede, *Arthrosphaera magna.* Ph.D. Thesis, Mangalore University, Mangalore, India.

Attems, C., 1936. Diplopods of India. Memoirs of the Indian Museum 11: 133-167.

Bailey, P.T. and Kovaliski, J. 1993. Summer quiescent behaviour of the millipede *Ommatoiulus moreleti* (Julida: Julidae). J. Zool. (London) 231: 523-532.

Bailey, A.P., Rehman, T., Park, J., Keatinge, J.D.H. and Tranter, R.B.1999. Towards the Method for the economic evaluation of environmental indicators for UK integrated arable farming systems. Agriculture, Ecosystems and Environment 72: 145-158.

Baker, G.H. 1980. The water and temperature relationships of *Ommetoiulus moreleti* (Diplopoda: Iulidae). J. Zool. 190: 97-108.

Baker, G.H. 1988. Recognising and responding to the influences of agriculture and other land-use practices on soil fauna in Australia. Appl. Soil. Ecol. 9: 303-310.

Bano, K. 1979. Some ecological studies on the millipede *Jonespeltis splendidus* in relation to soil humification. Ph.D. Thesis, Bangalore University, Bangalore, India.

Bano, K. and Krishnamoorthy, R.V. 1985. Reproductive strategy and life history of *Jonespeltis splendidus* (Verhoeff) (Diplopoda: Polydesmida), J. Soil Biol. Ecol. 5: 48-57.

Bano, K., Bagyaraj, D.J. and Krishnamoorthy, R.V. 1976. Feeding activity of the millipede, *Jonespeltis splendidus* Verhoeff and soil humification. Proc. Indian Acad. Sci. 83B:1-11.

Bernier, N. and Ponge, J.-F. 1994. Humus form dynamics during the sylvogenetic cycle in a mountain spruce forest. Soil Biol. Biochem 26: 183-220.

Bignell, D.E. 1984. The arthropod gut as an environment for microorganisms. In: Anderson, J.M., Rayner, A.D.M. and Walton, D.W.H.(eds.), Invertebrate: Microbial Interactions, British Mycological Society Symposium No. 6, Cambridge University Press, pp. 205-227.

Bignell, D.E. 1989. Relative assimilation of ^{14}C-labelled microbial tissue and ^{14}C-plant Fibre ingested with leaf litter by the millipede *Glomeris marginata* under experimental conditions. Soil Biol. Biochem. 21: 819-827.

Blower, J.G. 1955. Millipedes and centipedes as soil animals. In: Kevan, D.K.McE. (ed.), Soil Zoology. Proceedings of University of Nottingham 1, pp. 138-151.

Blower, J.G. 1956. Some relations between millipedes and the soil. Sixth International Congress on Soil Science 3: 169-176.

Blower, J.G. 1970a. The millipedes of a Cheshire wood. J. Zool 160: 455-496.

Blower, J.G. 1970b. Notes on the life histories of some British Iuidae. Bulletin du Muséum National d'Historie Naturelle, Série II, 41: 19-23.

Blower, J.G. 1974. Food consumption and growth in a laboratory population of *Ophyiuls pilosus* (Newport). Symposia of the Zoological Society of London 32: 527-551.

Blower, J.G. 1985. Millipedes. Linnean Society Synopses of the British Fauna (New Series), Number 35, E.J. Brill/Dr. W. Backhuys, London.

Bocock, K.L. 1963. The digestion of food by *Glomeris*. In: Doeksen, J. and van der Drift, J. (eds.), Soil Organisms. Elsevier-North Holland, Amsterdam, pp. 85-91.

Bocock, K.L. and Heath, J. 1966. Feeding activity of the millipede *Glomeris marginata* (Villers) in relation to its vertical distribution in the soil. In: Graff, O. and Satchell, J.E. (eds.), Progress in Soil Biology Friedrich Vieweg and Shon, Berlin, pp. 233-240.

Bonham, K.J., Mesibov, R. and Bashford, R. 2002. Diversity and abundance of some ground-dwelling invertebrates in plantation vs. native forests of Tasmania, Australia. Forest Ecology and Management 158: 237-247.

Bonkowski, M., Scheu, S. and Schaefer, M. 1998. Interactions of earthworms (*Octolasion lacteum*), millipedes (*Glomeris marginata*) and plants (*Hordelymus europaeus*) in a beech wood on a basalt hill: implications for litter decomposition and soil formation. Appl. Soil Ecol. 9: 161-166.

Branquart, É., Kime, R.D., Dufrêine, M., Tavernier, J. and Wauthy, G. 1995. Macroarthopod-habitat relationships of oak forests of South Belgium. 1. Environments and communities. Pedobiologia 39: 243-263.

Bueno-Villegas, J., Sierwald, P. and Bond, J.E. 2003. Diplopoda. In:. Bousquets, J.L. and Morrone, J.J. (eds.), Biodiversidad, Taxonomia y Biogeografia de Artrópodos de México 569-599.

Bultman, T.L. and Mathews, P.L. 1996. Mycophagy by a millipede and its possible impact on an insect-fungus mutualism. Oikos 75: 67-74.

Burgess, N.D. 1998. Coastal forests of eastern Africa: status, endemism patterns and their potential causes. Biol. J. Linn. Soc. 64: 337-367.

Byzov, B.A., Claus, H., Tretyakova, E.B., Zvyagintsev, D.G. and Filip, Z. 1996. Effects of soil invertebrates on the survival of some genetically engineered bacteria in leaf litter and soil. Biol. Fertl. Soils 23: 221-228.

Byzov, B.A., Kurakov, A.V., Tretyakova, E.B., Thanh, V.N., To Luu, N.D. and Rabinovich, Y.M. 1998. Principles of the digestion of microorganisms in the gut of soil millipedes: specificity and possible mechanisms. Appl. Soil Ecol. 9: 145-151.

Cárcamo, H.A., Abe, T.A., Prescott, C.A., Holl, F.B. and Chanway, C.P. 2000. Influence of millipedes on litter decomposition, N mineralization and microbial communities in a coastal forest in British Columbia, Canada. Can. J. For. Res 30: 817-826.

Cárcamo, H.A., Abe, T.A., Prescott, C.A., Holl, F.B. and Chanway, C.P. 2001. Do soil fauna increase rates of litter breakdown and nitrogen release in forests of British Columbia, Canada? Can. J. For. Res. 31: 1195-1204.

Carrel, J.E. 1990. Chemical defence in the pill millipede *Glomeris marginata*. In: Minelli, A. (ed.), Proceedings of the 7th International congress of Myriapodology. E.J.B. Brill, Leiden, pp. 157-164.

Carter, A. and Grag, J.B. 1976. Concentrations and standing crops of calcium, magnesium, potassium and sodium in soil and litter arthropods and their food in an aspen woodland ecosystem in the Rocky Mountains (Canada). Pedobiologia 16: 379-388.

Chowdaiah, B.N. 1962. Cytological Studies of Some Indian Diplopoda (Myriapoda). Ph.D. Thesis, University of Mysore, India.

Chowdaiah, B.N. 1966. Chromosome studies in two species of pill-millipedes (Diplopoda- Myriapoda). Caryologia 19: 135-141.

Chowdaiah, B.N. 1969. A note on the occurrence of chromatid bridges in the male of *Arthrosphaera* sp., a pill-millipede. Current Science 7: 169-170.

Chowdaiah, B.N. and Kanaka, R. 1974. Cytological studies in six species of pill-millipedes (Diplopoda-Myriapoda). Caryologia 27: 55-64.

Cloudsley-Thompson, J.L. 1988. Evolution and Adaptation of Terrestrial Arthropods. Springer-Verlag, Berlin.

Cole, L.C. 1946. A study of the cryptozoa of an Illinois woodland. Ecological Monographs 16 : 49-86.

Coûteaux, M.-M., Aloui, A. and Kurz-Besson, C. 2002. *Pinus halepensis* litter decomposition in laboratory microorganisms as influenced by temperature and a millipede, *Glomeris marginata*. Appl. Soil Ecol. 20: 85-96.

Crawford, C.S. 1976. Feeding-season production in the desert millipede *Orthoporus ornatus* (Girard) (Diplopoda). Oecologia 24: 265-276.

Crawford, C.S., Bercovitz, K. and Warburg, M.R. 1987. Regional environments, life history patterns and habitat use of Spirostreptid millipedes in arid regions. Zool. J. Linn. Soc. London 89, 63-88.

Cromack, K., Sollins, P., Tood, R.L., Crossley, D.A., Fender, W.M., Fogel, R. and Todd, A.W. 1977a. Soil microorganisms-arthropod interactions: Fungi as major calcium and sodium sources. In: Mattson, W.J. (ed.), The Role of Arthropods in Forest Ecosystems Springer-Verlag, Berlin, pp. 78-84.

Cromack, K., Sollins, P., Tood, R.L., Fogel, R., Todd, A.W., Fender, W.M., Crossley, M.E. and Crossley, D.A.Jr. 1977b. The role of oxalic acid and bicarbonate in calcium cycling by fungi and bacteria – Some possible implications for soil animals. In: Lohm, U. and Persson, T. (eds.), Soil Organisms as components of Ecosystems Ecological Bulletin 25.

Curry, J.P. 1994. Grassland invertebrates: Ecology, Influence on soil Fertility and Effects on Plant Growth. Chapman and Hall, London.

Dangerfield, J.M. 1990. Abundance, biomass and diversity of soil macrofauna in savanna woodland and associated managed habitats. Pedobiologia 34: 141-150.

Dangerfield, J.M. 1993a. Soil animals and soil fertility: a critical component of woodland productivity? In: Piearce, G.D. and Gumbo, D.J. (eds.), The Ecology and Management of Indigenous Forest in Southern Africa Zimbabwe Forestry Commission and SAREC, pp. 209-215.

Dangerfield, J.M. 1993b. Ingestion of mineral soil/litter mixture and faecal pellet production in the southern African millipede *Alloporus uncinatus* (Attems). Pedobiologia 37: 159-166.

Dangerfield, J.M. 1994. Ingestion of leaf litter by millipedes: The accuracy of laboratory estimates for predicting litter turnover in the field. Pedobiologia 38: 262-265.

Dangerfield, J.M. 1995. Effects of starvation and mechanical manipulation of leaf litter on faecal pellet production and assimilation in some millipedes from southern Africa: Implications for feeding strategies. Eur. J. Soil Biol. 31: 199-206.

Dangerfield, J.M. and Kaunda, S.K. 1994. Millipede behaviour in a savanna woodland habitat in south-east Botswana. Afr. J. Ecol. 32: 337-341.

Dangerfield, J.M. and Milner, A.E. 1993. Ingestion and assimilation of leaf litter in some tropical millipedes. J. Zool. (London) 229: 683-693.

Dangerfield, J.M. and Milner, A.E. 1996. Millipede faecal pellet production in selected natural and managed habitats of southern Africa: implications for litter dynamics. Biotropica 28:113-120.

Dangerfield, J.M. and Telford, S.R. 1989. Are millipedes important for soil fertility? Zim. Sci. News 23: 66-68.

Dangerfield, J.M. and Telford, S.R. 1991. Seasonal activity patterns of Julid millipedes in Zimbabwe. J. Trop. Ecol. 7: 281-285.

Dangerfield, J.M. and Telford, S.R. 1992. Species diversity of Julid millipedes: Between habitat comparisons within the seasonal tropics. Pedobiologia 36: 321-329.

Dangerfield, J.M. and Telford, S.R. 1993. Aggregation in the tropical millipede *Alloporus uncinatus*. J. Zool. (London) 230: 503-511.

Dangerfiled, J.M., Milner, A.E. and Mathews, R. 1992. Seasonal activity pattern and behaviour of juliform millipedes in south-eastern Botswana. J. Tropical Ecol. 8: 451-464.

Davenport, D., Wooten, D.M. and Cushing, J.E. 1952. The biology of the Sierra luminous millipede *Luminodesmus sequoiae* Loomis and Davenport. Biological Bulletin 102: 100-110.

David, J.-F. and Celerier, M.-L. 1996. Effect of yeast on the growth and reproduction of the saprophagous millipede *Polydesmus angustus* (Diplopoda, Polydesmidae). Biol. Fertl. Soils 24: 66-69.

David, J.-F., Devernay, S., Loucougarry, G. and Le Floc'h, E. 1999. Below ground biodiversity in a Mediterranean landscape: relationships between saprophagous macroarthorpod communities and vegetation structure. Biodiversity and Conservation 8: 753-767.

Devi, D.S. and Prabhoo, N.R. 1990. Studies on food and feeding habits, food preference and feeding mechanism in the millipede *Jonespeltis splendidus* Verhoeff in captivity. Uttar Pradesh J. Zool. 10: 48-56.

Didham, R.K., Ghazoul, J., Stork, N.E., Davis, A.J. 1996. Insects in fragmented forests. TREE 11: 255-260.

Dzik, J. 1975. Sprobolid millipedes from the late Cretaceous of the Gobi desert, Mongolia. Palaeontologia Polonica 33: 17-24.

Eaton, T.H.Jr. 1943. Biology of the mull forming millipede *Glomeris*. Amer. Midl. Nat. 29: 713-723.

Edwards, C.A. 1967. Relationship between weights, volumes and numbers of soil animals. In: Graft, U. and Satchell, J.E. (eds.), Progress in Soil Biology. North Holland, Amsterdam, pp. 585-594.

Edwards, C.A. 1974. Macroarthropods. In: Biology of Plant Litter Decomposition (Ed. Dickenson, C.H. and Pugh, G.J.F.), Academic Press, New York., pp. 533-553.

Edwards, C.A. and Bohlen, P.J. 1995. Biology of Earthworms. Chapman and Hall, New York.

Edwards, C.A., Reichle, D.E. and Crossley, D.A.Jr. 1968. Experimental manipulation of soil invertebrate populations for trophic studies. Ecology 50: 495-498.

Edwards, C.A., Reichle, D.E. and Crossley, D.A.Jr. 1970. The role of soil invertebrates in turnover of organic matter and nutrients in ecological studies. In: Reichle, D.E.(ed.), Analysis and Synthesis, Volume 1. Springer-Verlag, New York, 147-171.

Enghoff, H. 1982a. The millipede genus *Cylindroiulus* on Madeiran instar species swarm (Diplopoda, Julida: Julidae). Entomologica Scandinavica Supplement 18: 1-142.

Enghoff, H. 1982b. The Zosteractinidae, a Nearetic family of millipedes (Diplopoda, Julida). Myriopodologica 1: 69-80.

Enghoff, H. 1990. The ground-plan of chilognathan millipedes (external morphology). In: Minelli, A. (ed.), Proceedings of the 7th International Congress of Myriapodology. E.J. Brill, Leiden, pp. 1-24.

Enghoff, H. and Báez, M. 1993. Evolution of distribution and habitat patterns in endemic millipedes of the genus *Dolichoiulus* (Diplopoda: Julidae) on the Canary Islands, with notes on distribution patterns of other Canarian species swarms. Biol. J. Linn. Soc. 49: 277-301.

Erwin, M. and Andreas, S. 1997. Distribution, seasonal activity and abundance of millipedes inforests of Voralerg (Australia). Berichte des Naturwissenschaftlich Medizinischen vereins in Innsbruck 84: 287-306.

Gebhardt, K., Schimana, J., Müller, J., Fiedler, H.-P., Kallenborn, H.G., Holzenkämpfer, M., Krastel, P., Zeeck, A., Vater, J., Höltzel, A., Schmid, D.G., Rheinheimer, J. and Dittner, K. 2002. Screening for biologically active metabolites with endosymbiotic bacilli isolated from arthropods. FEMS Microbiol. Lett. 217 : 199-205.

Geoffroy, J.J. 1984. *Opisthoceiron canayerensis* (Diplopoda: Craspedosornida): repartition de l'epèce et variation de la pigmentation. Mémoires de Biopéology 11: 295-302.

Gere, G. 1956. Examination of the feeding biology and humification function of Diplopoda and Isopoda. Acta Biol. Hung. 6: 257-271.

Giller, P.S. 1996. The diversity of soil communities, the 'poor man's tropical rainforest'. Biodiversity and Conservation 5: 135-168.

Golovatch, S.I., Hoffman, R.L. and de Morais, J.W. 1995. Identification plate for the millipede orders populating the neotropical region south of Central Mexico (Myriapoda, Diplopoda). Studies on the Neotropical Fauna and Environment 30, 159-164.

Grelle, C., Fabre, M.-C., Leprêtre, A. and Descamps, M. 2000. Myriapod and isopod communities in soils contaminated by heavy metals in northern France. Eur. J. Soil Sci. 51, 425-433.

Greyling, M.D., van Aarde, R.J. and Ferreira, S.M. 2001. Seasonal changes in habitat preferences of two closely related millipede species. African Journal of Ecology 39: 51-58.

Gromysz-Kalkowska, K., Szubartowska, E. and Bienko, M. 1994. The role of millipedes (Diplopoda) in nature and agriculture. Przeglad Zoologiczny 38: 25-34.

Haacker, U. and Fuchs, S. 1972. Tree climbing in pill-millipedes. Oecologia 10: 191-192.

Hanlon, R.D.G. 1981. Some factors influencing microbial growth on soil animal faeces, II. Bacterial and fungal growth on soil animal faeces. Pedobiologia 21: 264-270.

Hannibal, J.T. and Feldmann, R.M. 1981. Systematics and functional morphology of Oniscomorph millipedes (Arthropoda: Diplopoda) from the Carboniferous of North America. Journal of Palaeontology 55: 730-746.

Harinikumar, K.M. and Bagyaraj, D.J. 1994. Potential of earthworms, ants, millipedes and termites for dissemination of vesicular-arbuscular mycorrhizal fungi in soil. Biology and Fertility of Soils 18: 115-118.

Heywood, V.H. and Watson, R.T. 1995. Global Biodiversity Assessment. United Nations Environment Programme, Cambridge University Press.

Hoffman, R.L. 1980. Classification of the Diplopoda. Muséum d'Histoire Naturelle, Genève, Switzerland.

Hoffman, R.L. 1993. Biogeography of East African montane forest millipedes. In: Lovett, J.C. and Wasser, S.K.(eds.), Biogeography and Ecology of the Rain Forests of Eastern Africa. Cambridge University Press, 103-115.

Hoffman, R.L. and Payne, J.A. 1969. Diplopods as carnivores. Ecology 50: 1096-1098.

Hoffman, R.L., Golovatch, S.I., Adis, J. and de Morais, J.W. 2002. Diplopoda. In: Satchell, J.E.(ed.), Arachnida and Myriapoda..), Pensoft, Sofia-Moscow, 501-533.

Hölldobler, B. and Wilson, E.O. 1990. The Ants. Springer-Verlag, Berlin.

Hopkin, S.P. and Read, H.J. 1992. The Biology of Millipedes. Oxford University Press, Oxford.

Hopkin, S.P., Watson, K., Martin, M.H. and Mould, J.L. 1985. The assimilation of heavy metals by *Lithobius variegatus* and *Glomeris marginata* (Chilopoda: Diplopoda). Bijdragen tot de Dierkunde 55: 88-94.

Humphreys, W.F. 1989. The status of relict cave fauna of Cape Range, Western Australia, especially the Schizomida. Report to the Australian National Parks and Wildlife Service, Canberra.

Humphreys, W.F. 1991. Experimental re-establishment of pulse-driven populations in a terrestrial troglobite community. J. Animal Ecol. 60: 609-623.

Iatrou, G.D. and Stamou, G.P. 1989. Seasonal activity patterns of *Glomeris balcanica* (Diplopoda, Glomeridae) in an evergreen-sclerophyllous formation in northern Greece. Revue d'Ecologie et de Biologie du Sol. 26 : 343-353.

Ishii, K. and Yamaoka, H. 1982. The species and number of symbiotic Penicillate millipedes in arboreal ant nests. Canadian Entomologist 114: 767-768.

Janardanan, K.P. and Ramachandran, P. 1983. Observations on two new species of cephaline gregarines (protozoa: Sporozoa) of the genus *Stenoductus*

Ramachandran, 1976 from sphaerotherid millipedes in Kerala, India. Archiv für Protistenkunde 127: 405-412.

Jocteur Monrozier, L. and Robin, A.M. 1988. Action de la fauna du sol sur une litière de feuillu: application de techniques pyrolytiques à l'*étude des* modifications subies par les feuilles de charme (*Carpinus betulus*) ingérées par *Glomeris marginata*. Revue d'Ecologie et de Biologie du Sol 25 : 203-214.

Kaneko, N. 1999. Effect of millipede *Parafontaria tonominea* Attems (Diplopoda: Xystodesmidae) adults on soil biological activities: A microcosm experiment. Ecological Research 14: 271-279.

Kelley, J. and Peterson, R. 1997. Crop residues as a resource – The use of fungi to upgrade lignocellulosic wastes. Biology International 35: 16-20.

Kheirallah, A.M. 1979. Behavioural preference of *Julus scandinavius* (Myriapoda) to different species of leaf litter. Oikos 33: 466-471.

Kime, R.D. and Golovatch, S.I., 2000. Trends in the ecological strategies and evolution of millipedes (Diplopoda). Biol. J. Linn. Soc. 69: 333-349.

Kime, R.D. and Wauthy, G. 1984. Aspects of relationships between millipedes, soil texture and temperature in deciduous forests. Pedobiologia 26: 387-402.

Köhler, H.R., Albeti, G. and Storch, V. 1991. The influence of mandibles of Diplopoda on the food – a dependence of fine structure and assimilation efficiency. Pedobiologia 35: 108-116.

Kollmansperger, F. 1956. Lumbriciden in humuden und ariden Gebieten un ihre Bedentung für die Fruchtbarkeit des Bodens. Rapp. VI Congr. Inst. Sci. Sol., Paris, pp. 293-297.

Kraus, O. 1974. On the morphology of Palaeozoic Diplopids. Symposia of the Zoological Society of London 32: 13-22.

Kremen, C., Colwell, R.K., Erwin, T.L., Murphy, D.D., Noss, R.F. and Sanjayan, M.A. 1993. Terrestrial arthropod assemblages: their use in conservation planning. Conserv. Biol. 7: 796-808.

Krivolutzky, D.A. and Pokarzhevsky, A.D. 1977. The role of soil animals in nutrient cycling in forest and steppe. Ecological Bulletin 25: 253-260.

Kubíková, J. and Rusek, J. 1976. Development of xerotermic rendzinas-A study in ecology and soil microstructure. Rozpravy ÈSAV, Mat. Pøír. 86: 1-78.

Kuhnelt, W. 1976. Soil Biology with Special Reference to the Animal Kingdom, Faber and Faber, London.

Kurcheva, G.F. 1960. Role of invertebrate in the decomposition of oak litter. Soviet Soil Sci. 4: 360-365.

Lavelle, P. 1983. The structure of earthworm communities. In: Satchell, J.E.(ed.), Earthworm Ecology – From Darwin to Vermiculture. Chapman and Hall, London, pp. 449-465.

Lavelle, P. 1988. Earthworm activities and the soil system. Biol. Fertl. Soils 6: 237-251.

Lawrence, J.M.1984. The Centipedes and Millipedes of Southern Africa: A Guide. A.A. Balkema, Cape Town and Roterdam.

Lawrence, J.M. and Samways, M.J. 2003. Litter breakdown by the Seychelles giant millipede and the conservation of soil process on Cousine Island, Seychelles. Biological Conservation 113: 125-132.

Lehtinen, P.T. and Terhivuo, J. 1996. Notes on records of five garden millipedes (Diplopoda, Iulida) in Finland. Memorenda Societatis Pro-Fauna et Flora Fennica 72: 6-12.

Lewis, J. G.E. 1971a. The life history and ecology of the millipede *Tymbodesmus falcatus* (Ploydesmida: Gomphodesmidae) in northern Nigeria with notes on *Sphenodesmus sheribongensis*. J. Zool 164: 551-563.

Lewis, J.G.E. 1971b. The life history and ecology of three paradoxosomatid millipedes (Diplopoda: Polydesmida) in northern Nigeria. J. Zool. 165: 431-452.

Loomis, H.F. 1972. Millipedes associated with ants of Washington State. Florida Entomologist 55, 145-151.

Loranger, G. 2001. Particular humus forms in a tropical semievergreen forest of Guadeloupe. Life Sciences 324: 725-732.

Loranger, G., Ponge, J.F. and Lavelle, P. 2003. Humus forms in two secondary semi-evergreen tropical forests. Eur. J. Soil Sci. 54: 17-24.

Lyford, W.H. 1943. Palatability of freshly fallen leaves of forest trees to millipedes. Ecology 24: 252-261.

Manton, S.M. 1977. The Arthropoda. Oxford University Press, Oxford.

Maraun, M., Martens, H., Migge, S., Theenhaus, A. and Scheu, S. 2003. Adding to 'the enigma of soil animal diversity'; fungal feeders and saprophagous soil invertebrates prefer similar food substrates. European J. Soil. Biol. 39: 85-95.

Maraun, M. and Scheu, S. 1996. Changes in microbial biomass respiration and nutrient status of beech (*Fagus sylvatica*) leaf litter processed by millipedes (*Glomeris marginata*). Oecologia 107: 131-140.

Marcus, E.R., Baransky, M., Kooker, de G.M. and Knox, C.M. 1987. Haemolymph composition of the Kalahari millipede *Triaenostreptus triodus* (Attems). Comparative Biochem. Physiol. 87A: 603-606.

Marcuzzi, G. 1970. Experimental observations on the role of *Glomeris* spp. (Myriapoda, Diplopoda) in the process of humification of litter. Pedobiologia 10: 401-406.

Márialigeti, K., Conteras, E., Barabás, Gy., Heydrich, M. and Szabó, I.M. 1985. True intestinal actinomycetes of millipedes (Diplopoda). J. Invert. Pathol. 45: 120-121.

Martens, H., Slphei, J., Schaefer, M. and Scheu, S. 2001. Millipedes and earthworms increase the decomposition rate of ^{15}N-labelled winter rape litter in an arable field. Isotopes Environ. Health Stud. 37, 43-51.

McBrayer, J.F. 1973. Exploitation of deciduous leaf litter by *Apheleria montana* (Diplopoda: Eurydesmidae). Pedobiologia 13, 90-98.

McGonigle, T.P. 1995. The significance of grazing on fungi in nutrient cycling. Can J. Bot. (Suppl. 1), S1370-S1376.

Meentenmeyer, V. 1978. Macroclimate and lignin control of litter decomposition rates. Ecology 59: 465-472.

Moss, S.T. and Taylor, J. 1996. Mycobionts in the guts of millipedes–the Eccrinales. Mycologist 10: 121-124.

Mwabvu, T. 1996a. Decomposition of litter and faecal pellets of the tropical millipede, *Alloporus unicinatus* (Diplopoda). J. Afr. Zoology 110: 397-401.

Mwabvu, T. 1996b. Soil in millipede diet: implications on faecal pellet stability and nutrient release. Pedobiologia 40: 495-497.

Nadkarni, N.M. and Longino, J.T. 1990. Invertebrates in canopy and ground organic matter in a neotropical montane forest, Costa Rica. Biotropica 22: 286-289.

Nakamura, A., Proctor, H. and Catterall, C.P. 2003. Using soil and litter arthropods to assess the state of rainforest restoration. Ecological Management and Restoration 4: S20-S28.

Neuhauser, E.F. and Hartenstein, R. 1978. Phenolic content and palatability of leaves and wood to soil isopods and diplopods. Pedobiologia 18: 99-109.

Nicholson, P.B., Bocock, K.L. and Heath, O.W. 1966. Studies on the decomposition of the faecal pellets of a millipede (*Glomeria marginata* (Villers)). J. Ecol. 54: 755-766.

Niijima, K. 1998. Effects of outbreak of the train millipede *Parafontaria laminata armigera* Verhoeff (Diplopoda:Xystodesmidae) on litter decomposition in natural beech forest in Central Japan. 1. Density and biomass of soil invertebrates. Ecological Research 13: 41-53.

Oldeman, R.A.A. 1990. Forests: Elements of Silvology. Springer, Berlin.

Oravecz, O., Nyiro, G. and Márialigeti, K. 2002. A molecular approach in the analysis of the faecal bacterial community in an African millipede belonging to the family Spirostreptidae (Diplopoda). European J. Soil Biol. 38: 67-70.

Petersen and Luxton, 1982. A comparative analysis of soil fauna populations and their role in decomposition processes. Oikos 39: 287-388.

Plowman, K.P. 1990. Forest-floor decomposers. In: Webb, L.J. and Kikkawa, J. (eds.), Australian Tropical Reinforests: Science-Values-Meaning. CSIRO Publications, East Melborne, pp. 98-104.

Pobozsny, M. 1985. Die bedeutung der diplopodenart *Chromatoiulus projectus* Verh. Bie der zersetzung von eichenstreu. Opuscula Zoologica Budapest 20: 91-98.

Pobozsny, M. 1986. Uber Struzersetzungsprotesse in Hainbuchen-Eichenwarden unter Berucksichtigung der Diplopoden. Opuscula Zoologica Budapest 22: 77-84.

Pobozsny, M. 1997. Vergleichende untersuchungen des konsums bei mannchen und weibchen von diplopoden. Opuscula Zoologica Budapest 29-30: 117-123.

Pocock, R.I. 1892. Report upon two collections of Myriapoda sent from Ceylon by Mr. E.E. Green and from various parts of Southern India by Mr. Edger Thurston of the Government Central Museum, Madras. Bombay Nat. Hist. Soc. 17: 131-175.

Pocock, R.I. 1899. A monograph of the pill-millipedes (Zephroniidae) inhabiting India, Cylon and Burma. J. Bombay Nat. Hist. Soc. 12: 269-285.

Ponge, J.-F. 1999a. Heterogeneity in animal communities and the development of humus forms. In: Rastin, N. and Bauhus, J. (eds.), Going Underground – Ecological Studies in Forest Soils Research Signpost, Trivandrum, India, pp. 33-44.

Ponge, J.-F. 1999b. Interaction between soil fauna and their environment. In: Rastin, N. and Bauhus, J. (eds.), Going Underground – Ecological Studies in Forest Soils. Research Signpost, Trivandrum, India, pp. 45-76.

Pramanik, R., Sraker, K. and Joy, V.C. 2001. Efficiency of detritivore soil arthropods in mobilizing nutrients from leaf litter. Tropical Ecology 42: 51-58.

Read, H.J. and Martin, M.H. 1990. A study of myriapod communities in woodlands contaminated with heavy metals. In: Minelli, A. (ed.), Proceedings of the 7th International Congress of Myriapodology. Brill, Leiden, pp. 289-298.

Reichle, D.E. 1968. Relation of body size to food intake, oxygen consumption and trace element metabolism in forest floor arthropods. Ecology 49: 538-542.

Reichle, D.E., Shanks, M.H. and Crossley, D.A. 1969. Calcium, potassium and sodium content of forest floor arthropods. Annals of the Environmental Society of America 62: 57-62.

Rettenmeyer, C.W. 1962. The behaviour of millipedes found with neotropical army ants. Journal of the Kansa Entomological Society 35: 377-384.

Rossi, W. and Weir, A. 1998. *Triainomyces*, a new genus of Laboulbeniales on the pill-millipede *Procyliosoma tuberculatum* from New Zealand. Mycologia 90: 282-289.

Roth, M. 1993. Investigations on lead in the soil invertebrates of a forest ecosystem. Pedobiologia 37: 270-279.

Saito, S. 1968. Productivity of high and low density populations of *Japonaria laminata armigers* (Diplopoda) in a warm temperate forest ecosystem. Res. Popul. Ecol Kyoto Univ. 9: 153-164.

Sakwa, W.N. 1974. A consideration of the chemical basis of food preference in millipedes. Symp. Zool. Soc. Lond. 32: 329-346.

Saulnier, L and Binche, A.F. 1986. Regeneration patterns of Mediterranean ecosystems after fire: soil arthropods (Myriapoda). Vie et Milieu 36: 191-204.

Scheu, S. and Wolters, V. 1991. Influence of fragmentation and bioturbation on the decomposition of ^{14}C-labelled beech leaf litter. Soil Biol. Biochem.23: 1029-1034.

Shear, W.A. 1969. A synopsis of the cave millipedes of the United States with an illustrated key to genera. Psyche 76: 126-143.

Shear, W.A. 1981. Two fossil millipedes from the Dominican amber (Diplopoda: Chytodesmidae, Siphonophoridae). Myriapodologica 1: 51-54.

Shear, W.A. 1984. Cave millipedes of the United States. III. Two new species from the western states (Diplopoda: Polydesmida, Chrodeumatida), Myriapodologica 351: 283-289.

Shelley, R.M. 2003. A revised, annotated, family-level classification of the Diplopoda. Arthropoda Sel. 11: 187-207.

Sibly, R.M. 1981. Strategies of digestion and defecation. In: Townsend C.R. and Calow, P. (eds.), Physiological Ecology: An Evolutionary Approach to Resource Use. Blackwell Scientific Publications, Oxford, pp. 109-139.

Sierwald, P. and Bond, J.E. 2007. Current status of the Myriapod Class Diplopoda (millipedes): Taxonomic diversity and phylogeny. Ann. Rev. Entomol. 52: 401-420.

Smit, A.-M. and van Aarde, R.J. 2001. The influence of millipedes on selected sol elements: a microcosm study on three species occurring on coastal sand dunes. Functional Ecol. 15: 51-59.

Stork, N.E. 1993. How many species are there? Biodiversity and Conservation 2: 215-232.

Striganova, B.R. 1971. Age changes in feeding activity of millipedes (Julidae). Zool. J. 50: 1472-1476.

Striganova, B.R. 1972. Effect of temperature on the feeding activity of *Sarmatiulus kessleri* (Diplopoda). Oikos 23: 197-199.

Striganova, B.R. and Prishutova, Z.G. 1990. Food requirements of diplopods in the dry steppe subzone of the USSR. Pedobiologia 34: 37-41.

Striganova, B.R. and Rachmanov, R.R. 1972. Comparative study of the feeding activity of Diplopoda in Lenkorn province of Azerbaijan. Pedobiologia 12: 430-433.

Swift, M.J., Heal, O.W. and Anderson, J.M. 1979. *Decomposition in Terrestrial Ecosystems.* Blackwell Scientific, Oxford.

Szabo, I.M., Jager, K., Contreras, E., Marialigetti, K. Dzingov, A., Barabas, G. and Pobozsny, M. 1983. Composition and properties of the external and internal microflora of millipedes (Diplopoda). In: NewTrends in Soil Biology. Lebrun, P., Andre, H.M. and Medts, A.De. (ed.), Budapest, Hungary, pp. 197-206.

Tajovský, K. 1992. Feeding biology of the millipede *Glomeris hexasticha* (Glomeridae, Diplopoda). Ber. Nat. med. Verein innsbruck Suppl. 10: 305-311.

Tajovský, K. 2001. Colonization of spoil heaps by millipedes (Diplopoda) and terrestrial isopods (Oniscidea) in the Sokolov region, Czech Republic. Restoration. Ecology 9: 365-369.

Tajovský, K., Mock, A. and Krumpál, M. 2001. Millipedes (Diplopoda) in birds' nests. *Eur.* J. Soil Biol. 37: 321-323.

Tajovský, K. Šanùètrková, H., Hánel, L., Balík and Lukešová. 1992. Decomposition of faecal pellets of the millipede *Glomeris hexasticha* (Diplopoda) in forest soil. Pedobiologia 36: 146-158.

Taylor, E.C. 1982. Fungal feeding preference by a desert millipede *Orthoporus ornatus* (Spirostreptidae). Pedobiologia 23: 331-336.

Taylor, E.C. and Crawford, C.S. 1982. Microbial gut symbionts and desert detritivores. *Sci.* Rev. Arid Zone Res. 1: 37-52.

Teuben, A. and Verhoef, H.A. 1992. Direct contribution by soil arthropods to nutrient availability through body and faecal nutrient content. Biol. Fertl. Soils 14: 71-75.

Thiele, H.U. 1956. Die Tiergesellschaften der Bodenstreu in den Verchiedenen Waldtypen des Niederbergischen Landes. Z. Angew. Ent. 39: 316-367.

Tian, G., Brussaard, L. and Kang, B.T. 1995. Breakdown of plant residues with contrasting chemical compositions under humid tropical conditions: effects of earthworms and millipedes. Soil Biol. Biochem. 27: 277-280.

Tracz, H. 1987. The role of *Proteroiulus fuscus* (Am Stein 1957) (Diplopoda) in the circulation of some elements in the fresh pine forest. Annals of Warsaw Agricultural University, Forestry and Wood Technology 36: 101-107.

Tretyakova, E.B., Dobrovolskaya, T.G., Byzov, B.A. and Zvyagintsev, D.G. 1996. Bacterial communities associated with soil invertebrates. Mikrobiologiya 65: 102-110.

Urbasek, F. and Tajovský, K. 1991. The influence of food and temperature on enzymatic activities of the millipede *Glomeris hexasticha* (Diplopoda) Rev. Ecol. Biol. Sol. 28: 155-163.

van Aarde, R.J., Ferriera, S.M. and Kritzinger, J.J. 1996a. Millipede communities in rehabilitating coastal dune forests in northern KwaZulu-Natal, South Africa.. J. Zool. (London) 238: 703-712.

van Aarde, R.J., Ferriera, S.M. and Kritzinger, J.J. 1996b. Successional changes in rehabilitating coastal dune communities in northern KwaZulu-Natal, South Arica. Landscape and Urban Planning 34: 277-286.

van Aarde, R.J., Ferreira, S.M., Kritzinger, J.J., van Dyk, P.J., Vogt, M. and Wassenaar, T.D. 1996c. An evaluation of habitat rehabilitation on coastal dune forests in Northern KwaZulu-Natal, South Africa. Restoration Ecology 4: 334-345.

van Aarde, R.J., Smit, A.-M. and Claasens, A.S. 1998. Soil characteristics of rehabilitating and unmined coastal duned at Richards Bay, KwaZulu-Natal, South Africa. Restoration Ecology 6: 102-110.

van der Drift, J. 1951. Analysis of the animal community in a beech forest floor. Tijdschrift voor Entomologie 94: 1-168.

van der Drift, J. 1963. The disappearance of litter in mull and mor in connection with weather conditions and the activity of the macrofauna. In: Doeksen, J. and van der Drift, J. (eds.), Soil Organisms. North-Holland Publishing Company, Amsterdam, Netherlands, pp. 125-133.

van der Drift, J. 1975. The significance of the millipede *Glomeris marginata* (Villers) for oak-litter decomposition and an approach of its part in energy flow. In: Vanek, J. (ed.), Progress in Soil Zoology. Academia, Prague, pp. 293-298.

Vohland, K. and Schroth, G. 1999. Distribution patterns of the litter macrofauna in agroforestry and monoculture plantations in central Amazonia as affected by plant species and management. Appl. Soil Ecol. 13: 57-68.

Voigtländer, K. and Düker, C. 2001. Distribution of species grouping of millipedes (Myriapoda, Diplopoda) in dry biotopes in Saxony-Anhalt/Eastern Germany. Eur. J. Soil. Biol. 37: 235-238.

Wardle, D.A. 1995. Impact of disturbance on detritus food webs in agro-ecosystems of contrasting tillage and weed management practices. Adv. Ecol. Res. 26: 105-185.

Warren, M.W. and Zou, X. 2002. Soil macrofauna and litter nutrients in three tropical tree plantations on a disturbed site in Puerto Rico. Forest Ecology and Management 170: 161-171.

Webb, D.P. 1977. Regulation of deciduous forest litter decomposition by soil arthropod faeces. In: Mattson, W.J. (ed.), The Role of Arthropods in Forest Ecosystem Springer-Verlag, Heidelberg, pp. 57-69.

Weiermans, J. and van Aarde, R.J. 2003. Roads as ecological edges for rehabilitating coastal dune assemblages in northern KwaZulu-Natal, South Africa. Restoration Ecology 11: 43-49.

Wilson, E.O. 1992. The Diversity of Life. Harvard University Press, Harvard.

Witkamp, M. and Crossley, D.A. 1966. The role of arthropods and microflora in the breakdown of white oak litter. Pedobiologia 6: 293-303.

WCMC (World Conservation Monitoring Centre) 1992. Global Biodiversity: Status of the Earth's Living Resources. Chapman and Hall, London.

Wooten, R.C. and Crawford, C.S. 1975. Food, ingestion rates and assimilation in the desert millipede *Orthoporus ornatus* (Girard) (Diplopoda). Oecologia 20: 231-236.

Zenova, G.M., Babkina, N.I., Polyanskaya, L.M. and Zvyagintsev, D.G. 1996. Actinomycetes in the intestinal tract of soil invertebrates fed with vermicompost or litter. Microbiology 65: 360-365.

Chapter 2

Phytomedicine for Respiratory Complaints

B. Sandhya Sri, S. Prasanthi and T.V.V. Seetharami Reddi

Department of Botany, Andhra University,
Visakhapatnam – 530 003

Introduction

Medicinal plants have been considered as an important therapeutic aid for alleviating ailments of humankind. Search for eternal health and longevity and to seek remedy to relieve pain and discomfort prompted the early man to explore his immediate natural surrounding to develop vareity of therapeutic agents using natural resources.

Herbs, glorified in the ancient medicinal science of Ayurveda have a gentle yet powerful effect in enhancing the health and beauty of man. These are capable of bringing the body into harmony and health. Plants having remarkable medicinal value have been rediscovered and the search is still on at global level. The fact that a very large part of Indian population has inadequate access or none to essential and life-saving medicines is of grave concern. It results in a vast loss of life and much suffering, especially among the under privileged poors. It is estimated that one half population in Asian countries lacks access to the most basic, essential medicinal remedies. A high proportion of deaths in India is due to illnesses which are in principle curable with medicines which currently exist, *e.g.*, pneumonia, tuberculosis etc.

Respiratory disease is the term for diseases of the respiratory system. These include diseases of the lung, pleural cavity, bronchial tubes, trachea, upper respiratory tract and of the nerves and muscles of breathing. Respiratory diseases range from mild and self-limiting such as the common cold to life-threatening such as bacterial pneumonia or pulmonary embolism. They are a common and important cause of illness and death. The study of respiratory disease is known as pulmonology. The symptoms of respiratory disease differ depending on the disease. Common symptoms are:

- General malaise
- Shortness of breath
- Loss of appetite
- Weight loss
- Cachexia
- Cyanosis, a bluish discoloration of the lips, tongue or fingers.

Treatment of respiratory disease depends on the particular disease being treated, the severity of disease and the patient. Regular exercise and healthy nutrition are important in preventing and treating respiratory disease. Vaccination can prevent some respiratory diseases.

Bronchitis is inflammation of the mucous membranes of the bronchi, the airways that carry airflow from the trachea into the lungs. Bronchitis can be classified into two categories, acute and chronic, each of which has unique etiologies, pathologies and therapies. *Acute bronchitis* is characterized by the development of a cough, with or without the production of sputum, mucus that is expectorated (coughed up) from the respiratory tract. Acute bronchitis often occurs during the course of an acute viral illness such as the common cold or influenza. Viruses cause about 90 per cent of cases of acute bronchitis while bacteria account for less than 10 per cent.

Chronic bronchitis, a type of chronic obstructive pulmonary disease, is characterized by the presence of a productive cough that lasts for 3 months or more per year for at least 2 years. Chronic bronchitis most often develops due to recurrent injury to the airways caused by inhaled irritants. Cigarette smoking is the most common cause, followed by air pollution and occupational exposure to irritants, and cold air. Chronic bronchitis is a chronic inflammation of the bronchi (medium size airways) in the lungs. It is generally considered one of the two forms of chronic obstructive pulmonary disease (COPD). It is defined clinically as a persistent cough that produces sputum (phlegm) and mucus.

Bronchitis may be indicated by an expectorating cough (also known as a productive cough, *i.e.* one that produces sputum), shortness of breath (dyspnea) and wheezing. Occasionally chest pains, fever and fatigue or malaise may also occur. Mucus is often green or yellowish green and also may be orange or pink depending on the pathogen causing the inflammation. Tobacco smoking is the most common cause. Pneumoconiosis and long-term fume inhalation are other causes.

Cough is a sudden and often repetitively occurring reflex which helps to clear the large breathing passages from secretions, irritants, foreign particles and microbes. The cough reflex consists of three phases: an inhalation, a forced exhalation against a closed glottis, and a violent release of air from the lungs following opening of the glottis, usually accompanied by a distinctive sound.

Frequent coughing usually indicates the presence of a disease. Many viruses and bacteria benefit evolutionarily by causing the host to cough, which helps to spread the disease to new hosts. Most of the time, coughing is caused by a respiratory tract infection but can be triggered by choking, smoking, air pollution, asthma, gastroesophageal reflux disease, post-nasal drip, chronic bronchitis, heart failure and medications such as ACE inhibitors.

Cough suppressants such as codeine or dextromethorphan are frequently prescribed, but have been demonstrated to have little effect. Other treatment options may target airway inflammation or may promote mucus expectoration. As it is a natural protective reflex, suppressing the cough reflex might have deleterious effects, especially if the cough is productive.

A cough can be classified by its duration, character, quality, and timing. The duration can be either acute (of sudden onset) if it is present less than three weeks, subacute if it is present between three to eight weeks and chronic when lasting longer than eight weeks. A cough can be dry or productive depending on whether sputum is coughed up. It may occur only at night, during both night and day or just during the day.

A cough can be the result of a respiratory tract infections such as the common cold, pneumonia, pertussis, or tuberculosis. In people with a normal chest X-ray, tuberculosis is a rare finding. Pertussis is increasingly being recognised as a cause of troublesome coughing in adults. Postinfectious cough typically consists of a dry, non-productive cough that produces no phlegm. Symptoms may include a tightness in the chest and a tickle in the lungs. This cough may often persist for weeks after an illness. The cause of the cough may be inflammation similar to that observed in repetitive stress disorders. The repetition of coughing produces inflammation which produces discomfort, which in turn produces more coughing creating a closed loop. Inflammation may increase sensitivity to other existing issues such as allergies, and treatment of other causes of coughs (such as use of an air purifier or allergy medicines) may help speed recovery. A bronchodilator, which helps open up the airways, may also help treat this type of cough.

The *common cold* (viral upper respiratory tract infection (VURI), acute viral rhinopharyngitis, acute coryza, or cold) is a contagious, viral infectious disease of the upper respiratory system caused primarily by rhinoviruses and coronaviruses. There is no cure; however, symptoms usually resolve spontaneously in 7 to 10 days, with some symptoms possibly lasting for up to three weeks.

The common cold is the most frequent infectious disease with an average two to four infections a year in adults and up to 6–12 months in children. Collectively, cold, influenza, and other infections with similar symptoms are included in the diagnosis of influenza-like illness. They may also be termed upper respiratory tract infections (URTI). Influenza involves the lungs while the common cold does not.

Symptoms are cough, sore throat, runny nose and nasal congestion; sometimes this may be accompanied by conjunctivitis (pink eye), muscle aches, fatigue, headaches, shivering, and loss of appetite. Fever is often present thus creating a symptom picture which overlaps with influenza. The common cold usually resolves spontaneously in 7-10 days, but some symptoms can last for up to three weeks. In children the cough lasts for more than 10 days in 35-40 per cent and continue for more than 25 days in 10 per cent.

The common cold is due to a viral infection of the upper respiratory tract. The most common implicated is rhinovirus (30-50 per cent), a type of picornavirus with 99 known serotypes. Others include: coronavirus (10-15 per cent), influenza (5-15 per cent), human parainfluenza viruses, human respiratory syncytial virus, adenoviruses, enteroviruses, and metapneumovirus. Coronaviruses are particularly implicated in adult colds. Due to the many different types of viruses and their tendency for continuous mutation, it is impossible to gain complete immunity to the common cold.

Pertussis, also known as *whooping cough*, is a highly contagious disease caused by the bacterium *Bordetella pertussis*. It is known to last for a duration of approximately 6 weeks before subsiding. The disease derives its name from the "whoop" sound made from the inspiration of air after a cough. A similar, milder disease is caused by *B. parapertussis*. This is generally the case with infected babies and children only, not adults.

Pertussis in infants or young children is characterized initially by mild respiratory infection symptoms such as mild coughing, sneezing, or runny nose (catarrhal stage). After one to two weeks, the coughing develops into uncontrollable fits, each with five to ten forceful coughs, followed by a high-pitched "whoop" sound as the patient struggles to breathe in afterwards. Coughing fits are sometimes followed by vomiting, which can lead to malnutrition. Fits can occur on their own or can be triggered by yawning, stretching, laughing, eating or yelling; they usually occur in groups, with multiple episodes every hour around the clock. This stage lasts two to eight weeks or sometimes longer. Common complications of the disease include pneumonia, encephalopathy, earache or seizures. Adults and adolescents are the primary reservoir for pertussis. Pertussis is spread by contact with airborne discharges from the mucous membranes of infected people, who are most contagious during the catarrhal stage.

Influenza, commonly referred to as the *flu*, is an infectious disease caused by RNA viruses of the family Orthomyxoviridae (the influenza viruses), that affects birds and mammals. The most common symptoms of the disease are chills, fever, sore throat, muscle pains, severe headache, coughing, weakness/fatigue and general discomfort. Sore throat, fever and coughs are the most frequent symptoms. In more serious cases, influenza causes pneumonia, which can be fatal, particularly for the young and the elderly. Influenza may produce nausea and vomiting, particularly in children.

Typically, influenza is transmitted through the air by coughs or sneezes, creating aerosols containing the virus. Influenza can also be transmitted by direct contact with bird droppings or nasal secretions or through contact with contaminated surfaces. Influenza viruses can be inactivated by sunlight, disinfectants and detergents. As the virus can be inactivated by soap, frequent hand washing reduces the risk of infection. Influenza can be spread in three main ways: by direct transmission (when an infected person sneezes mucus directly into the eyes, nose or mouth of another person); the airborne route (when someone inhales the aerosols produced by an infected person coughing, sneezing or spitting) and through hand-to-eye, hand-to-nose, or hand-to-mouth transmission, either from contaminated surfaces or from direct personal contact such as a hand-shake.

Influenza spreads around the world in seasonal epidemics, resulting in the deaths of between 250,000 and 500,000 people every year, up to millions in some pandemic years. Three influenza pandemics occurred in the 20th century and killed tens of millions of people, with each of these pandemics being caused by the appearance of a new strain of the virus in humans. Often, these new strains appear when an existing flu virus spreads to humans from other animal species, or when an existing human strain picks up new genes from a virus that usually infects birds or pigs. An avian strain named H5N1 raised the concern of a new influenza pandemic, after it emerged in Asia in the 1990s, but it has not evolved to a form that spreads easily between people. In April 2009, a novel flu strain evolved that combined genes from human, pig and bird flu, initially dubbed "swine flu" and also known as influenza A/H1N1, emerged in Mexico, the United States, and several other nations.

Symptoms of influenza can start quite suddenly one to two days after infection. Usually the first symptoms are chills or a chilly sensation, but fever is also common early in the infection, with body temperatures ranging from 38-39 °C (approximately 100-103 °F). Many people are so ill that they are confined to bed for several days, with aches and pains throughout their bodies, which are worse in their backs and legs. Symptoms of influenza may include:

- Fever and extreme coldness (chills shivering, shaking (rigor)
- Cough
- Nasal congestion
- Body aches, especially joints and throat
- Fatigue
- Headache
- Irritated, watering eyes
- Reddened eyes, skin (especially face), mouth, throat and nose
- In children, gastrointestinal symptoms such as diarrohea and abdominal pain, (may be severe in children with influenza B)

Tuberculosis or *TB* (short for tubercles bacillus) is a common and often deadly infectious disease caused by various strains of mycobacteria, usually *Mycobacterium tuberculosis* in humans. Tuberculosis usually attacks the lungs but can also affect other parts of the body. It is spread through the air, when people who have the disease cough, sneeze, or spit.

The classic symptoms are a chronic cough with blood-tinged sputum, fever, night sweats and weight loss. Infection of other organs causes a wide range of symptoms. Diagnosis relies on radiology (commonly chest X-rays), a tuberculin skin test, blood tests, as well as microscopic examination and microbiological culture of bodily fluids. Treatment is difficult and requires long courses of multiple antibiotics. Antibiotic resistance is a growing problem in (extensively) multi-drug-resistant tuberculosis. Prevention relies on screening programs and vaccination, usually with Bacillus Calmette-Guérin vaccine.

A third of the world's population are thought to be infected with *M. tuberculosis*, and new infections occur at a rate of about one per second. In 2007 there were an estimated 13.7 million chronic active cases, 9.3 million new cases, and 1.8 million deaths, mostly in developing countries. In addition, more people in the developed world are contracting tuberculosis because their immune systems are compromised by immunosuppressive drugs, substance abuse, or AIDS. The distribution of tuberculosis is not uniform across the globe; about 80 per cent of the population in many Asian and African countries test positive in tuberculin tests, while only 5-10 per cent of the US population test positive.

When the disease becomes active, 75 per cent of the cases are pulmonary TB, that is, TB in the lungs. In the other 25 per cent of active cases, the infection moves from the lungs, causing other kinds of TB, collectively denoted extrapulmonary tuberculosis. This occurs more commonly in immunosuppressed persons and young children.

When people suffering from active pulmonary TB cough, sneeze, speak, or spit, they expel infectious aerosol droplets 0.5 to 5 µm in diameter. A single sneeze can release up to 40,000 droplets. Each one of these droplets may transmit the disease, since the infectious dose of tuberculosis is very low and inhaling less than ten bacteria may cause an infection. People with prolonged, frequent, or intense contact are at particularly high risk of becoming infected, with an estimated 22 per cent infection rate. A person with active but untreated tuberculosis can infect 10–15 other people per year. Others at risk include people in areas where TB is common, people who inject drugs using unsanitary needles, children exposed to adults in high-risk categories, patients immunocompromised by conditions such as HIV/AIDS, people who take immunosuppressant drugs, and health care workers serving these high-risk clients.

Asthma is a common chronic inflammatory disease of the airways characterized by variable and recurring symptoms, airflow obstruction, and bronchospasm. Symptoms include wheezing, cough, chest tightness, and shortness of breath. Although asthma is a chronic obstructive condition, it is not considered as a part of chronic obstructive pulmonary disease as this term refers specifically to combinations of bronchiectasis, chronic bronchitis, and emphysema. Unlike these diseases, the airway obstruction in asthma is usually reversible; however, if left untreated, asthma can result in chronic inflammation of the lungs and irreversible obstruction.

Asthma exists in two states: the steady-state of chronic asthma, and the acute state of an acute asthma exacerbation. The symptoms are different depending on what state the patient is in. Common symptoms of asthma in a steady-state include:

night time coughing, shortness of breath with exertion but no dyspnea at rest, a chronic 'throat-clearing' type cough and complaints of a tight feeling in the chest. Severity often correlates to an increase in symptoms.

Asthma is caused by environmental and genetic factors. It affects 7 per cent of the population of the United States, 6.5 per cent of British people and a total of 300 million worldwide. Medicines such as inhaled short-acting beta-2 agonists may be used to treat acute attacks. Attacks can also be prevented by avoiding triggering factors such as allergens or rapid temperature changes and through drug treatment such as inhaled corticosteroids. Leukotriene antagonists are less effective than corticosteroids, but have fewer side effects. Monoclonal antibodies, such as mepolizumab and omalizumab, are sometimes effective.

Pneumonia is an abnormal inflammatory condition of the lung. It is often characterized as including inflammation of the parenchyma of the lung (that is, the alveoli) and abnormal alveolar filling with fluid (consolidation and exudation). The alveoli are microscopic air-filled sacs in the lungs responsible for gas exchange. Pneumonia can result from a variety of causes, including infection with bacteria, viruses, fungi, or parasites, and chemical or physical injury to the lungs. The most common causes of pneumonia are viruses and bacteria. Less common causes of infectious pneumonia are fungi and parasites.

Pneumonia is a common illness which occurs in all age groups, and is a leading cause of death among the elderly and people who are chronically and terminally ill. Additionally, it is the leading cause of death in children under five years old worldwide. Vaccines to prevent certain types of pneumonia are available. People with infectious pneumonia often have a cough producing greenish or yellow sputum, or phlegm and a high fever that may be accompanied by shaking chills. Shortness of breath is also common, as is pleuritic chest pain, a sharp or stabbing pain, either experienced during deep breaths or coughs or worsened by them. People with pneumonia may cough up blood, experience headaches or develop sweaty and clammy skin. Other possible symptoms are loss of appetite, fatigue, blueness of the skin, nausea, vomiting, mood swings, and joint pains or muscle aches.

Catarrh is a thick exudate of mucus and white blood cells caused by the swelling of the mucous membranes in the head in response to an infection. It is a symptom usually associated with the common cold and chesty coughs, but can also be found in patients with infections of the adenoids, middle ear, sinus or tonsils. The catarrh may either discharge or cause a blockage which may become chronic.

Croup is a group of respiratory diseases that often affect infants and children under age 6. It is characterized by a barking cough; a whistling, obstructive sound as the child breathes in and hoarseness due to obstruction in the region of the larynx. It may be mild, moderate or severe, and severe cases, with breathing difficulty, can be fatal if not treated. Another type of croup is known as spasmodic croup. People with spasmodic croup first catch a cold, rarely with fever, and then the croupy cough begins. In some cases spasmodic croup may begin suddenly without any preceding cold symptoms. Unlike viral croup, spasmodic croup usually recurs and can occur in older children, and rarely even in adults.

Croup is most often caused by parainfluenza virus, primarily types 1 and 2. Approximately 75 per cent of cases are caused by parainfluenza virus. Influenza A and B, measles, adenovirus and respiratory syncytial virus (RSV) are other viruses that sometimes cause croup. It is most common in the autumn and winter but can occur year-round.

Sore throat or *throat pain* is a common physical symptom usually caused by acute pharyngitis or throat inflammation, however occurs from a number of other situations such as post trauma. It can be extremely painful. A sore throat is usually from irritation or inflammation. The most common cause is acute viral pharyngitis (80 per cent) which is a viral infection of the throat. Other causes include other infections, trauma and tumors.

The present review aims at discussing the plant wealth available to treat respiratory disorders and includes the works of Gamble (1916-1935), Dastur (1962), Krishnam Nambiar *et al.* (1982), Agarwal and Ghosh (1985), Thammanna and Narayana Rao (1990), Bhattarai (1993), Varier (1994-1995), Yoganarasimhan (1996), Jain (1997), Pal and Jain (1998), Christopher *et al.* (1999), Bhattacharjee (2001), Pullaiah (2002), Kirtikar and Basu (2003), Seetharami Reddi *et al.* (2005), Patil (2005) and Suneetha *et al.* (2009).

ENUMERATION

1. *Abelmoschus esculentus* (L.) Moench (Malvaceae)
Syn: *Hibiscus esculentus* L.

VN: Benda S: Bhenda H: Bhindi E: Lady's finger

Bronchitis and pneumonia: Decoction of leaves is used.

Tuberculosis: Roots with leaves of *Andrographis paniculata* and *Azadirachta indica* are taken in equal quantities and ground. Two spoonfuls of paste mixed in a glass of cow's milk is administered daily twice for 30 days.

2. *A. manihot* (L.) Medic. (Malvaceae)
Syn: *Hibiscus manihot* L., *H. pungens* Roxb.

VN: Peda benda

Tuberculosis: A spoonful of root extract along with a glass of hot water is administered daily once for a month.

3. *Abies spectabilis* (D. Don) Mirb. (Pinaceae)
Syn: *A. webbiana* Lindl.

VN: Talisapatri S: Talisam H: Talisapatra E: East Himalayan silver fir

Whooping cough: About ten g of the fresh leaf tops are boiled in 400 ml of water, until it is boiled down to half this amount. The resultant decoction with a little sugar or salt is administered orally warm twice a day for 3-5 days.

4. *Abroma angusta* L. f. (Sterculiaceae)
H: Kumal E: Devil's cotton

Tuberculosis: One glassful decoction of root is taken twice a day for a month.

5. *Abrus precatorius* L. (Fabaceae)

Syn: *Glycine abrus* L.

VN: Guruvinda S: Gunja H: Gunj E: Crab's eye

Asthma: A spoonful of leaf juice is administered daily twice for 3 days.

Roots along with the roots of *Azima tetracantha* and black pepper are pounded and 3 spoonfuls of powder is given per day for a period of 3 days.

Young leaves are chewed on empty stomach.

Leaves along with those of *Cymbopogon giganteus*, palm oil and water are boiled for few minutes and decanted. This liquid is taken orally twice a day.

Half teaspoonful of root powder is taken every morning with a cup (50 ml) of water till cure.

Roots are boiled in water and the decoction is administered orally till cure.

Cough: Decoction of the root is given.

About 5 ml of leaf juice is given to children in the morning and evening for 3 days.

Roasted leaves are made into powder which is taken with honey twice a day.

Cold and cough: Leaves along with young leaves of *Psidium guajava* and powdered seeds of black pepper is effective.

6. *Abutilon indicum* (L.) Sweet (Malvaceae)

Syn: *Sida indica* L.

VN: Tuttura benda S: Atibala H: Tepari E: Indian abutilon

Bronchitis: Decoction of leaves is used.

Asthma: Two to five drops of extract of leaves with honey is given thrice a day.

Cold and cough: Seeds along with leaves of *Adiantum caudatum* and young leaves of *Psidium guajava* in the form of decoction is given.

7. *Acacia catechu* (L.f.) Willd. (Mimosaceae)

Syn: *Mimosa catechu* L. *f.*

VN: Podali manu S: Khadirah H: Khair E: Cutch tree

Cough: Two to three teaspoonfuls of decoction of stem (without bark) is given twice a day.

Cough and throat infection: About ten g of the mature wood is boiled in 500 ml of water, until it is boiled down to half this amount. The warm decoction is orally administered 2-3 times a day for 2-3 days to relieve throat infection and cough.

Warm decoction of mature wood (about 50 g boiled in a cup of water) is taken 2-3 times a day for 3-4 days.

8. *A. leucophloea* (Roxb.) Willd. (Mimosaceae)

Syn: *Mimosa leucophloea* Roxb.

VN: Tella tumma S: Arimedah H: Safed babul E: White babool

Cold and cough: Bark decoction is used.

Stem bark is squeezed and inhaled.

9. *A. nilotica* (L.) Willd. ex Del. ssp. *indica* (Benth.) Brenan (Mimosaceae)
Syn: *A. arabica* auct. (Lam.) Willd.

VN: Nalla thumma S: Babbuka H: Babul E: Black babool

Cold and whooping cough: Fifty to hundred g of bark is made into decoction in water with a small quantities of salt and 'Gur' (25 g of solidified sugarcane juice) and one teaspoonful of it is given twice a day.

10. *A. pennata* (L.) Willd. (Mimosaceae)
Syn: *Mimosa pennata* L.

VN: Korinta, Teega Indiga S: Valli khadira H: Agla-bel

Cold and cough: Two spoonfuls of stem bark paste mixed with a spoonful of honey is administered daily twice for 3 days.

Asthma: Fresh stem sap is sucked daily once till cure.

Croup: Leaves with tender leaves of *Bauhinia vahlii* and *Asparagus racemosus* are taken in equal quantities and ground. Paste is tied with cloth and bandaged around the hip till cure.

Whooping cough: Pulp of 2 ft tendril is mixed with a half glass of water and administered once a day for 5 days.

11. *A. sinuata* (Lour.) Merr. (Mimosaceae)
Syn: *Acacia concinna* (Willd.) DC.

VN: Sikaya S: Saptala H: Satala E: Soapnut-acacia

Asthma: Juice of green bark is given in small doses daily for few days as an effective remedy.

12. *A. torta* (Roxb.) Craib (Mimosaceae)
Syn: *Mimosa torta* Roxb.

VN: Korintha chekka

Whooping cough: Ten g of stem sap with 2 spoonfuls of mother's milk is administered to children twice a day till cure.

13. *Acalypha indica* L. (Euphorbiaceae)
VN: Kuppinta S: Haritamanjari H: Kuppikhokli E: Indian Acalypha

Asthma: Ten ml of plant juice mixed with 5 ml of old ghee is massaged gently over the chest of children to get relief.

About four teaspoonfuls of whole plant juice is administered orally. In the case of children the dose is reduced to 1 teaspoonful till cure.

Fresh juice of the leaves is given orally till cure.

Asthma and bronchitis: One g powder of whole plant with a few drops of honey is given twice daily for 4-5 days (or till required) for cure in children.

Asthma, bronchitis, pneumonia and whooping cough: Decoction of leaves is administered. It accelerates removal of mucous from the bronchial tubes.

Cold and cough: Decoction of the plant is mixed with jaggery and used till cure.

Cough: One cup extract of leaves is taken once in a day for 2 days.

Flowers with leaves of *Piper betle* are taken in equal quantities and ground. A spoonful of paste is administered daily twice for 3 days.

Whooping cough: Plant is crushed and juice is extracted. Nearly 10 ml is administered twice a day along with honey.

14. *Achyranthes aspera* L. (Amaranthaceae)
VN: Uttareni S: Apamargah H: Circita E: Prickly chaff flower plant

Asthma and cough: Five ml of whole plant juice is used twice daily for 2-3 days in children.

Five ml of whole plant juice is used twice daily to cure.

Curry prepared from crushed seeds is given orally.

Asthma: The ash of burnt dried plants is given with honey twice a day for 7 days.

Powder of roasted seeds is taken orally with warm water.

About two g powder of leaves and roots along with 4-5 fruits of black pepper is taken orally.

Whole plant ash is mixed with the seed powder of *Calotropis gigantea* and made into small pills with lemon juice and are given 2 times daily.

Dried leaf vapours are inhaled.

Two to six g of plant ash along with honey is given till cure.

Cough: Half teaspoon each of roasted plant powder and honey is taken with warm water at bedtime.

One to two teaspoonfuls of root ash is given twice a day for 5 days to children.

Ten to fifteen g of root ash is given to children.

Tuberculosis: About ten g of fresh root with 100 g onion are fried in 200 ml of mustard oil and peanut size pills are made. Two pills are given twice a day.

Whooping cough: Root ash is suspended in water and decanted. Water soluble portion is dried and about 2 g of it is taken.

15. *A. bidentata* Bl. (Amaranthaceae)
Whooping cough: Ash of the plant mixed with honey is given.

16. *Aconitum violaceum* Jacq. ex Stapf (Ranunculaceae)
Cold: One to two spoonfuls of the root extract is very beneficial.

17. ***Acorus calamus*** **L. (Araceae)**

VN: Vasa S:Vacha H: Bach E: The Sweet flag

Asthma: Five ml decoction of rhizome is used twice daily till cure in children.

Powdered root with a cup of hot water is used.

About 10 ml of whole plant infusion is administered orally for every 2-3 hours till cure.

Rhizome along with roots of *Momordica dioica* and *Solanum virginianum* are crushed and the juice is administered orally till cure.

Cough: Roots are hung around the neck.

Root along with that of *Boerhavia diffusa* are taken in equal quantities and ground and then boiled with ghee. A spoonful of paste is administered daily once for 3 days in children.

Cough and sore throat: Small pieces of rhizome are chewed for instant cure.

Two grams of the rootstocks are boiled in 250 ml of water ratio. with a little salt and the decoction is orally administered twice a day, in the morning and before bed-time warm for 3 or more days.

A decoction is prepared from 1 g of the rootstock with 5 g of the crushed leaves of neem with salt and drunk twice a day in the morning and before bedtime for 2 days.

Small pieces of rhizome are chewed for instant cure.

Lung infection: Rhizome paste is used.

Cold: Small cut pieces of the rhizome together with the bulbs of *Allium sativum* are made into chain and prescribed to child to wear as garland on neck.

Rhizome is roasted and powdered and mixed with honey and a pinch of it is taken.

Asthma and pneumonia: Five ml of decoction of rhizome is used twice daily till cure.

18. ***Actiniopteris radiata*** **(Sw.) Link. (Actiniopteridaceae)**

Cough: Plants are soaked overnight in a glass of water and taken orally in morning for control of tuberculosis. Plants are dried and one teaspoonful powder is taken orally, once a day for four days.

Tuberculosis: Leaves are made into paste along with *Teramnus labialis* and jaggery in equal quantity. This is made into pills of bengal gram seed size. These are given twice daily for about 10-15 days.

Whole plant is soaked in water for 24 hrs, filtered and one glass of filtrate is given orally early in the morning on empty stomach.

19. ***Adenanthera pavonina*** **L. (Mimosaceae)**

VN: Bandigurvinda S: Tamrakah H: Baragunci E: Coral wood tree

Asthma: Root bark with seeds of *Withania somnifera* and tuberous roots of *Asparagus racemosus* are taken in equal quantities and ground. Two spoonfuls of paste mixed with a spoonful of honey is administered daily twice for 5 days.

20. *Adhatoda zeylanica* Medic. (Acanthaceae)

E: Malabar Nut H: Arusha S: Vasaka

Cold and cough: Two teaspoonfuls decoction of leaves and *Ocimum sanctum* leaves mixed in 2:1 ratio is given thrice a day for 3-5 days.

Dried powder of leaves with honey twice a day is an excellent remedy.

Leaf powder is mixed with water and taken orally in the morning.

Leaf juice with common salt and honey (2:1:2) is given as cure for children.

Leaf juice of 5 ml is administered twice daily.

Asthma: Leaves with the roots *of Solanum surattense* and fruits of *Piper longum* are taken in equal proportions and made into powder and it is taken with honey in doses of one spoonful twice a day for about a week.

About 20 g of the leaves are crushed and boiled in 500 ml of water, until it is reduced to half the amount. The warm decoction is drunk twice a day for a week or longer.

Half Kg juice of this plant, ½ Kg jaggery, ¼ litre honey are mixed and cooked in an earthen pot, reduced to ½ litre and again ½ Kg jaggery is added and boiled for 4 hrs. One teaspoonful of the paste is given orally twice a day till cure.

Decoction of roots and leaves is taken in doses of 10-15 ml 3-4 times daily till cure.

Half teaspoonful of leaf juice is administered daily once till cure.

Twenty ml of leaf decoction is taken along with honey and long pepper till cure.

Dried leaf or heartwood vapours are inhaled till cure.

Leaves are smoked as cigarettes.

Leaf paste is taken orally to cure asthma for about 40 days.

Dried inflorescence is taken and burnt. The vapours are inhaled till cure.

About 10 ml of the leaf juice is given with equal amount of honey twice a day for a week.

About 20 g of the leaves are crushed and boiled in 500 ml of water till it is reduced to half. The warm decoction is drunk twice a day for a week.

A decoction prepared from about 5 g of the flowers and tender leaves is drunk warm with honey regularly till cure.

A decoction of about 2 Kg is prepared from flowers and leaves by boiling in 250 ml water. To this 50 g of ammonium chloride and 50 ml leaf extract of *Datura metel* are added. About 12 ml of this drug thus obtained is given thrice a day for 5 days.

Tender shoot is boiled in water and the concentrated solution is given till cure.

Two teaspoonfuls of juice of leaf or root mixed with one spoonful of ginger juice is given orally for few days.

Flowers and shoots are crushed along with leaves of *Ocimum sanctum* in equal quantities and the extract juice mixed with equal quantity of honey is given in doses of two teaspoonfuls twice daily for a week.

Ten to twenty g of root bark powder or leaf powder is mixed with honey and is administered orally once a day.

One tablespoonful leaf juice or decoction mixed with black pepper powder is administered orally for every half an hour.

Leaf and wood ash mixed with honey is taken internally till cure.

Juice of a mixture of fresh leaves and ginger with salt is used once daily for 5-20 days.

Roots are ground with dried ginger and tablets of 10 g each are prepared. These are administered orally twice a day till cure.

Shade dried leaf powder is mixed with equal proportions of honey and is given 3-4 times a day.

Warm decoction of leaves (about 5 leaves crushed and boiled in a cup of water) is given twice a day for a month.

About two teaspoonfuls of leaf juice is taken with honey twice a day for a week as expectorant.

Cough, cold and catarrh: Leaf juice, long pepper powder and 2 teaspoonfuls of honey is taken on empty stomach for 2-3 days.

Bronchitis: A decoction prepared from about 5 g of the flowers and tender leaves is drunk warm with honey regularly to treat.

Asthma, bronchitis and cough: Five to fifteen ml of decoction of the leaves and roots is taken orally 2-3 times daily.

Cough. Decoction is prepared by adding leaves with sugar and ginger. It is taken orally thrice a day to cure cough.

One teaspoonful of extract of leaves with half the quantity of honey is given thrice a day.

Catarrh and cough: Vapours emerging from the decoction of the leaves and root is inhaled deeply for 2-3 minutes in the morning and evening.

Asthma and cough: Hundred g of dried leaf powder smoked as cigarette twice a day for 15 days.

Tuberculosis: Decoction of leaves along with jaggery and water are kept in an earthen pot for a month. This is taken orally in a sufficient quantity till cure.

21. ***Adiantum capillus–veneris*** **L. (Adiantaceae)**

E: Maiden-hair Fern H: Hansraj S: Brahmadani

Cough: Two small cups of plant decoction is administered per day till cure.

22. ***A. incisum*** **Forsk. (Adiantaceae)**

Bronchial diseases: Infusion of young fronds is used.

Cold and cough: One teaspoonful decoction of rhizomes is given thrice a day for 7 days.

23. ***Aegle marmelos*** **(L.) Correa (Rutaceae)**

Syn: *Crataeva marmelos* L.

VN: Meredu S: Vilvah H: Bel E: Bael tree

Asthma: Decoction of leaves help to remove mucous from the bronchial tubes and relieves asthma.

Four tender shoots are chewed and the sap is swallowed daily once for 40 days.

Fresh leaf juice is mixed with half spoon of honey and is taken internally till cure.

Leaf decoction of about 2-5 ml or leaf juice of about 2-3 ml is given till cure.

Cold and cough: Leaf juice is given along with honey.

24. ***Agave americana*** **L. (Agavaceae)**

E: Century plant H: Kantala S: Kalakantala

Pulmonary tuberculosis: Sap obtained from the plant is drunk warm twice a day for a week or still longer till relief.

25. ***Ageratum conyzoides*** **L. (Asteraceae)**

VN: Pumpullu S: Visamustih H: Visadodi E: Goat weed

Catarrh: Two to three drops of the leaf juice are used as a nasal drops 2-3 times a day for 2-3 days to treat catarrh. It is said to increase the expulsion of phlegm. Based upon the severity of the ailment, the juice may be prescribed for 5 or more days.

Dried leaves are powdered and inhaled as snuff 4-6 times a day for 3 or more days.

Catarrh and cold: About 500g of leaves are ground with 50 g of jaggery and made into pills of bengal gram seed size. Six pills are administered at a time twice a day for 2 days. Take a fibre cloth bed and cover the bed with leaves of this plant. A stove with burning coals is arranged under the bed and patient is made to sleep on that bed for 30 minutes to get relief.

26. ***Ailanthus excelsa*** **Roxb. (Simarubaceae)**

VN: Pedda manu S: Mahanimba H: Arua E: Tree of heaven

Bronchitis: Powder prepared from the resin is mixed with milk and administered.

Asthma and cough: Decoction of stem bark is given.

Asthma: Thirty to fifty ml of bark infusion is administered orally twice daily till cure.

27. *Ajuga bracteosa* Wall. ex Benth. (Lamiaceae)

Bronchitis and pneumonia: Infusion of leaves is used.

28. *Alangium salvifolium* (Linn. f.) Wang. (Alangiaceae)

Syn: *A. lamarkii* Thw., *A. decapetalum* Lam.

VN: Uduga chettu S: Ankolah H: Angol E: Sage-leaved alangium

Tuberculosis: Ten to thirty ml bark decoction is given once a day for 10-15 days.

29. *Albizia amara* (Roxb.) Boivin (Mimosaceae)

Syn: *Mimosa amara* Roxb.

VN: Chikireni, Nalla regi S: Krishna sirish

Asthma: Flowers and seeds are consumed till cure.

30. *A. lebbeck* (L.) Benth. (Mimosaceae)

Asthma: Flowers are given till cure.

31. *A. odoratissima* (L. f.) Benth. (Mimosaceae)

Syn: *Mimosa odoratissima* L. f.

VN: Chinduga S: Bhusirisah H: Kalasiris E: Black siris

Cough: A spoonful of leaf juice is administered daily thrice for 2 days.

32. *Allium cepa* L. (Liliaceae)

VN: Ullipaya H: Pyaj S: Palanduh E: Onion

Asthma and bronchitis: Raw onion bulb is consumed.

Cold: Three bulbs are mashed after covering with a cotton cloth and smelled.

33. *A. hookeri* Thw. (Liliaceae)

Cold and cough: A pounded bulb mixed with oil is applied on the throat and chest.

34. *A. hypsis* Stearn (Liliaceae)

Catarrh: About 300 ml of decoction prepared from 0.5 g of the dried leaves and 3 g of the leaves of *Ocimum sanctum* with common salt is drink twice a day for 3-4 days.

Catarrh and cough: About 500 ml of decoction is prepared from the leaves of *Allium hypsis*, stem bark of *Cinnamomum zeylanicum* and common salt about 1 g of each and the rhizomes of *Curcuma domestica* and *Zingiber officinale* and fruits of *Zanthoxylum armatum* about 2 g of each. The resultant decoction is drunk warm, inhaling the vapour through mouth and nostrils as well. It is administered before bedtime for 3 or more days to cure catarrh and cough, to control excessive flow of mucous through the nostrils, and to loosen the sputum to facilitate expectoration. Occasionally, the stem bark of *Cinnamomum tamala* is substituted for that of *Cinnamomum zeylanicum*.

35. *A. sativum* L. (Liliaceae)

VN: Tella gadda S: Lasunah H: Lasun E: Garlic

Inflammation of respiratory tract: Decoction of bulb and coconut oil together are used.

Cough: About ten ml of juice extracted from the bulb is mixed with equal amount of honey and taken twice a day for 2 or more days as an expectorant.

Bulb is fried in mustard oil and the oil is used for massaging newborns and is also taken orally.

Bulb is also taken with hot milk.

Cold and cough: Bulb paste is applied on a cloth piece which is folded so that the paste is covered from both the sides. Now this piece of cloth with paste inside is tied around neck before bed time.

Asthma: A bulb of garlic is peeled, crushed and boiled in 100 ml of vinegar. After cooling, an equal quantity of honey is mixed and preserved in a clean bottle. Two teaspoonfuls of syrup is taken in the evening and before bed time till cure.

Three cloves are boiled in milk and administered every night till cure.

Ten cloves are boiled in 30 ml of milk and this mixture is administered daily once till cure.

Steaming tea with 2 minced cloves is taken in the morning and evening till cure.

36. *Aloe vera* (L.) Burm. f. (Liliaceae)

Syn: *A. barbadensis* Mill.

VN: Kala bandha S: Ghratkumari H: Guarpatha E: Indian aloe

Asthma and cough: Twenty g of mucilaginous jelly from leaves is eaten twice a day for 2 days.

37. *Alpinia calcarata* Rosc. (Zingiberaceae)

Bronchial catarrh: Young shoots are cooked as vegetable and about 10 g rhizome in powdered form is given.

38. *A. sericeum* Roxb. (Zingiberaceae)

Asthma: About 10 g of seeds are taken twice daily.

39. *Alysicarpus bupleurifolius* (L.) DC. (Fabaceae)

Asthma: Ten to thirty ml of root extract is given twice a day for 7-9 days.

40. *A. monilifer* DC. (Fabaceae)

Cold and cough: A decoction of the leaves is said to be beneficial.

41. *Ammania baccifera* L. (Lythraceae)

VN: Agnivendrapaku H: Kuranta S: Kurandika E: Blistering ammania.

Cough: Leaf decoction is given.

42. ***Amorphophallus commutatus*** **Engl. (Araceae)**

Syn: *A. sylvaticus* Dalz. & Gibs.

S: Surana H: Vanasuran

Asthma: Tuberous chips are dried and made into powder and given orally till cure.

43. ***A. paeoniifolius*** **(Dennst.) Nicolson var.** ***campanulatus*** **(Decne.) Sivadasan (Araceae)**

Syn: *A. campanulatus* (Decne.) Sivadasan

VN: Siri kanda S: Suranah H:Suranah E: Elephant foot yam

Asthma: One teaspoonful ash of the root with one table spoonful of honey is administered once a day in early morning for one month.

44. ***Ananas comosus*** **(L.) Merr. (Bromeliaceae)**

Syn: *Bromelia comosa* L., *Ananas sativus* Schult. f.

VN: Anas pandu S: Anamnasam H: Ananas E: Pineapple

Asthma: About 15-20 ml decoction of the fruits and of the stalk of *Colocasia esculenta* is mixed with sugar and taken till cure.

Fruit juice mixed with honey and crushed bulb of *Allium sativum* is given thrice daily until cure.

One glass of leaf decoction is taken orally twice a day till cure.

45. ***Andrographis paniculata*** **(Burm. f.) Wall. ex Nees (Acanthaceae)**

Syn: *Justicia paniculata* Burm.f., *Andrographis subspathulata* Clarke

VN: Nelavemu H: Kalamegh S: Kirtatiktah E: Green chiretta

Asthma: Stem is mixed with that of *Gymnema sylvestre* and *Justicia adhatoda* leaves, ground and the infusion is given orally till cure.

Decoction is made from whole plant parts of this plant, *Evolvulus alsinoides* and *Mollugo pentaphylla* taken in equal proportions. This is administered 2 or 3 times a day till cure.

A spoonful of whole plant (leaf, stem and flower) paste mixed in half glass of hot water is administered daily twice for 2 days.

Cold and cough: Leaf powder is administered internally.

Shoot along with stem bark of *Mimusops elengi* is boiled in water and decoction is given orally.

Decoction of leaf and seed is diluted in water and filtered. The filtrate with a little honey added is taken thrice a day.

Cough: Whole plant is dried and burnt in a closed clay container and the ash is used with honey.

Crushed leaves mixed in warm water is administered.

46. *Angelica glauca* Edgew. (Apiaceae)

Acute bronchitis: About one teaspoonful dried powder of the root with hot water is given orally twice a day for a week.

47. *Anisochilus carnosus* (L.f.) Wall. ex Benth. (Lamiaceae)

VN : Kapuravalli H: Panjirikapat

Cold and cough: Juice of the fresh leaves mixed with sugar candy is used.

48. *Anisomeles indica* (L.) O. Kuntze (Lamiaceae)

Syn: *A. ovata* R. Br., *Nepeta indica* L.

VN: Ada beera S: Vaikunthah H: Gopali E: Devil fraghtener

Cold: Dried leaves are rubbed and smoked.

Cold and catarrh: Leaves are pounded with turmeric powder and camphor and this mixture is put on fire, the smoky fumes of it are inhaled in case of male children.

49. *Annona squamosa* L. (Annonaceae)

VN: Seethapalam S: Krishnabija H: Sitaphal E: Custard apple

Asthma: Bark juice mixed with a glass of water is taken orally twice a day till cure.

Ten leaves and 30 g of bark are boiled in a glass of water till it reduces to half glass and is given orally once a day till cure.

50. *Anogeissus latifolia* (Roxb. ex DC.) Wall. ex Guill. & Perr. (Combretaceae)

Syn: *Conocarpus latifolia* Roxb. ex DC.

VN: Chirumanu S: Dhavah H: Bankli E: Button tree

Cough: Two tablespoons of bark extract with the internodes of *Cissus quandragularis* are administered for a week.

Bark extract is taken in doses of 3 spoonfuls thrice a day for 3 days.

About one teaspoonful of juice of stem bark is given thrice a day for about a week.

Dry cough: Decoction of the bark is pounded with black pepper and is given.

51. *Anthocephalus chinensis* (Lam.) A. Rich ex Walp. (Rubiaceae)

Cough: Stem bark and leaves of *Andrographis paniculata* are taken in equal quantities and ground. Two spoonfuls of juice is administered daily twice for 3 days.

52. *Ardisia humilis* A. DC. (Myrsinaceae)

Asthma: Leaves and flowers are powdered along with black pepper and garlic and infusion is prepared. This is taken in doses of one spoonful daily in the morning till cure.

Root bark paste with dried flower powder of *Bauhinia purpurea* with goat milk in the ratio of 4:3:4 is administered orally till cure.

Whooping cough: Infusion of leaves and flowers is drunk.

53. *Areca catechu* L. (Arecaceae)

VN: Vakka H: Supari S: Pugah E: Betelnut tree.

Cough: Seed is semi-roasted on a gentle fire then cut into small pieces and chewed after each meal regularly.

54. *Argemone mexicana* L. (Papaveraceae)

VN: Balla rakasi S: Svarnaksiri H: Bhatbhamt E: Mexican poppy

Whooping cough: Five ml of plant juice is used once a day for 3 days.

55. *Aristolochia bracteolata* Lam. (Aristolochiaceae)

VN: Gadidagadapaku S: Kitamari H: Kiramar E: Bracteated birthwort

Asthma: Leaves together with roots of *Melothria heterophylla* and the fruits of *Solanum nigrum* are ground properly and mixed with water are given orally twice a day.

56. *A. indica* L. (Aristolochiaceae)

VN: Nalla eswari S: Isvari H: Isarmul E: Indian birthwort

Dry cough: Dried seed powder with honey (3:1) is used.

Asthma: Three mature leaves are chewed on empty stomach for 21 days and about 3-4 fresh leaves are chewed for atleast 15 days.

57. *Artemisia dubia* Wall.ex Besser (Asteraceae)

E: Mug-wort

Asthma: Five pieces of root (each about 5 cm long) are boiled in two cups of water and warm decoction taken twice a day for a week.

Five g of the root are boiled in 400 ml of water until reduced to half this amount then drunk warm twice a day for a week or longer.

Cold and cough: Crushed leaves are smeared on throat and at the same time, 2-3 drops of the leaf juice are given with milk 3-4 times each with an interval of 3-4 hrs.

58. *A. gmelinii* Web. ex Stechm. (Asteraceae)

Cold and cough: Extract of leaves and flowers is given in small doses for a week.

59. *A. nilgirica* (Clarke) Pamp. (Asteraceae)

Syn: *A. vulgaris* auct. non L.

VN: Masipattri H: Davana S: Damanah E: Fleabane

Cold and cough: The warm infusion of leaves is tied around the chest with the help of long cloth or bandage.

Asthma and bronchitis: Inhalation of smoke from leaves and flowering twigs is effective.

Asthma: Stem and flowering tops are used as bitter tonic after putting in hot water.

60. *Artocarpus heterophyllus* Lam. (Moraceae)
Syn: *A. integrifolius* auct. non L.

VN: Panasa S: Panasah H: Kathal E: Jack fruit tree

Asthma: Root decoction is used.

Tuberculosis: Edible part of the fruit is taken and kept in a mud pot with jaggery in alternate layers, then it is sealed with a thick cloth and is kept in sunlight for 21 days and later it is pound into a paste. Two spoonfuls of paste is given once a day on an empty stomach for 21 days.

61. *Asparagus racemosus* Willd. (Liliaceae)
VN: Pilli theegalu S: Satavari H: Satavar E: Wild asparagus

Cold and cough: Tuber extract along with ginger is given.

62. *Astragalus hosackioides* Benth. (Fabaceae)
Chest infection and whooping cough: Decoction of the root is given.

63. *Azadirachta indica* A. Juss. (Meliaceae)
Syn: *Melia azadirachta* L.

VN: Vepa S: Nimbha H: Nimb E: Margosa tree

Asthma: Seed oil is given in doses of 10 ml orally till cure.

Cold and cough: Leaves along with tubers of *Dioscorea pentaphylla* are taken in equal quantities and ground. Two spoonfuls of paste mixed with a spoonful of honey is administered daily once for 3 days. Meanwhile paste is soaked in hot water and is inhaled daily once for 3 days.

Cough: Paste of one Kg leaves is soaked in water and applied on the body at night for 2-3 days.

64. *Azima tetracantha* Lam. (Salvadoraceae)
S: Lavanee H: Kantagurkamii

Asthma: Root along with honey is made into paste or its powder mixed honey is administered orally till cure.

65. *Bacopa monnieri* (L.) Pennell (Scrophulariaceae)
Syn: *Lysimachia monnieri* L., *Gratiola monnieri* (L.) L., *Monniera cuneifolia* Michx.
VN: Sambranichettu S: Brahmi H: Barami E: Thyme leaved gratiola

Asthma: A spoonful of whole plant juice with a spoonful of seed powder of *Terminalia chebula* mixed in a glass of hot water is administered twice a day before meals.

Half spoon of decoction of leaf is administered daily twice for one week in children.

Dried plant powder is given internally.

Cough: A poultice of boiled plant is applied externally on child's chest.

A spoonful of whole plant juice with a spoonful of seed powder of *Terminalia chebula* mixed in a glass of hot water is administered twice a day before meals.

Bronchitis: Leaf juice is given in doses of 1 teaspoonful till cure.

Cold: Plant juice of 5 ml is given with half cup of lukewarm cow's milk twice daily for 2-3 days.

66. *Balanites aegyptiaca* (L.) Del. (Balanitaceae)

Syn: *B. roxburghii* Planch.

S: Indugi H: Hingan E: Hingan

Asthma: About 3 spoonfuls of bark paste is administered orally along with honey thrice a day for 3 days.

67. *Balsamodendron myrrha* Bal. (Burseraceae)

Asthma: Gum from the bark is used as a stimulating expectorant.

68. *Bambusa arundinacea* (Retz.) Roxb. (Bambusaceae)

Syn: *Bambusa bambos* (L.) Voss., *Arundo bambos* L.

VN: Veduru S: Tavakshiri H: Kanta baus E: Bamboo

Cough: Leaf juice (with water) 10 ml with a little honey is used twice daily 5-6 days in children.

69. *B. vulgaris* Schrad. (Poaceae)

E: Bamboo H:Udha-bans S: Bahupallava

Respiratory diseases: Consumption of young shoots is beneficial.

70. *Barleria cristata* L. (Acanthaceae)

S: Ragaprasava H: Tadrelu E: Crested purple nail dye

Bronchitis and bronchial asthma: One teaspoonful of leaf juice (about 20 leaves crushed by mortar and pestle) is taken with water twice a day for 15 days.

Asthma: About 5 ml of the leaf juice is diluted in water and drunk twice a day regularly till cure.

Four to five g of leaves are dried and powdered and is administered daily once a day till cure.

Cough: Infusion of roots is given.

Whooping cough: Small quantity of ash of the plant with honey and 'Desi ghee' is given.

Tuberculosis: Three tablespoons of whole plant decoction given twice a day for six months.

71. *B. prionitis* L. (Acanthaceae)

VN: Mullagorinta S: Kantakuranta H: Kat-sareya E: Thronnail dye

Asthma: A spoonful of leaf juice with powder of 3 seeds of *Piper nigrum* mixed in a glass of hot milk is administered daily once for a week.

Leaves and roots are powdered and given with milk till cure.

About 4-5 g of leaves are dried and powdered and taken once a day till cure.

Whooping cough: Dried stem bark powder is given to children.

Cough: About 5 g of root powder is boiled with 200 ml water till it is reduced to 50 ml. Five ml of decoction is given three times a day with honey.

Decoction of the bark mixed with honey or jaggery is administered orally.

Catarrh: Juice of the leaves is generally administered with a little honey or sugar in children.

Bronchitis: Root decoction is taken in doses of 3 teaspoonfuls twice daily for about a week.

Cold and cough: Two spoonfuls of leaf paste mixed with a glass of water and kept for overnight. It is administered daily once for 3 days.

72. *Barringtonia acutangula* (L.) Gaertn. (Barringtoniaceae)
E : Indian Oak H: Samudraphal S: Hijjala

Congestion of respiratory tract: Fruit paste prepared in water after rubbing is applied on chest.

73. *Basella rubra* L. (Basellaceae)
VN : Allabatsalla H: Bonpoi S : Apodika E: Indian spinach

Asthma: Twenty ml of leaf juice mixed with goat's milk (15 ml) and honey (10 ml) is taken orally 2-3 times daily for a week.

74. *Bauhinia manandra* Kurz (Caesalpiniaceae)
Cold: Decoction prepared from crushed leaves is taken.

75. *B. purpurea* L. (Caesalpiniaceae)
Syn: *Phanera purpurea* (L.) Benth.

VN: Bhodinti koora S: Kanchan H: Lal kanchan E: Pink Bauhinia

Cold and cough: About 10 ml root decoction is given with one glass of luke warm water twice a day for 5-7 days.

76. *B. racemosa* Lam. (Caesalpiniaceae)
Syn: *Piliostigma racemosa* (Lam.) Benth.

VN: Chinna aare S: Ashamantaka H: Mawal E: Sweta kanchanara

Asthma: Paste of the stem bark is eaten till cure.

77. *B. variegata* L. (Caesalpiniaceae)
S: Kanchana H: Barial E: Variegated bauhinia

Asthma: One teaspoon of flower paste is administered daily twice for a fortnight.

Tuberculosis: Twenty ml decoction of bark is given twice a day for 15 days.

Cough: Decoction of the buds is used.

78. *Benincasa hispida* (Thunb.) Cogn. (Cucurbitaceae)

Syn: *B. cerifera* Savi.

VN: Budidha gummadi S: Kusmandah H: Petha E: Ash gourd

Asthma: Five ml of infusion of the root is administered 3-4 times daily.

Juice of fruit pulp is taken till cure.

Cough and asthma: Old ripe fruits are selected, divided into two halves and the pulp is scraped out in thin flakes with iron comb or scraper. The watery juice that oozes out abundantly during this process is preserved. The pulp is boiled in the juice until it becomes soft, then tied up tightly in a cloth and the fluid portion is allowed to drain away. Then 50 g of this prepared pulp is fried in 16 g of clarified butter and again boiled in the juice of the fruit, till reduced to the consistence of honey. To this, 50 g of refined sugar and the whole is heated over a gentle fire till the mass assumes such a consistence as to adhere to the ladle. The pot is now removed from the fire and add pepper, ginger, cumin seeds, cardamoms, cinnamon and coriander to the confection. One to two g of it is prescribed.

79. *Benkara malabarica* (Lam.) Tirv. (Rubiaceae)

Asthma: Stem bark along with that of *Syzygium cumini* each 50 g is crushed with 3-4 pepper seeds and mixed in a litre of water. About 40-50 ml of this extract is administered daily once till cure.

80. *Bergenia ciliata* (Haw.) Sternb. (Saxifragaceae)

Cold and cough: About 10 ml of the juice obtained from the crushed rhizome is taken with warm saline water twice a day regularly.

81. *B. ligulata* (Wall.) Engl. (Saxifragaceae)

VN : Tela nurupindi H: Pakanabheda S: Ash mabhedaka

Pulmonary infections: Five to ten ml of decoction of the root is administered 3-5 times daily.

82. *B. stracheyi* (Hook. f. & Thoms.) Engl. (Saxifragaceae)

Cold and cough: Decoction of the roots is used.

83. *Bidens biternata* (Lour.) Merr. & Sherff. (Asteraceae)

Cough: Poultice made from leaf juice is used.

84. *Biophytum nervifolium* Thw. (Oxalidaceae)

Asthma: A decoction of the leaves is given.

85. *Blumea balsamifera* DC. (Asteraceae)

Asthma and cough: Ten ml of decoction of the leaves is administered 3-4 times daily.

Cold: Five ml of decoction of the stem and root is administered 2-3 times daily.

86. *Boerhavia diffusa* L. (Nyctaginaceae)

S: Rakta punarnava H: Pimura E: Spreading hog-weed

Asthma: About 2 teaspoons of root powder is administered with quarter spoon of black pepper powder daily once for a fortnight.

Root paste is administered orally in doses of 2 spoonfuls thrice a day for 7 days.

Dried plant powder is smoked to get relief.

Cough: Root powder with equal amount of sugar candy is taken.

87. *B. erecta* L. (Nyctaginaceae)
Cold and cough: Leaf juice or leaf decoction is used.

88. *B. repens* L. (Nyctaginaceae)
VN : Atikamamidi H : Sant S : Bhauma E: Hog weed

Whooping cough: Root powder with sugar is taken.

89. *Bombax ceiba* L. (Bombacaceae)
Syn: *B. malabaricum* DC., *Salmalia malabarica* (DC.) Schott. & Endl.

VN: Buruga S: Moca H: Semal E: Red silk cotton tree

Cough: Bark decoction is used.

90. *Boswellia serrata* Roxb. ex Colebr. (Burseraceae)
S: Sallaki H: Salai E: The Indian olibanum tree

Cough: Bark decoction is used.

Asthma: Gum is given orally till cure.

Cold: Powdered resin is sprayed on the burning charcoal and the smoke is inhaled especially for children.

91. *Botrychium lonuginosum* Wall. ex Hook. *et* Grev. (Botrychiaceae)
E : Moon wort

Catarrh and pneumonia: Boiled rhizomes are used.

92. *Brassica juncea* L. (Brassicaceae)
E: Mustard H:Rai, Sarson S: Sarsapah

Bronchitis: Seed paste is applied externally.

93. *B. napus* L. var. *napus* L. (Brassicaceae)
E : Cole seed

Cold: Seed oil is smeared on the body as a protective measure particularly to the newly-born babies and postnatal mothers.

94. *B. nigra* (L.) Koch. (Brassicaceae)
Syn: *Sinapsis nigra* L.

VN: Varnavalu S: Mudhurika H: Kall rai E: Tree mustard

Asthma: Oil mixed with a little camphor and is massaged over the back of the chest till cure.

95. *Breynia vitis-idaea* (Burm. f.) Fischer (Euphorbiaceae)

Bronchitis and cold: Leaf after shade drying is smoked till cure.

Cold and cough: Fruit powder is burnt on fire and the smoke is inhaled.

96. *Bridelia retusa* (L.) Spreng. (Euphorbiaceae)

VN : Koramanu S: Ekavira H: Kattian E: Feather foil

Asthma: Stem bark is crushed with that of *Terminalia bellirica* in equal proportions and the paste of redgram seed size is administered daily once for 3 days.

97. *B. scandens* Willd. (Euphorbiaceae)

Asthma: Smoke of dried plant powder is inhaled to get relief.

98. *Bruguiera sexangula* Poir. (Rhizophoraceae)

Cough: Leaf decoction of about 5 ml is used twice daily for 2-3 days.

99. *Bryophyllum pinnatum* (Lamk.) Oken (Crassulaceae)

Cold and Cough: One teaspoonful infusion of leaf is given to infants.

100. *Buchanania lanzan* Spreng. (Anacardiaceae)

Syn: *B. latifolia* Roxb.

VN: Chinna morlu S: Tapasya-priya H: Piyar E: Almondette tree

Bronchitis and asthma: Roots of plants are harvested and cut into small pieces of 1-2 inches length. About 100 g of roots are boiled in 8-10 litres of water for a period of about 6-8 hrs till the volume reduces to one-fourth. The roots are crushed and kept overnight and in the morning the extract is filtered and decoction is stored in dry, cool place. Five to ten ml of decoction is given 2-3 times a day at an interval of 3-4 hrs for a period of about 8-10 days.

101. *Butea monosperma* (Lam.) Taub. (Fabaceae)

S: Palashah H: Palas E: Bastard teak

Asthma: Seed paste is taken orally along with fresh cow's milk early in the morning for 20 days.

Flower paste is applied over the chest as a plaster till cure.

102. *Byttneria herbacea* Roxb. (Sterculiaceae)

VN: Konda chikkudu dumpa.

Bronchitis: Root mixed with *Piper longum*, black peppers and *Trachyspermum ammi* in equal proportions are ground and about 5 g of this powder is given twice daily till cure.

103. *Caesalpinia bonduc* (L.) Roxb. (Caesalpiniaceae)

E: Fever Nut H: Karanju S: Karanj

Asthma: Seed powder is used.

104. *C. cristata* Lam. (Caesalpiniaceae)

VN : Gachakaya S: Kuberakshi H: Kantkarej E: Bonduc nut

Asthma: Roasted seeds in doses of 10-15 g are eaten till cure.

Seed powder is mixed with black pepper powder and sealed in a corked glass bottle. This is administered in doses of 5–20 g depending on the age twice or thrice a day.

105. *C. pulcherrima* L. (Caesalpiniaceae)
E: Peacock Flower H: Gulutora S: Ratangandhi

Asthma and bronchitis: Infusion of flowers is taken orally till cure.

106. *Cajanus cajan* (L.) Millsp. (Fabaceae)
Syn: *Cytisus cajan* L., *Cajanus indicus* Spreng.

VN: Kandhi S: Adhaki H: Tuvar E: Red gram

Asthma: Dried leaf powder is smoked along with the petals of *Nymphaea nouchali* in 4:1 ratio till cure.

Bronchitis: Decoction of green fruits is taken internally.

107. *Calamus viminalis* Willd. (Arecaceae)
Asthma and cough: About one cup of leaf decoction along with half spoonful of black salt and one spoonful of honey is given.

108. *Calotropis gigantea* (L.) R. Br. ex Ait. (Asclepiadaceae)
Syn: *Asclepias gigantea* L.

VN: Nalla jilledu S: Arka H: Mudar E: Gigantic swallow wort

Cough: One spoonful of powdered root/flowers is taken twice a day with lukewarm water.

Dried powder of flowers (1-3 g) in honey is given thrice a day.

Asthma: Flowers are mixed with black pepper seeds and made into a paste. Pills of soapnut seed size are made from this paste and dried under shade. These are given once daily in the morning.

Roasted leaf ash is taken till cure.

Dried leaf powder along with the powder of black peppers in 3:1 ratio is administered till cure.

Flowers are burnt and the ash is mixed with honey and given orally thrice a day till cure.

About 10 g of seed powder mixed with the ash of *Achyranthes aspera* is made into pills with lemon juice and is given 2 times daily till cure.

Flowers are fried in cow ghee and one flower is taken each time twice a day for 30 days.

109. *C. procera* (Ait.) R. Br. (Asclepiadaceae)
Syn: *Asclepias procera* Ait.

VN: Tella jilledu S: Arka H: Safed E: Swallow-wort

Asthma: Smoke of the burnt leaves is inhaled till cure.

Honey is mixed in the powder of roasted roots and given orally before meals.

Root powder is mixed with latex and tamarind juice and is given internally twice a day till cure.

Whole plant is uprooted from the soil. In the hole left in the soil, a dried piece of *Curcuma longa* and a lump of salt are put for 7 days. The salt melts and the turmeric piece swells up. The turmeric piece is taken out and roasted on fire and ground. This is given to the patient (Oil, jaggery, sour food should not be taken with it).

Catarrh: Leaves are crushed and warmed and the juice is dropped in the nose.

110. *Camphora officinarum* L. (Campanulaceae)
VN : Chandramu S: Karpura H: Kapur E : Camphor

Asthma: One pill of 3 or 4 mg of camphor is given with an equal quantity of asafoetida till cure.

111. *Cananga odorata* (Lam.) Hook. f. & Thoms. (Annonaceae)
E: Ylang-ylang

Asthma: Paste made from fresh flowers is applied on the chest till cure.

112. *Canna edulis* Ker.-Gawl. (Cannaceae)
Cold and cough: Roots and corns of maize are crushed and made into a tonic and given in a small quantity before and after food. A few spoons of it is given to children.

113. *Cannabis sativa* L. (Cannabinaceae)
VN: Bangiaku S: Madana H: Ganja E: Indian hemp

Asthma: Dried leaf powder of about 1-2 g is taken internally till cure.

114. *Capsicum annuum* L. (Solanaceae)
VN: Mirapa S: Katuirah H: Lalmirca E: Chillies

Cold and cough: Fruits are dried and powdered. One spoonful of powder mixed with honey is consumed 4 times a day.

Asthma: The paste of a ripe fruit with pulp of three fruits of *Tamarindus indica* and a pinch of common salt is given twice or thrice daily till cure.

115. *Careya arborea* Roxb. (Barringtoniaceae)
VN: Kummadi S: Kumbhi H: Kamba E: Patana oak

Cold and cough: Fresh flower paste with honey (4:1) is given to children.

116. *Carica papaya* L. (Caricaceae)
VN: Boppayi S: Eranda H: Papaya E: Papay tree

Asthma: Pulp of ripened fruit along with seeds is a good remedy.

Influenza: Infusion of flowers with honey is used.

Dry cough: Fruits are cut open at one end, seeds are taken out and filled with common salt and tightly plugged with bamboo pins. This is burned and the charred material is then cooled and mixed in water and filtered. Filtrate obtained is administered till cure.

117. *Carissa carandas* L. (Apocynaceae)
VN: Pedda vaka S: Karamardah H: Karunda E: Karanda

Cough: Decoction of the roots is used.

Whooping cough: Leaf juice is taken with honey.

118. *C. spinarum* L. (Apocynaceae)
VN: Chinna vaka S: Bahudala H: Karaunda E: Karanda

Cough: Immature fruits and flowers are taken early in the morning with an empty stomach for about 9 days.

119. *Carlemania griffithii* Benth. (Rubiaceae)
Cough: Palm decoction is given thrice a day.

120. *Carthamus tinctorius* L. (Asteraceae)
VN: Kusuma H: Kussum S: Kusumbha E: African saffron

Asthma: Half teaspoon of powder of the dry seeds mixed with a tablespoon of honey is taken once or twice a day till cure.

An infusion of 5 g of flowers mixed with one tablespoon of honey is taken once daily till cure.

121. *Cascabela thevetia* (L.) Lipp. (Apocynaceae)
Syn: *Cerbera thevetia* L., *Thevetia neriifolia* Juss. ex Steud.

VN: Paccha ganneru S: Ashvarodhaka H: Pnila Tree E: Lucky nut tree

Asthma: Flowers and jaggery are taken in equal proportions and they are kept layer by layer in alternation with each other in a glass jar and it is exposed to sun and moon lights for about 45 days. Later on the contents in the jar are mixed well and it is taken in doses of 1 spoonful once a day for 45 days.

122. *Cassia alata* L. (Caesalpiniaceae)
VN: Mettatamara S: Dadrugna H: Dadmurdan E: Ringworm shrub

Asthma: Decoction of the leaves is given many times a day till cure.

Decoction of flowers in combination with the decoction of leaves is administered internally till cure.

123. *C. fistula* L. (Caesalpiniaceae)
VN: Rela S: Aragvadha H: Amaltas E: Indian laburnum

Cough: Ash of the pod is given with honey to the children.

Asthma: Roots are macerated and the extract is drunk 3 times a day till cure.

Mucilaginous substance is pasted on midrib of tobacco leaf and then smoked.

Ten ml of leaf juice mixed with honey is taken orally 2-3 times daily.

Whooping cough: Ash of fruit along with honey (2-3 g) is given.

Sore throat: Usually, 100 g of the crushed bark are boiled in 1 litre of water until it remains only 500 ml. This solution along with common salt is used for gargling. The treatment is prescribed twice a day for 2-3 days.

124. *C. occidentalis* L. (Caesalpiniaceae)
Syn: *Senna occidentalis* Roxb.

VN: Kasintha, Peddha vempali S: Kasamandah H: Kasondi E: Negro coffee

Asthma and cough: Decoction made from roots and leaves is used.

Cough: Seeds are boiled in water, strained and about 4 teaspoons of decoction is given thrice a day.

125. *C. sophera* L. (Caesalpiniaceae)
VN: Konda kasindha S: Kasamanda H: Kasaundi E: Senna sophera

Asthma, bronchitis, cold and cough: Whole plant juice in honey or infusion or decoction is administered till cure.

126. *C. tora* L. (Caesalpiniaceae)
VN: Tottemu S: Chakramardah H: Chakunda E: Foetid cassia

Asthma: Stem and seeds are taken in equal quantities, boiled in water and filtered. About 100 ml of filtrate is taken orally twice a day for 5-10 days.

Tea is prepared from seeds and given.

Bronchial asthma: Leaf juice is administered internally.

Cold and cough: Two spoonfuls of seed powder mixed in a cup of tea is administered daily twice for 2 days.

Cough: Seed powder of 5-10 mg mixed with tea is used two or three times daily.

About one teaspoonful of crushed seed is taken with tea 2-3 times daily for a week to cure cough.

Tuberculosis: Seed oil is given.

127. *Catunaregam spinosa* (Thunb.) Tirv. (Rubiaceae)
Syn: *Gardenia spinosa* Thunb., *Randia dumetorum* (Retz.) Poir.

VN: Manga chettu S: Madana H: Main phal E: Emitic nut

Asthma: Fruit extract is given orally till cure.

128. *Celastrus paniculatus* Willd. (Celastraceae)
VN: Jyothishmathi S: Jyothismati H: Jamrasi E: Black oil plant.

Bronchitis and asthma: Decoction of seeds is administered.

Asthma, cold and pneumonia: Oil obtained from seeds is used internally in children.

Tuberculosis: Three to four drops of seed oil are taken with one teaspoonful of water once a day on empty stomach for a month.

129. ***Celosia argentea*** **L. (Amaranthaceae)**
VN: Gunugu S: Vitunna H: Safed murgha E: Quail grass

Tuberculosis: A spoonful of root paste mixed with a spoonful of leaf paste of *Andrographis paniculata* is administered daily once for 21 days.

Cough and asthma: Plant ash is mixed with honey and given internally.

130. ***C. trigyna*** **L. (Amaranthaceae)**
Asthma: Leaves are made into curry with chicken and is taken twice a day till cure.

131. ***Centaurea depressa*** **M. Bieb. (Asteraceae)**
Cold and cough: Whole plant is boiled in water and the decoction is taken in small doses for a week or more.

132. ***Centella asiatica*** **(L.) Urban (Apiaceae)**
Syn: *Hydrocotyle asiatica* L.

VN: Saraswathi aaku S: Mandukaparni H: Brahmi E: Indian pennywort

Dry cough: Leaf juice is mixed with ginger juice and powder of black pepper and half tea spoon of it is taken twice daily for 7 days.

Cough: About 10 ml of leaf juice mixed with 5-7 g of cardamom and seeds of *Piper longum* each is given twice a day.

Roots are crushed, powdered and the filtrate is given.

Syrup resulting from boiling of the leaves with small ginger and black pepper is taken.

Five ml of honey and plant juice is mixed and given once for a week.

133. ***Ceratopteris thallictroides*** **(L.) Brong. (Parkeriaceae)**
Cough: Decoction of the whole plant is used.

134. ***Chamaesyce hirta*** **Millsp. (Euphorbiaceae)**
Asthma: Tincture of root is given.

135. ***Cheilanthes farinosa*** **Kaulf. (Sinopteridaceae)**
Tuberculosis: Plant juice mixed with honey is given after meals.

136. ***Chenopodium botrys*** **L. (Chenopodiaceae)**
Catarrh and asthma: Plant extract is given twice a day.

137. ***Chlorophytum arundinaceum*** **Bak. (Liliaceae)**
H: Safed musli

Asthma: One glassful of decoction of root is taken twice a day for a month.

138. ***Chloroxylon swietenia*** **DC. (Flindersiaceae)**
VN: Billa chettu S: Bhillotaka H: Bhera E: Satin wood tree

Asthma: Ten ml of the stem bark is administered daily once for a week.

139. ***Choerospondias axillaris*** **(Roxb.) Burth ex Hill (Anacardiaceae)**
Cold and cough: About 10 ml of the pulp of the ripe fruit is boiled in 150 ml of water with salt and the warm decoction is taken before bedtime regularly.

140. ***Christella parasitica*** **(L.) Lev. (Thelypteridaceae)**
Asthma: Lukewarm rhizome juice mixed with honey is given once daily for a fortnight.

141. ***Chrysanthemum indicum*** **L. (Asteraceae)**
VN: Chamunti E: Chrysanthemum H: Guldandi S: Sevanti

Bronchitis: Decoction of entire plants is used.

142. ***Cicer arientinum*** **L. (Fabaceae)**
E: Chickpea H: Chana S: Chanaka

Bronchial catarrh: Parched seeds are consumed at night followed by milk to get relief.

143. ***Cinnamomum camphora*** **(L.) Nees & Eberm. (Lauraceae)**
VN: Karpuramu S: Karpurah H: Kapur E: Camphor tree

Cold and cough: An ointment prepared by mixing equal amounts of camphor and cow's butter is warmed and used to massage on the chest and throat before bed time regularly as a remedy.

144. ***C. glanduliferum*** **Meissn. (Lauraceae)**
Cough, bronchitis and pneumonia: Stem bark is crushed and 5 ml of juice is given thrice daily.

Juice of young leaves is also given orally.

145. ***C. tamala*** **Nees & Eberm. (Lauraceae)**
VN : Tallishapatri H: Barahmi S: Ankusha E: Cassia lignea

Cough and cold: One glassful decoction of leaves is taken twice a day for 3 days.

Cough: Bark is boiled and decoction is taken.

146. ***C. zeylanicum*** **Garc. ex Bl. (Lauraceae)**
VN: Dalchina chekka S: Darusita H: Dalchini E: Cinnamon tree

Asthma, cough and tuberculosis: Bark is dried, made into powder and is then smoked along with tobacco.

Cough: Leaves are boiled and the decoction is taken.

Cold: Root bark with rhizome of *Curcuma longa* is taken in equal quantity and ground. A spoonful of paste is administered daily twice for 5 days and this paste is also inhaled along with hot water.

147. ***Cissampelos pareira*** **L. (Menispermaceae)**
Syn: *C. hirsuta* Buch.-Ham. ex DC.

VN: Chiru boddhi S: Patha H: Akanadi E: Velvet-leaf

Cold: About 30 ml extract of fresh roots is given twice or thrice daily for 3 days.

Asthma: Dried root is ground with dried root of *Rauvolfia serpentina* (1:1). A spoonful of powder along with hot water is administered daily thrice for 2 days.

Cold and cough: Root decoction mixed with honey is consumed thrice daily.

148. *Cissus quadrangularis* (L.) Mant. (Vitaceae)
Syn: *Vitis quadrangularis* (L.) Wt. & Arn.

S: Vajravalli H: Hadjora E: The edible stemmed vine

Asthma: Roots (after burning the shoots) are dried and powdered, made into paste with honey and given orally.

Stem along with root bark of *Agave cantula* is taken in equal proportions with a pinch of salt and made into paste. This paste is then made into pills of bengal gram seed size. These pills are administered in doses of 2 pills twice a day for 30 days.

Cough: Chutney is made out of the twigs, cumin seeds and pepper and eaten.

149. *Citrullus colocynthis* (L.) Schrad. (Cucurbitaceae)
VN: Chittipapara H: Ghorumba S : Atmaraksha E: Bitter apple

Asthma: Ash of fruits is taken orally with water.

150. *Citrus aurantifolia* (Chr.) Sw. (Rutaceae)
Syn: *Limonia aurantifoia* Chr.

VN: Narinja S: Nimbuka H: Nimbu E: Acid Lime

Asthma: The juice of one lemon diluted in a glass of water is taken with meals till cure.

151. *C. medica* L. (Rutaceae)
VN: Dabba S: Jambirah H: Jamiri nimbu E: Lemon

Asthma and cough: Two spoonfuls of fruit juice and a spoonful of honey are ground with 10 g of sugar candy and administered with a glass of water once a day for 4 days.

Cold: Leaf decoction is used for bathing newborns and infants.

152. *Cleistanthus collinus* (Roxb.) Benth. ex Hook. f. (Euphorbiaceae)
Syn: *Cluytia collina* Roxb.

VN: Gavine S: Indrayava H: Gorrar E: Rafterwood

Cold and cough: Dried bark powder along with dried rhizome of *Polygala crotalarioides* and 10 black peppers is consumed in the manner of raw tobacco.

153. *Clematis gouriana* Roxb. ex DC. (Ranunculaceae)
VN: Budia maal E: Indian Traveller's joy

Cough: Smoke obtained by burning the root is inhaled.

154. ***Clerodendrum indicum* (L.) Kuntze (Verbenaceae)**
E: Indian tube flower

Asthma: Root paste of about 5 g is taken with water once a day for 12-15 days in children.

Dried root powder is used for smoking till cure.

Cough: Leaf paste with honey is taken.

Asthma and bronchitis: Cigarette made by dried leaves or semi-dried stem is smoked.

155. ***C. serratum* (L.) Moon (Verbenaceae)**
Syn: *Volkameria serrata* L.

VN: Bommala marry S: Bhriguja H: Bharangi E: Beetle Killer

Asthma: About 6 g of leaf powder is mixed with sufficient quantity of ginger and the paste is administered daily once till cure.

Two to three spoons of root decoction mixed with black pepper and salt is given till cure.

Cold and cough: Boiled extract of young parts is given.

156. ***C. siphonanthus* R. Br. (Verbenaceae)**
VN: Chiruteka bharangi H: Barangi S: Bhargi

Cold and cough: Juice of leaf along with a little honey is diluted in lukewarm water and a glassful of the liquid is taken thrice a day.

157. ***C. viscosum* Vent. (Verbenaceae)**
Syn: *C. infortunatum* Gamble

VN: Bokkudu S: Bhandirah H: Titabhant E: Glory tree

Cough: Decoction of the boiled leaves is given orally twice a day.

158. ***Clitoria ternatea* L. (Fabaceae)**
VN: Sanku pushpamu S: Aparijitha H: Aparjit E: Butter fly pea

Chronic cough: Ten g of root powder mixed with little hen blood is dried for seven days in sun; honey is added to this mixture and 45 pills are made. These are taken thrice a day for 15 days.

Root juice of 10 ml mixed with honey (10 ml) is taken orally thrice daily for a week.

Chronic bronchitis: Two tolas of the root juice is given in cold milk to remove the phlegm.

159. ***Coccinia indica* Wt. et Arn. (Cucurbitaceae)**
VN: Kakidonda S: Bimbi H: Kanduri E: Koval fruti

Cough and respiratory tract infection: Decoction of the leaves, fruits and stem of 5-10 ml is given thrice daily.

Fruit is also eaten.

160. *Cochlospermum religiosum* (L.) Alston (Cochlospermaceae)
Syn: *Bombax religiosum* L., *Cochlospermum gossypium* DC.

VN: Konda buruga S: Peethakarpass H: Gabdi E: Golden silk cotton

Asthma: Thirty g of gum of stem is taken thrice a day for 2 days.

Cough: A spoonful of stem bark gum mixed with a glass of hot water is administered daily twice for 3 days.

161. *Cocos nucifera* L. (Arecaceae)
VN: Kobbari chettu S: Narikela H: Nariel E: Coconut

Bronchitis: Decoction of roots is advised.

162. *Colebrookea oppositifolia* Sm. (Lamiaceae)
Asthma: Two teaspoons infusion of root is given twice daily for 21 days.

Cough: Leaf is chewed and its juice is swallowed.

163. *Colocasia affinis* Schott. (Araceae)
Cough and tuberculosis: One or two spathe(s) and inflorescence are taken twice.

164. *Commiphora caudata* (Wight & Arn.) Engl. (Burseraceae)
Syn: *Protium caudatum* Wight & Arn.

VN: Konda mamidi E: Hill mango

Cold and cough: A spoonful of stem bark powder mixed in a glass of water is administered daily twice for 3 days.

165. *Corallocarpus epigaeus* (Rottl. & Willd.) Clarke (Cucurbitaceae)
Syn: *Bryonia epigaea* Rottl. & Willd.

VN: Naga donda S: Katunahi H: Akasgaddha E: Bitter apple.

Asthma: One teaspoonful of tuber paste is administered daily once till cure.

166. *Corchorus aestuans* L. (Tiliaceae)
Syn: *C. acutangulus* Lam.

VN: Nela beera S: Cuncuh E: White jute

Asthma: Mixture of root powder (10 g), *Asparagus racemosus* rhizome powder (5 g) and *Ocimum tenuiflorum* leaves (5 g) (50 : 50) is given in doses of one teaspoon thrice a day for about 15 days.

167. *Cordia dichotoma* Forst. f. (Boraginaceae)
Syn: *C. obliqua* Willd.

VN: Banka nakkeri S: Shleshmanthaka H: Lasura E: Sebesten plum

Cold and Cough: Two spoonfuls of leaf decoction with a pinch of salt is administered daily twice for 3 days.

A decoction of fresh leaves with the leaves of *Ocimum sanctum* is taken thrice a day.

Leaf decoction with common salt (3:1) is used.

168. ***C. myxa*** **L. (Boraginaceae)**
Cough: Decoction of the dried fruits is effective.

169. ***Costus speciosus*** **(Koen.) Smith (Zingiberaceae)**
Syn: *Banksia speciosa* Koen.

VN: Chengalva kostu S: Canda H: Kebu E: Spiral ginger

Cough: Young shoot is crushed and boiled with a pinch of salt and 1-2 cups of the extract is taken orally twice a day for 15 days.

Cold, cough and asthma: Fresh rhizome is crushed and extract is mixed with a glass of boiled milk and taken thrice a day for five days.

170. ***Crinum asiaticum*** **L. (Amaryllidaceae)**
VN: Pedda chalava dumpa S: Nagadamani H: Pindar E: Poison bulb.

Chest pain due to Cough: The underground stem is crushed, fried in mustard oil and is used for massaging the chest.

171. ***C. defixum*** **Ker.-Gawl. (Amaryllidaceae)**
VN: Kesarchettu

Asthma: One teaspoonful of extract of leaves with equal quantity of honey is given thrice a day.

172. ***Crocus sativus*** **L. (Iridaceae)**
VN: Kunkumapuvvu S: Kesaram H: Kesar E: Saffron

Asthma: One teaspoonful of honey (older the better) is mixed with 60 g of liquor made from this plant. The mixture is put in a bottle which is then kept in a drum containing grains of *Triticum aestivum*. After 7 days, about 11 g of this mixture is taken in the morning after brushing the teeth. It is said that if the honey is 5 years old, the patient will not suffer from for 5 years.

Cough: Dried stigma powder is used.

173. ***Crossandra infundibuliformis*** **(L.) Nees (Acanthaceae)**
VN ; Gobbi

Cold and Cough: Leaf decoction is given.

174. ***Crotalaria orixensis*** **Willd. (Fabaceae)**
Tuberculosis: Ten g of crushed root mixed with equal quantity of *Bambusa arundinacea* root is taken with a glass of milk once a day for one month.

175. ***Croton tiglium*** **L. (Euphorbiaceae)**
VN: Nepalum S: Jayapala H: Jeypal E: Purging cotton

Asthma: Seeds are burnt on fire and the smoke is inhaled.

176. ***Cryptolepis elegans*** **Wall. ex G. Don (Periplocaceae)**
Cough: Two spoonfuls of root extract mixed with cow's milk is administered once a day till cure.

177. *Curculigo orchioides* Gaertn. (Hypoxidaceae)
Syn: *C. malabarica* Wight, *C. brevifolia* Dryan.

VN: Chukka gadda S: Talamuli H: Mushali E: Black muscale

Asthma: Tuber is washed, dried, powdered and about 5 g powder is rolled to balls with coconut milk. This is swallowed along with a glass of lukewarm water.

Two spoonfuls of tuber extract is administered orally twice a day till cure.

Roots are boiled in water and the decoction is administered internally till cure. A little jaggery is taken while drinking the decoction.

Rhizome powder along with sugar is given orally till cure.

Cough: Half teaspoon tuber powder mixed with honey is given twice a day for 3 days.

Cold and Cough: One teaspoonful of rhizome powder along with 2 teaspoonfuls of *Trachyspermum ammi* seed powder is boiled in 2 teaglasses of water and taken.

178. *Curcuma amada* Roxb. (Zingiberaceae)
VN: Mamidi allam S: Karpuraharidra H: Ama-haldi E: Mango-ginger

Bronchitis and Asthma: About 5 ml of rhizome decoction is given twice a day.

Cold and Cough: Decoction of rhizome with common salt (3:2) is given.

Two spoonfuls of rhizome decoction with a pinch of salt is administered daily twice for 3 days.

179. *C. angustifolia* Roxb. (Zingiberaceae)
VN : Ararutgaddalu H: Tavakhira S: Govayodbhava E: East Indian arrow root

Asthma: About 10 ml of rhizome decoction is administered orally twice a day.

180. *C. aromatica* Salisb. (Zingiberaceae)
VN: Kasthuri dumpa, Pandava dumpa, Pandava cheetakam

S: Aranya harida H: Jangli-Haldi E: Wild turmeric

Cough and cold: Root decoction is taken in doses of one cupful twice a day for 3 days.

181. *C. caesia* Roxb. (Zingiberaceae)
VN: Manupasupu H: Kalihaldi E: Black zedoary

Asthma: Decoction of the rhizome is used.

Roasted rhizome is eaten during bed time daily till cure.

182. *C. longa* L. (Zingiberaceae)
Syn: *C. domestica* Valeton

VN: Pasupu S: Haridra H: Haldi E: Turmeric.

Bronchitis: Decoction of dried rhizome is advised to take internally.

Asthma: Rhizome is ground with crab and 15 ml of the decoction is taken thrice daily.

A teaspoon of the rhizome powder with a glass of milk is administered twice or thrice a day on empty stomach till cure.

Rhizome paste along with honey and milk is administered daily.

Rhizomes are chewed for relief.

Cold: Rhizome powder mixed with warm milk is said to be beneficial.

Throat infection: A decoction prepared from the powdered rhizome and common salt, 2 g each in 300-400 ml of water is drunk warm before bedtime for 3-4 days. It is also used as an expectorant.

Cough: Rhizome paste along with half teaspoonful of fresh turmeric powder is mixed in 30 ml of warm milk is administered daily till cure.

183. *C. neilgherrensis* Wight (Zingiberaceae)
Cough and cold: Fifty g of crushed roots is boiled in a glass of water and taken once a day for 2 days.

184. *C. zeodaria* (Christ.) Rosc. (Zingiberaceae)
VN: Kachoram H: Kachura S: Dravida E: Zeodary

Asthma: Rhizome decoction is given.

Cold and Cough: Rhizome paste crushed with water is used for taking bath.

Decoction of rhizome is taken in doses of one cupful twice a day for 4 days.

185. *Cuscuta reflexa* Roxb. (Cuscutaceae)
VN: Sitasavaramu H: Akashabela S: Akashabhavana E: Dodder

Cough: Plant juice mixed with coconut water is taken in the early morning for 2 weeks.

Whooping Cough: One to one and a half teaspoonfuls of the plant juice is drunk with 100 ml of boiled warm water thrice a day, each with an interval of 20-25 minutes for 3-4 days.

Cold and Cough: Whole plant along with latex of *Calotropis procera* in 1:3 ratio are burnt in a clay pot to prepare ash. About 10 g of ash in honey is given thrice a day.

186. *Cycas pectinata* Griff. (Cyacadaceae)
Asthma: Two to three g powder of microsporophylls is taken orally with a little water once in a day.

187. *Cynodon dactylon* (L.) Pers. (Poaceae)
Syn: *Panicum dactylon* L.

VN: Garika gaddi S: Durva H: Dub E: Bahama grass

Asthma: Extract prepared from pollen grains is used.

188. *Cyperus rotundus* L. (Cyperaceae)
Syn: *C. subcapitatus* Clarke, *C. tuberosus* Rottl.

VN: Tunga gaddi S: Hima H: Nagarmotha E: Nut grass

Bronchitis and bronchial Asthma: A decoction prepared from 10 g of the crushed tubercles in 100 ml of water with a little salt is given twice a day for 2-5 days.

189. *C. scariosus* R. Br. (Cyperaceae)
VN: Thungagaddalaveru H: Nagarmotha S: Chakranksha

Cold and Cough: About 100 ml decoction of whole plant with 5 g of *Piper nigrum* is given twice or thrice daily for 4-5 days.

190. *Dactyloctenium aegyptium* (L.) P. Beauv. (Poaceae)
VN: Muttengapillu H: Makra

Asthma: Culm decoction with root of *Diplazium esculentum* in equal quantities is given in doses of 20 ml thrice daily for two to three months in children.

191. *Dalbergia paniculata* Roxb. (Fabaceae)
VN: Pachari H: Dhobein

Cold and Cough: Fifty ml leaf juice is taken internally twice a day for a week to cure.

192. *Datura innoxia* Mill. (Solanaceae)
S: Dhattura H: Safed datura E: Thorn apple

Asthma: Leaves are dried and burnt. The smoke is inhaled to get relief during early stages.

Hundred g of root is ground along with a pinch of salt and made into pills of red gram seed size. Two pills are administered with a glass of water once a day for a week.

Hundred g of leaf powder is smoked as cigarette twice a day for a month.

193. *D. metel* L. (Solanaceae)
Syn: *D. fastuosa* L., *D. alba* Nees

VN: Nalla umetha S: Dhustura H: Kaladatura E: Downy datura

Asthma: Leaf powder is used as snuff.

Leaves of this plant and *Calotropis gigantea* mixed with common salt and honey are made into paste. One spoon of it is taken orally twice a day.

Dried leaves and stems are smoked. At first 10-20 g are smoked and this quantity is gradually increased to 30 g till cure.

Dried leaf powder is inserted into a cigar and the smoke is inhaled till cure.

Root powder is mixed with honey and is given which acts as an anticongesant and gives quick relief.

Seeds of unripe fruits are removed from the plant while the fruits are on the plant. In the place of seeds, cloves are inserted. After 24 h, cloves are taken out roasted and the drug is obtained. Two cloves are given at each time with an interval of 8 h for 2 days.

Bronchitis and asthma: Hundred g of powdered leaves are smoked as bidis twice a day for a month.

Acute bronchial infestation and asthma: Leaves paste is gently applied 2-3 times a day on chest parts and lungs for a period of 3-4 days. The paste is applied regularly for 8-10 days to chronic patients and for 2-3 days in acute case. The paste gives immediate relief in inflammation of lungs and restores respiratory distress.

Cough: Leaves are dried and powdered, mixed with powder of black pepper and with honey or jaggery to alleviate constant coughing.

194. *Dendrophthoe falcata* (L. f.) Etting (Loranthaceae)
S: Bandaka H: Banda E: Honey suckle-mistletoe

Cough: Five tender leaves are eaten daily once for 3 days.

Tuberculosis: About one teaspoonful of the aqueous extract of the whole plant is taken thrice a day for 3 days

195. *Desmodium biarticulatum* (L.) F. v. Muell. (Fabaceae)
Syn: *Hedysarum biarticulatum* L., *Dicerma biarticulatum* (L.) DC.

Asthma: Whole plant powder is burnt and the smoke is inhaled daily twice till cure.

196. *D. gangeticum* (L.) DC. (Fabaceae)
Syn: *Hedysarum gangeticum* L.

VN: Bhumi ippa S: Prsniparni H: Salpan E: Geethanaram

Cough: Two spoonfuls of root paste mixed with a spoonful of honey is administered daily twice.

Asthma: Half cup of root decoction is administered orally daily once for 3 days.

197. *D. heterocarpon* (L.)DC. (Fabaceae)
Cough: Decoction of the plant is considered to be a tonic.

198. *D. triflorum* (L.) DC. (Fabaceae)
Syn: *Hedysarum triflorum* L.

VN: Muntha mandhu S: Tripadi H: Kudaliya

Cold and Cough: Leaf juice with honey (2:1) is given to children.

199. *Dicanthium pertusum* (L.) Clayton (Poaceae)
Cough: Paste of inflorescence with decoction of long peppers (3:1) is given to children.

200. *Dioscorea bulbifera* L. (Dioscoreaceae)
Syn: *D. sativa* Thunb.

VN: Adavi dumpa S: Brahmaputri H: Bhirvolikanda E: Bulb bearing yam

Cold and Cough: About 10 g of fresh rhizome is chewed twice a day for 3 or more days.

About one fourth of fresh tuber is chewed twice a day for 2-3 days.

201. *Diospyros malabarica* (Desr.) Kostel. (Ebenaceae)
Syn: *Garcinia malabarica* Desr., *Diospyros embryopteris* Pers.,

VN: Naramamidi S: Tindukah H: Gabh E: Gaub persimon

Cough: Four to five g of dried rind of the fruit is smoked till cure.

Unripe fruit is eaten for relief from nagging cough.

202. *Dolichos trilobus* L. (Fabaceae)
Cold and Cough: Twenty five to fifty g seed powder is taken with a cup of water twice a day for 3 days. The paste is also applied on forehead for 3 days at night.

203. *Dracocephalum heterophyllum* Benth. (Lamiaceae)
Cold and Cough: Decoction of flowers and leaves boiled in water is given twice a day.

204. *Drimia indica* (Roxb.) Jessop (Liliaceae)
E: Indian Squill H: Jangli-Piyaz S: Vanpalandu

Cough, chronic bronchitis and asthma: The dry outer coats of the bulbs are sliced and dried and is used as an expectorant.

205. *Drypetes roxburghii* (Wall.) Hurusawa (Euphorbiaceae)
Cold: A decoction of the leaves and fruits is given.

206. *Eclipta prostrata* (L.) L. Mant. (Asteraceae)
Syn: *Verbesina prostrata* L.

VN: Choda ganti S: Bhringaraja H: Bhangra E: Trailing eclipta

Cold and Cough: Leaf decoction along with a little honey is diluted in lukewarm water and a glassful of the liquid is taken thrice a day.

Leaves are mixed with the leaves of *Enicostemma axillare, Phyla nodiflora, Ocimum tenuiflorum* and *Coleus amboinicus* and roasted. The juice is extracted and 2-3 teaspoonfuls of this juice is given 3-4 times a day.

About 70 g flowers are boiled in half litre of water till the water is reduced to about 250 ml. To this decoction add dry grapes and sugar candy.

Cough and bronchitis: About two teaspoonfuls juice of the crushed aerial parts (about 5 branches crushed by mortar and pestle) with equal amount of honey is taken orally twice a day for a week.

Catarrh: Two drops of fresh leaf juice with eight drops of honey is given to new born child.

Cough: Whole plant crushed and mixed with *Ocimum sanctum* leaf juice and honey and is given.

207. ***Echinops echinatus*** **Roxb. (Asteraceae)**
H: Gokhru S: Kantalu

Chronic cough and asthma: Decoction of roots is given 4-5 times.

208. ***E. nivens*** **Wall. (Asteraceae)**
Cough and asthma: Root bark powder mixed with honey is taken.

209. ***Elsholtzia blanda*** **(Benth.) Benth. (Lamiaceae)**
Cold and Cough: Juice of about 20 leaves is applied for two days on the back and chest of a child as a remedy.

210. ***Elytraria acaulis*** **(L. f.) Lindau (Acanthaceae)**
Syn: *Justicia acaulis* L. f., *Elytraria crenata* Vahl

VN: Doodh Chettu, Pokka sunga

Cold and Cough: Plant is squeezed and inhaled.

211. ***Embelia ribes*** **Burm. f. (Myrsinaceae)**
VN: Vidangamu H: Haruch S: Amogha

Bronchitis: Fruits are used.

Cough: Root infusion is given.

212. ***Enhydra fluctuans*** **Lour. (Asteraceae)**
H: Haruch S: Achari

Cough: Leaf juice mixed with honey is given thrice a day.

Leaf is heated with little water and pressed, the juice is mixed with honey and taken orally thrice daily.

213. ***Ephedra gerardiana*** **Wall. ex Stapf (Ephedraceae)**
H: Asmania E: Ephedra

Asthma: Dried leaf vapours are inhaled till cure.

One teaspoonful powder of roots, stems and branches is taken twice a day.

About 20 g of the aerial parts are boiled in 500 ml of water, until it is boiled down to half this amount. The warm decoction is orally administered twice a day regularly.

Powder of the herb in doses of 0.5 to 2 g is swallowed with water till cure.

214. ***Erigeron acer*** **L. var.** ***multicaulis*** **(Wallich ex DC.) C.B. Clarke (Asteraceae)**
Breathing trouble: Smoke of whole plant with immature buds is often inhaled.

Cold and Cough: Decoction of whole plant is used.

215. ***Ervatamia heyneana*** **(Wall.) Cooke (Apocynaceae)**
Cough: Fruits are tied with thread and put around the neck in children.

216. ***Erythrina variegata*** **L. (Fabaceae)**
Syn: *E. indica* Lam.

VN: Baradapu, Pardha S: Paribhadrah H: Pharhad E: Indian coral tree

Whooping cough: Leaf juice is given to children till cure.

217. *Eucalyptus globulus* Labill. (Myrtaceae)
VN: Euklis chettu S: Nilaniryasa H: Yukeliptas E: Eucalyptus

Cold: Leaves are pounded and boiled in water and the concentrated decoction is applied externally.

Infections of the upper respiratory tract, chronic bronchitis and asthma: Oil is obtained from leaves by steam distillation and is used. Internally it is used as a stimulating expectorant.

218. *Euphorbia heyneana* Spreng. (Euphorbiaceae)
Dry cough: Whole plant mixed with common salt, turmeric, garlic and pepper ground into paste, given orally with mother's milk for 3 days.

219. *E. hirta* L. (Euphorbiaceae)
Syn: *E. pilulifera* Hook. f.

VN: Nela jilledu S: Dugdhika H: Lal dudhi E: Snake weed

Cough and asthma: Two teaspoonfuls paste of a plant is taken twice a day for a week.

Cough: Stem, leaves of *Leucas aspera* and seeds of *Piper nigrum* are taken in equal quantity and ground. Two spoonfuls of paste is administered daily twice for 3 days.

Asthma: Twenty ml of leaf decoction is administered orally for 10 days.

220. *E. ligularia* Roxb. (Euphorbiaceae)
Syn: *Euphorbia neriifolia* L.

VN: Akujemudu H: Sehund S: Sehunda, Sudha

Bronchitis: Young branches are first roasted and then juice is sqeezed out. It is administered with molasses for producing vomiting in bronchitis in children.

Cough: Juice of roasted leaf and juice of 5-7 leaves of *Ocimum sanctum* are mixed with crushed rhizome of *Zingiber officinale* (5 g). It is warmed on metallic pot with 2-3 ml of honey and 5 ml of it is given orally 2-3 times daily to children.

Leaves boiled to make decoction and 10 ml of decoction is taken internally as tonic once daily before bed.

221. *E. nivulia* Buch.- Ham. (Euphorbiaceae)
VN: Boggu jilledu S: Patrasnchi H: Thor E: Leafy milk hedge

Cough: The plant is heated on fire and the juice is extracted by compression which is taken for severe cough especially in children.

222. *E. thymifolia* L. (Euphorbiaceae)
VN: Bidurunanabiyam S: Laghudugdhika

Cough: Stem, bulbs *of Allium sativum* and seeds of *Piper nigrum* are taken in equal quantity and ground. A spoonful of paste is administered daily twice for 3 days.

223. ***E. tirucalli*** **L. (Euphorbiaceae)**
VN: Kada jamudu S: Vajraduma H: Sehund E: Round milk hedge

Cough: Stem and seeds *of Piper nigrum* (3:1) are taken and ground. Two spoonfuls of paste mixed with a pinch of salt is administered daily twice for 3 days.

Two drops of latex mixed in a half glass of seed decoction of *Piper nigrum* is administered thrice a day.

224. ***Evolvulus alsinoides*** **(L.) L. (Convolvulaceae)**
Syn: *Convolvulus alsinoides* L.

VN: Neelam puvvu S: Vishnu kranth H: Syama kantha

Chronic bronchitis and Asthma: Dried leaves powder is smoked as cigarettes in chronic bronchitis and asthma, twice a day for a month.

Hundred ml of plant decoction is prescribed once daily.

Bronchitis: Whole plant (10 g) with 10 g borax powder, whole plant of *Bacopa monnieri* and 50 ml of honey is mixed together and pills are made; two pills are given twice a day.

Cold and cough: Whole plant decoction is administered 2-3 times a day.

225. ***Fagonia cretica*** **L. (Zygophyllaceae)**
VN: Chittigara H: Damahan S: Ajabhakshya

Asthmatic bronchitis: Decoction of whole plant is given.

226. ***Feronia limonia*** **L. (Rutaceae)**
E: Wood Apple H: Kowit S: Kapittha

Acute bronchitis: Warmed poultice of leaves is applied.

227. ***Ferula asafoetida*** **L. (Apiaceae)**
H: Hing E: Asafoetida

Asthma: Three to six hundred g of gum mixed with 2 teaspoons of honey, a quarter teaspoon of white onion juice and one teaspoon of betel leaf juice is administered thrice daily till cure.

228. ***Ficus carica*** **L. (Moraceae)**
E: Common Fig H: Anjir S: Anjira

Bronchitis: Fruit pulp mixed with vinegar and sugar is consumed.

229. ***F. glomerata*** **Roxb. (Moraceae)**
E: Cluster Fig H: Gular S: Udumber

Asthma and bronchitis: Decoction of stem bark is administered till cure.

230. ***F. heterophylla*** **L. f. (Moraceae)**
VN: Buroni S:Trayamani

Asthma: Root bark powder along with coriander seeds is given orally till cure.

231. *F. mollis* Vahl (Moraceae)
Asthma: Four leaves are boiled in a glass of water till half of it remains and this is administered daily in the morning for 40 days.

232. *F. racemosa* L. (Moraceae)
Syn: *F. glomerata* Roxb.

VN: Karuku bodda S: Sadaphala H: Domoor E: Gular fig

Bronchial asthma: Bark paste is applied on chest, lungs and thoracic region 2-3 times a day to patients suffering from acute inflammation. The application of paste gives immediate relief and restores the choked lungs.

233. *F. religiosa* L. (Moraceae)
Syn: *Urostigma religiosum* (L.) Gasp.

VN: Raavi S: Aswatha H: Pipal E: Peepul tree

Asthma: Fruits are dried and powdered and is given in doses of 3 g twice a day.

Dried fruit is pulverized and taken in water for a fortnight.

Five ml of fruit juice is given orally 3-5 times daily till cure.

Mature figs are dried and powdered. About 5 g of this powder is taken with warm milk before breakfast for a month or longer till relief.

Cold and cough: Stem bark is crushed and a teaspoonful of this paste is taken with equal amount of honey thrice a day for 2-3 days.

234. *F. rumphii* Blume (Moraceae)
H: Gagjaira

Asthma: Fruit paste with decoction of black peppers in 5:2 ratio is administered orally till cure.

235. *Flacourtia indica* (Burm. f.) Merr. (Flacourtiaceae)
Syn: *Gmelina indica* Burm. f., *Flacourtia ramontchi* L' Herit.

VN: Mandiakodi S: Aghori H: Kancu E: Governor's plum

Bronchitis: Root extract is taken in doses of 3 spoonfuls twice a day for 3 days.

236. *Garcinia pedunculata* Roxb. (Clusiaceae)
Cough: One fruit boiled in ½ litre of water is given to drink twice daily.

237. *Garuga pinnata* Roxb. (Burseraceae)
VN: Garuvu manu S: Golika H: Ghogar E: Garuga

Asthma: Juice obtained from the leaves mixed with honey is administered till cure.

Leaves are crushed along with those of *Justicia adhatoda* and *Vitex negundo* in equal quantity and made into a paste. Pigeon pea sized paste is taken twice a day for a month.

Leaves are boiled in a glass of water till half glass of water remains and this is taken daily in the morning for 40 days.

238. ***Gaultheria fragrantissima*** **Wall. (Ericaceae)**
E: Fragrant wintergreen

Cold and cough: About 3 teaspoonfuls of the young leaf paste is orally administered. The usual course is 2-3 times a day for 3 days or longer.

239. ***Gentiana algida*** **Pallas. var.** ***nubigena*** **(Edgew.) Kush. (Gentianaceae)**
Bronchitis and cough: Powdered flowers are taken twice a day for a week.

240. ***G. prolata*** **Balf. f. (Gentianaceae)**
Cough: About a teaspoonful of the plant juice diluted in water is taken 2-3 times a day. The medicament, if necessary, may be repeated for one more day.

241. ***Geranium wallichianum*** **D. Don (Geraniaceae)**
Cold and cough: Decoction of root is taken till cure.

242. ***Globba bulbifera*** **Roxb. (Zingiberaceae)**
Cough and asthma: Ten g of root or 3 cm long root and 10 g of *Piper nigrum* are made into paste and tablets are prepared out of it. One tablet is given twice daily for 15 days in the early morning and at bed time.

243. ***G. clarkei*** **Baker (Zingiberaceae)**
Cough: Rhizome extract is mixed with leaf extract of *Justicia adhatoda* and taken till cure.

244. ***Gloriosa superba*** **L. (Liliaceae)**
VN: Vanka vajram S: Langali H: Karkari E: Malabar glory lilly

Asthma: Leaf paste is heated and applied on the forehead and neck for 7 days in children.

Cough: Tuber is made into powder and a pinch of it mixed with juice of wild *Zingiber officinale* and given at night for 3 days.

245. ***Glycosmis mauritiana*** **(Lam.) Tanaka (Rutaceae)**
Cough and asthma: Juice of rhizome with black pepper is given twice a day till cure.

Root bark is ground and the exract is mixed with jaggery and is administered orally for children.

246. ***G. pentaphylla*** **(Retz.) DC. (Rutaceae)**
Syn: *Limonia pentaphylla* Retz.

VN: Konda gilugu S: Pithari H: Ban-Nimbu

Cold: Root bark extract is prepared and mixed with jaggery. One spoon of it is given to infants once a day for 3 days.

Cough: Root bark paste is mixed with jaggery (5:1) and administered daily twice for 3 days in children.

247. ***Glycyrrhiza glabra*** **L. (Fabaceae)**
VN: Atimadhuram H: Jethimadh S: Jalayashti E: Liquorice

Tuberculosis: 5 g paste of rhizome is consumed with milk.

248. *Gmelina arborea* Roxb. (Verbenaceae)
VN: Konda gummadu chettu S: Gamhari H: Gamhar E: Candahar tree

Cough: Ten ml of leaf juice is used orally thrice daily.

Barks facing east and west are ground with human urine, one pearl and brassica seeds and the juice is administered orally till cure.

249. *Gossypium hirsutum* L. (Malvaceae)
VN: Patti S: Karpasa H: Kapas E: Asiatic cotton

Asthma: Dried twigs of the plant are smoked as cigarette twice a day for one month.

250. *Grewia disperma* Rottl. ex Sprengl. (Tiliaceae)
Cough: About 2 teaspoonfuls of the juice from the stem bark is taken 2-3 times a day as an expectorant.

251. *Gymnema sylvestre* (Retz.) R. Br. ex Schult. (Asclepiadaceae)
Syn: *Periploca sylvestris* Retz.

VN: Podapathri S: Meshasringi H: Kavali E: Periploca of the wood

Asthma: Twelve g of leaves are powdered along with 5-6 black pepper seeds and 300 mg of this powder is given till cure.

252. *Gynura conyza* Cass. (Asteraceae)
Asthma: Five to fifteen ml decoction of the leaves is given orally 3-4 times daily till cure.

253. *Hedychium coccineum* Sm. (Zingiberaceae)
Asthma: Young shoots are eaten raw.

254. *Hedyotis corymbosa* (L.) Lam. (Rubiaceae)
Asthma: Leaves are boiled in water and the decoction is drunk till cure.

Two to four teaspoonfuls of decoction is prepared from the roots and leaves of *Justicia adhatoda*. This is administered orally twice a day for a month.

Tuberculosis: Plant juice along with goat's milk is used till cure.

255. *Helianthus annuus* L. (Asteraceae)
VN: Suryakanti S: Adityabhakta H: Surajmukhi E: Common sunflower

Bronchitis, asthma, whooping cough and pulmonary diseases: Tea prepared from flowers and dried leaves is used till cure.

Bronchitis: A tincture of flowers and leaves is recommended in combination with balsams.

Cough: Two spoonfuls of seed paste mixed with a spoonful of honey is administered daily twice for 3 days.

256. *H. lanceolatus* (Wall.) Brandis (Asteraceae)
Cough: Six g of leaf paste is given internally twice a day.

257. ***Helicteres isora*** **L. (Sterculiaceae)**
VN: Chamala S: Awartaki H: Jonkaphal E: East Indian screw tree

Cough: One spoonful of fruit decoction is taken once a day till cure. Dried fruit powder with fruit powder of *Phyllanthus emblica* is used till cure.

258. ***Hemidesmus indicus*** **(L.) R. Br. (Periplocaceae)**
Syn: *Periploca indica* L.

VN: Aavu sugandhi paala S: Ananthamula H: Barri Sugandhi E: Indian sarsaparilla

Cough: Tuber is washed, crushed and the juice is used.

Roots along with flowers of *Calotropis procera* and leaves of *Gymnema sylvestre* are taken in equal quantity and powdered. A spoonful of this powder along with half spoon of honey is administered daily twice for about 3 days.

A spoonful of leaf juice is administered twice a day.

Chronic cough: Watery extract of dried leaves when hot is taken with sugar and milk. This acts as an alternative and tonic.

259. ***H. indicus*** **(L.) R. Br. var.** ***pubescens*** **(Wight & Arn.) Hook. f. (Periplocaceae)**
Syn: *H. pubescens* Wight & Arn.

VN: Gede sugandha pala S: Saribha H: Sugandipal E: Indian sarsa

Cough: Two spoonfuls of root decoction mixed in a glass of milk is administered daily twice for 3 days.

Cold: A dosage of 20 ml of root juice is consumed once a day for 4 days.

260. ***Hibiscus rosa-sinensis*** **L. (Malvaceae)**
VN: Mandara S: Aruna H: Jasum E: Chinese rose

Bronchitis, tuberculosis and whooping cough: Petals of fresh flower are given to chew thrice a day till cure.

Pure red coloured petals of the double variety, half cup of butter milk, 2-3 black peppers and a pinch of salt are crushed and the paste is eaten as a chutney daily once for a month.

Throat infection: Three to four flowers are chewed for 4-6 times each with an interval of 3-4 hrs.

Bronchial catarrh: Decoction of the flowers is given.

261. ***Holarrhena pubescens*** **(Buch.–Ham.) Wall. ex G. Don (Apocynaceae)**
Syn: *Echites pubescens* Buch.- Ham.,

VN: Kodisa pala S: Kutaja H: Karchi E: Ivory tree

Cold and cough: Powder of stem bark is given orally along with pure honey to children.

Cough: Stem is soaked in water for whole night. Two spoonfuls of this water is administered daily once for about 3 days.

Bark decoction along with powder of *Piper nigrum* is taken twice daily for 3-5 days.

Asthma: Half to one teaspoonful of bark powder is given orally till cure.

262. *Holoptelea integrifolia* (Roxb.) Pl. (Ulmaceae)
Syn: *Ulmus integrifolius* Roxb.

VN: Nemali chettu S: Cirabilva H: Karanji-pipal E: South Indian elm

Asthma: Stem bark is crushed with garlic and the extract is administered in doses of 2 spoonfuls twice a day till cure.

263. *Hydrolea zeylanica* (L.) Vahl (Hydrophyllaceae)
S: Langali

Cough: Entire plant decoction is drunk thrice a day.

264. *Hyoscyamus niger* L. (Solanaceae)
VN: Kurashanivanam H: Khurasaniajvayan S: Dipya E: Belene

Asthma: Seeds are made into powder and given in small doses to subside pain due to asthma.

Whooping cough and asthma: Decoction made from leaves and fruits is given.

265. *Hypericum perforatum* L. (Hypericaceae)
H: Bassant E: Amber

Asthma and whooping cough: Leaves and flowers are used as an expectorant.

266. *Hyssopus officinalis* L. (Lamiaceae)
E: Hyssop

Asthma: Whole plant infusion is given in doses of a wine glassful 2 or 3 times a day till cure.

267. *Indigofera cassioides* Rottl. ex DC. (Fabaceae)
Syn: *I. pulchella* sensu Gamble

Asthma: Roots are crushed with black pepper and the extract is administered in doses of 2 spoonfuls twice a day for 5 days.

268. *I. tinctoria* L. (Fabaceae)
Syn: *I. sumatrana* Gaertn.

VN: Neeli mokka S: Nilini H: Nil E: Indian indigo

Asthma: Root juice is extracted and is administered in doses of 2 spoonfuls twice a day for about 20 days.

Root (50 g) crushed with black pepper (5 g) is taken in doses of one spoon twice a day for 21 days.

269. ***Inula grandiflora*** **Willd. (Asteraceae)**
Cough, cold and throat irritation: About 10 g root powder with lukewarm water is given twice a day.

270. ***I. racemosa*** **Hk. f. (Asteraceae)**
Bronchial Asthma: Roots are made into powder mixed with *Saussurea lappa* in equal quantities and is given in doses of one teaspoonful thrice a day for a month.

271. ***Ipomoea obscura*** **(L.) Ker-Gawl. (Convolvulaceae)**
VN: Nallakokkita S: Vachagandha

Cough: Juice of the leaves is given to children.

272. ***Ixora pavetta*** **Andrews (Rubiaceae)**
Whooping Cough: Flowers are pounded in milk and given.

273. ***Jasminum bignoniaceum*** **Wall. ex DC. (Oleaceae)**
Syn: *J. humile* L.

VN: Hema pushpika H: Malto S: Gandadhya E: Golden Jasmine

Throat infection and sore throat: About 10 fresh leaves are chewed and ingested thrice a day for 2-3 days.

274. ***Kaempferia galanga*** **L. (Zingiberaceae)**
VN: Kachoram H: Chandramula S: Chandramulika

Cough: Powder of rhizome mixed with honey is given.

275. ***Kalanchoe pinnata*** **(Lamk.) Pers. (Crassulaceae)**
Syn: *Cotyledon pinnatum* Lam., *Bryophyllum calycinum* Salish.

VN: Ranapala S: Asthi bakshi H: Zakhamhaiyat

Cough: Single raw leaf is eaten for 7 days to get relief.

276. ***Kyllinga nemoralis*** **(J.R & G. Forst) Danty ex Hutch. & Dalziel (Cyperaceae)**
Asthma: Two teaspoonfuls of whole plant extract given thrice a day.

277. ***Lannea coromandelica*** **(Houtt.) Merr. (Anacardiaceae)**
Syn: *Dialium coromandelicum* Houtt., *Odina wodier* Roxb.

VN: Gumpena manu S: Jingini H: Jingan E: Indian ash tree

Asthma: One cup of decoction of the bark is given daily for 3 days.

278. ***Lantana camara*** **L. (Verbenaceae)**
Syn: *Lantana aculeata* L.

VN: Deva ganneru S: Caturangi H: Lantana E: Wild sage

Cold and cough: Decoction of the leaf is given till cure.

279. ***Lavatera kashmiriana*** **Comb. (Malvaceae)**
Cold and respiratory ailments: The petals of the flowers are fermented with sugar and given to children.

280. ***Leontopodium jacotianum*** **Beauv. (Asteraceae)**
Cold: Dried whole plant is boiled in water for a long time and then used for washing feet and hands to relieve cold.

281. ***Lepidagathis cuspidata*** **Nees (Acanthaceae)**
Whooping cough: The ash obtained by burning the leaves is mixed with little honey and taken in a very small dose twice a day.

282. ***Leptadenia reticulata*** **(Retz.) Wt. & Arn. (Asclepiadaceae)**
VN: Kalasa H: Dori S: Arkapushpi

Asthma and cough: Leaves are chewed and the sap is swallowed.

Cough due to cold: Latex from the young stem is used as a nasal stimulant.

283. ***Leucas aspera*** **(Willd.) Link (Lamiaceae)**
Syn: *Phlomis aspera* Willd., *Leucas plukenetii* (Roth) Spreng.

VN: Tummi S: Drona pushpi E: Chhota-itakusa E: Thumbe

Cough and cold: Flowers are administered in the form of syrup or given with honey.

Two teaspoons of leaf juice is given orally thrice till cure.

Cold: Leaf juice mixed with turmeric (5:1) is heated on the live coal and 3 spoonfuls of paste is administered daily twice for 3 days.

Cough and asthma: Leaf juice and honey is taken twice a day till cure.

Cough: Leaves those that of *Ocimum tenuiflorum, Pergularia daemia*, bulbs of *Allium sativum* and seeds of *Piper nigrum* are taken in equal quantity and ground. Paste is made into pills of peanut seed size. Two pills are administered daily twice for 3 days.

Four to five flowers mixed with one teaspoonful of honey is given to chidren thrice daily for 3-4 days.

284. ***L. cephalotes*** **(Roth) Spreng. (Lamiaceae)**
VN: Pedda tummi H: Deldona S: Chhatruka

Cough: Five g of flower powder is given daily till cure.

Cold: Leaves are crushed on palm and the aroma is inhaled daily twice for 3 days.

285. ***Ligularia fischeri*** **(Ledeb.) Turecz. (Asteraceae)**
Cold: Leaves are crushed and used as snuff.

286. ***Limonia acidissima*** **L. (Rutaceae)**
Syn: *Feronia limonia* (L.) Sw., *F. elephantum* Correa

VN: Velaga S: Kapittha H: Kavita E: Elephant apple

Cough: Two teaspoonfuls of bark extract is given thrice a day till recovery.

287. *Lippia nodiflora* (L.) Rich. (Verbenaceae)
VN: Bokkena H: Bhuiokra S: Agnijvala

Cough and cold: Squeezed plant is inhaled.

288. *Litsea cubeba* (Lour.) Pers. (Lauraceae)
Chronic bronchitis and Asthma: Dry fruits with candy sugar are used.

289. *Lobelia nicotianaefolia* Heyne (Campanulaceae)
VN: Adavipogaku H: Nala S:Bibhishna E: Wild tobacco

Asthma: Smoke of leaves is inhaled 2-3 times a day till cure.

Croup: Leaves with tender leaves of *Bauhinia vahlii* and *Asparagus racemosus* are taken in equal quantities and ground. Paste is tied with cloth and bandaged around the hip till cure.

Whooping cough: Pulp of 2 ft tendril is mixed with a half glass of water and administered once a day for 5 days.

290. *Lobelia pyramidalis* Wall. (Campanulaceae)
Asthma and chronic bronchitis: Leaf decoction (about 10 leaves boiled in a cup of water) is given twice a day for 7 days.

291. *Ludwigia prostrata* Roxb. (Onagraceae)
Whooping Cough: Stem juice is taken thrice a day till cure.

292. *Luffa acutangula var. amara* Clarke (Cucurbitaceae)
H: Jhimni S: Ghantali

Asthma: Decoction of the plant is used till cure.

293. *L. tuberosa* Roxb. (Cucurbitaceae)
Asthma: Root extract is taken orally till cure.

294. *Madhuca indica* Gmel. (Sapotaceae)
Syn: *M. longifolia* (Koen.) Macbr. var. *latifolia* Chev.

VN: Ippa S: Madhuca H: Mohva E: South Indian mahua tree

Tuberculosis: Two teaspoons of pickle prepared from flowers is given twice a day for six months.

Asthma: Five flowers are boiled in a glass of water until it is reduced to half and is administered orally once a day for 5-10 days.

Pneumonia and cold: Oil along with mustard oil, *Pongamia pinnata* oil and cow's ghee mixed together and is put in a water container and churned. The butter like substance that emerges is applied on the chest for 7 days.

295. *M. longifolia* (Koen.) Macbr. (Sapotaceae)
E: Butter Tree H: Janglimoha S: Atavimadhuka

Bronchitis: Decoction of flowers is taken orally till cure.

296. *Maerua oblongifolia* (Forsk.) A. Rich. (Capparaceae)
Syn: *Capparis oblongifolia* Forsk.

VN: Bhuchakram S: Murahari

Asthma: Two teaspoonfuls of leaves are pounded with cumin and onion mixed with cow's milk and the mixture is given in early morning for 5 days.

297. *Maranta arundinacea* L. (Marantaceae)
VN: Palagunda E: Arrow root plant

Asthma: Rhizome powder along with milk is used till cure.

298. *Marsilea minuta* L. (Marsileaceae)
Cough and bronchitis: The decoction of the leaves with ginger is used.

299. *Martynia annua* L. (Martyniaceae)
Syn: *M. diandra* Gloxin

VN: Thelukondi S: Kakanasa H: Hatha-jori E: Tiger's claw

Cold: Flowers are shade dried, powdered and stored. One spoonful of powder with honey is consumed thrice a day for cold.

Sore throat: Leaf juice is used as a gargle.

300. *Melastoma malabathricum* L. (Melastomaceae)
VN: Gulapi S: Tinisah H: Phutki E: Indian Rhododendron

Cough: Leaves are pounded with garlic and ginger and taken till cure.

301. *Melia azedarach* L. (Meliaceae)
VN: Turaka vepa S: Himadruma H: Drek E: Persian lilac

Respiratory complaints: About 500 g of leaves are crushed with 3 spoonfuls of water and filtered. The extract is taken orally in the morning for 3 successive Saturdays.

302. *Melissa parviflora* Benth. (Lamiaceae)
Bronchitis and Tuberculosis: Leaf decoction is used till cure.

303. *Mentha spicata* L. (Lamiaceae)
Syn: *M. viridis* L.

VN: Pudina E: Spear mint

Cold and cough: About 5 spoonfuls of leaf decoction is administered daily twice for 3 days.

Throat infection: Three to four grams of the leaves are boiled in 300 ml of water with some common salt and the decoction is drunk twice a day warm for 3-5 days.

304. *Mesua nagassarium* (Burm. f.) Kost. (Clusiaceae)
Cold: Leaves are used in the form of poultice which is applied to head.

305. *Mimosa pudica* L. (Mimosaceae)
VN: Atti patthi S: Lajjalu H: Lajwanti E: Touch me-not

Asthma: About 20 ml of plant decoction is given with honey twice daily for a month.

Plant juice is mixed with coconut milk and used internally till cure.

Asthma and cough: About two teaspoonfuls paste of root (about 5 cm long 10 pieces of fresh root ground by mortar and pestle) is used twice a day for a week.

306. *Mimusops elengi* L. (Sapotaceae)
VN: Bogada S: Bakula H: Bakul E: Spanish cherry

Tuberculosis: Stem bark along with seeds of *Phoenix dactylifera* are powdered and given orally.

307. *Mitragyna parvifolia* (Roxb.) Korth. (Rubiaceae)
Syn: *Nauclea parvifolia* Roxb., *Stephegyna parvifolia* Korth.

VN: Bandari, Pulagara S: Vitanah H: Kaddam E: Kaim

Cold and cough: Stem bark mixed with leaf is squeezed and inhaled.

308. *Momordica charantia* L. (Cucurbitaceae)
VN: Kakara S: Karavellam H: Karela E: Bitter gourd

Asthma: A teaspoon of the root paste mixed with an equal quantity of honey or juice of *Ocimum* leaves is given once every night for a month.

309. *Morinda pubescens* J.E.Smith (Rubiaceae)
Syn: *M. tinctoria* Roxb.

VN: Togaru S: Papahnah H: Ach E: Togari wood of Madras

Asthma: Twenty to twenty five ml of stem bark extract is administered daily once for a fortnight.

Fruit is baked and is taken till cure.

310. *Moringa concanensis* Nimmo ex Dalz. & Gibs. (Moringaceae)
VN: Adavi munaga H: Sajana S: Bahulada

Cough: Stem bark extract (2 spoonfuls) is administered with sugar for 4 days.

311. *Moringa oleifera* Lam. (Moringaceae)
Syn: *M. pterigosperma* Gaertn.

VN: Munaga chettu S: Madhugunjana H: Soujna E: Indian horse radish

Cold and cough: Fruit is eaten as vegetable.

Acute cold: Fresh bark is inhaled to clear nasal congestion to get relief.

Asthma: Bark of this plant and seeds of *Piper nigrum* with sugar are ground and made into pills. These are taken twice a day till cure.

Fresh root juice is given with milk till cure.

Root decoction is taken along with asafoetida and salt till cure.

A handful of leaves are boiled in 180 ml of water for 5 minutes. After being cooled, a little salt, pepper and limejuice are added. This soup is administered once daily till cure.

312. ***Morus alba*** **L. (Moraceae)**
VN: Adavi munaga H: Chinni E: Mulberry

Throat inflammation: A decoction of the leaves is used as a gargle.

313. ***Mucuna pruriens*** **(L.) DC. (Fabaceae)**
Syn: *Dolichos pruriens* L., *Mucuna prurita* Hook., *M. utilis* Wight

VN: Dula dama S: Kapi kacchu H: Kivach E: Cowitch

Tuberculosis: Roots along with apical bud of *Plumeria rubra* are taken in 1:1 ratio and made into paste and 20 g of paste is given per day for 3 days.

314. ***Murraya koenigii*** **(L.) Spreng. (Rutaceae)**
Syn: *Bergera koenigii* L.

VN: Karivepaku S: Surabinimba H: Kari patha E: Curry leaf tree

Cough: Leaf extract mixed with honey is given 5 times.

315. ***Musa paradisiaca*** **L. (Musaceae)**
Syn: *M. sapientum* L.

VN: Arati S: Vanalakshmi H: Kela E: Plantain banana

Chronic cough: Flower is eaten as a medicine.

Asthma: Juice obtained from the plant is homogenized with honey and taken orally till cure.

Cough: After peeling the rinds, fruit is dried and ground. The powder (10-15 g) with water or with honey is taken twice a day for 3-4 days for difficulty in breathing.

316. ***Mussaenda laxa*** **(Hook. f.) Hutch. ex Gamble (Rubiaceae)**
Cough: Decoction of the stem is used as a remedy.

317. ***Myrica esculenta*** **Buch.–Ham. ex D. Don (Myricaceae)**
E: Box myrtle

Asthma: Five to ten ml of bark juice is given orally 3-5 times daily till cure.

318. ***Naravelia zeylanica*** **(L.) DC. (Ranunculaceae)**
Syn: *Atragene zeylanica* L.

VN: Konda mudusu teega S: Dhanavalli

Cold: Leaf paste is inhaled daily thrice through a cloth piece.

Cough and cold: Crushed whole plant is inhaled for quick relief.

319. ***Nasturtium officinale*** **R. Br. (Brassicaceae)**
Tuberculosis: Soup prepared from leaves and twigs is given to patients.

320. ***Nelumbo nucifera*** **Gaertn. (Nymphaeaceae)**
Syn: *Nymphaea nelumbo* L., *Nelubium speciosum* Willd.

VN: Thella thamara S: Pankaj H: Kamal E: Sacred lotus

Cold and cough: Roasted seed powder with black pepper and honey is said to be efficacious.

Asthma: Flower juice is taken orally 3 times a day till cure.

321. ***Nepeta floccosa* Benth. (Lamiaceae)**
Cough: Leaf decoction is taken twice a day till cure.

322. ***Nephrolepis cordifolia* (L.) Presl (Davalliaceae)**
Cough: About 4 teaspoons of root juice is given thrice a day till cure.

323. ***Neuracanthus sphaerostachys* (Nees) Dalz. (Acanthaceae)**
Asthma: A pinch of ash of the whole plant is mixed with either jaggery or honey and given orally twice a day for two weeks.

324. ***Nicotiana tabacum* L. (Solanaceae)**
VN: Pogaku S: Tamakhu H: Tamaku E: Tobacco

Cough: About 2 spoonfuls of leaf paste mixed with a spoonful of honey is administered twice daily for 3 days.

Cold: Flowers are boiled in milk and given into stomach.

Ten to twelve g of flowers are burnt to ash by heating in an air tight earthen pot and 2 g of this ash is given with *Piper betel* leaf once a day for about a month.

325. ***Nyctanthes arbor-tristis* L. (Oleaceae)**
VN: Parijatamu H: Harsinghar S: Atyuha E: Coral Jasmine

Bronchitis, asthma and whooping cough: Young leaf juice along with honey mixed with hot cow's milk is given to children twice daily for 2 weeks.

Whooping cough: Leaf juice of 3-5 ml with a pinch of common salt is given twice daily for 3-4 days in children.

Cough and cold: About 5 g of the bruised leaves are boiled in 500 ml of water and sieved. The resultant decoction is drunk thrice a day for 3 or more days.

Dried leaves with tobacco are smoked and aqueous extract of fresh leaf is taken once a day in the morning.

Grounded fruits are given with warm water.

Cold: Leaf juice (10 ml) mixed with equal portion of honey is taken orally 5 times a day for 2-3 days.

326. ***Ochna obtusata* DC. var. *pumila* (DC.) Kanis (Ochnaceae)**
Asthma: Dried root powder is smoked.

A decoction of the root is also used till cure.

327. ***Ochna squarrosa* L. (Ochnaceae)**
VN: Jammi S: Kanaka champa E: Golden champak

Tuberculosis: Root paste is made into pills and two pills are taken twice a day for one month.

328. ***Ocimum americanum* L. (Lamiaceae)**
Syn: *O. canum* Sims.

VN: Bhuthulasi S: Asurasa H: Ban Tulsi E: Common basil

Respiratory problems: Chopped stem pieces are worn around neck.

Whooping Cough: Equal proportions of the herb, ginger and honey mixed together is administered.

329. ***O. basilicum* L. (Lamiaceae)**
VN: Sabja S: Barbari H: Babuyitulsi E: Sweet basil

Whooping cough: Decoction of leaves aids to remove mucous from the bronchial tubes.

Cough and cold: Leaf juice with honey is given to children till cure.

Cough and bronchitis: Whole plant is pounded into paste and taken orally.

330. ***O. gratissimum* L. (Lamiaceae)**
VN: Ramatulasi S: Ramatulasi H: Ramtulsi E: Shrubby basil

Cough and cold: One teaspoonful of leaf juice is mixed with equal quantity of honey and a pinch of fruit powder of black pepper. This mixture is administered internally twice a day for a week.

Asthma: Dried leaf powder is smoked as a cigarette.

Four leaves are eaten early in the morning daily once for 3 days.

331. ***O. sanctum* L. (Lamiaceae)**
VN: Tulasi S: Surasah H: Kalatulsi E: Holy basil

Cough: Decoction is made by adding leaves of this plant with dried ginger and a pinch of black pepper and taken orally thrice a day.

Two spoonfuls of whole plant paste mixed with a spoonful of honey is administered twice daily for 3 days.

A tea prepared with the leaves is used.

Leaf juice of 5 ml with 2-3 drops of honey is given to children once daily for 2-3 days.

Ten to fifteen leaves are chewed and eaten in early morning for 4-5 days.

Influenza: About 10 ml of the leaf juice is added to 150 ml of boiled water with sugar or molasses and drunk 2-3 times a day for 2-5 days.

Cough and bronchitis: About 5 ml of the leaf juice is given with 1 ml of the juice obtained from the crushed rhizome of ginger and two teaspoonfuls of honey twice a day for 2-4 days for expectoration.

Tuberculosis: Powder of leaves along with sugar candies in milk is taken.

332. ***Odina wodier* Roxb. (Anacardiaceae)**
S: Jhingini H:Jingan E: Wodier

Asthma: Leaf or root or whole plant decoction is given internally till cure.

333. ***Olax scandens* Roxb. (Olacaceae)**
VN: Bapana mushti H: Dheniani

Cough and cold: Tender leaves are used as vegetable.

334. ***Oldenlandia umbellata* L. (Rubiaceae)**
VN: Cherivelu H: Chirval E: Chaya root

Bronchitis and asthma: Decoction of roots is administered till cure.

335. ***Oligocheta ramosa* (Roxb.) Wagenitz (Asteraceae)**
Cold and cough: Two g paste of equal quantities of stem and leaves is consumed twice a day for 4 days.

336. ***Opuntia dillenii* (Ker-Gawl.)Haw. (Cactaceae)**
Syn: *Cactus dillenii* Ker-Gawl., *C. stricta* (Haw.) Haw., *C. indicus* Roxb.

VN: Naaga jemudu S: Bahu shala H: Nag phana E: Slipper thorn

Cough: A spoonful of milky latex of stem is mixed with a glass of hot water and administered daily twice for 3 days.

Asthma: Fruit is baked or made into a syrup and is given in doses of about one teaspoonful 3 or 4 times daily till cure.

Whooping cough: Decoction of green fruits with paste of ginger (3:2) is given to children.

337. ***O. monacantha* (Willd.) Haw. (Cactaceae)**
VN: Nagajemudu

Whooping cough and asthma: Fruits are crushed to extract juice. Two to three teaspoonfuls of the juice are given 2-4 times a day for a week or longer.

338. ***Oroxylum indicum* Vent. (Bignoniaceae)**
E: Trumpet Flower H: Arlu S: Shyonaka

Asthma and bronchitis: Decoction of root bark and fruits is employed.

Cough: Bark, leaves and pods are boiled together in water and the decoction is drunk.

339. ***Osbeckia zeylanica* L. f. (Melastomataceae)**
Cold and Cough: Half a spoonful of plant paste is consumed with milk once a day for 3 days.

340. ***Oxalis corniculata* L. (Oxalidaceae)**
VN: Balusu koora, Senchulam S: Cangeri H: Amrul E: Indian sorrel

Cold: Two teaspoonfuls of leaf juice mixed with breast milk in equal quantities and given thrice a day till recovery to infants.

Juice of aerial shoots is mixed and warmed with equal volume of mustard oil and gently massaged on chest to treat congestion due to cold.

Whooping cough: Plant juice with goat milk (2 : 1) is given to children.

341. *Oxytropis cachemiriana* Cambess. (Fabaceae)
Cold and Cough: Decoction of leaves is used till cure.

342. *Paederia foetida* L. (Rubiaceae)
VN: Savirela H: Bakuchi S: Bala

Cold: Decoction is made from leaves along with those of *Murraya koenigii* in equal quantities with a little long pepper. About 15 ml is given twice or thrice daily till cure.

Ten to twenty ml decoction of the young shoot is taken orally 2-6 times daily till cure.

343. *Pandanus fascicularis* Lam. (Pandanaceae)
S: Ketaki H: Kevada E: Screw pine

Asthma: Two to five g of flower petals are rolled as a cigarette and smoked.

344. *Passiflora edulis* Sims. (Passifloraceae)
Asthma: Roots are dried, mixed with tobacco and smoked in a pipe till cure.

345. *Pavonia odorata* Willd. (Malvaceae)
VN: Chitti benda H: Sughandhabala S: Barhishtha

Cough and cold: Leaf powder or decoction is taken till cure.

346. *Pergularia daemia* (Forssk.) Chiov. (Asclepiadaceae)
Syn: *Asclepias daemia* Forsk., *Daemia extensa* (Jacq.) R. Br.

VN: Juttipaku S: Kurutakah H: Utaran E: Whitlow plant

Asthma: Ten to fifteen fresh flowers are crushed with 2-3 bulbs of garlic. This mixture is given orally early in the morning daily till it is cured.

Paste of leaves is given orally twice a day for a week.

Cold: Stem with fruits of *Phyllanthus emblica, Solanum surattense* and seeds of *Piper nigrum* are taken in equal quantities and ground. A spoonful of paste is administered daily twice for 3 days.

Juice of the leaves mixed with salt and turmeric powder is given.

Fresh shoot tip is introduced into nose results in continuous sneezing due to cold.

347. *Phlogacanthus curviflorus* Nees (Acanthaceae)
Cough: Red flowers mixed with fish curry is used.

348. *P. thyrsiflorus* Wall. (Acanthaceae)
Cough: Leaf decoction along with a little honey is taken.

Juice of 2-3 g flowers or leaves are given with water once daily for a week.

Leaf decoction with little common salt and honey is prescribed.

349. *P. tubiflorus* Nees (Acanthaceae)
Cough: Boiled leaf juice is given till cure.

350. ***Phoenix sylvestris*** **(L.) Roxb. (Arecaceae)**
S: Khar jura H: Khejur E: Wild date palm

Asthma: Root tuber is pounded with those of *Momordica dioica* and *Coccinia grandis* and the paste of one spoonful is administered once in a day for 3 days. Tamarind should be abstained during the treatment.

351. ***Phyla nodiflora*** **(L.) Greene (Verbenaceae)**
Syn: *Verbena nodiflora* L., *Lippia nodiflora* (L.) A. Rich.

VN: Bokkenaaku S: Jalapippali H: Chota-okra E: Purple lippia

Cold and cough: Whole plant is shade dried and the powder is given orally (2 g) along with milk.

Squeezed plant juice is inhaled.

352. ***Phyllanthus amarus*** **Schum. & Thonn. (Euphorbiaceae)**
Syn: *P. niruri* L., *P. fraternus* Webster

VN: Nela usiri S: Bhumyamalaki H: Bhu-Amla E: Phyllanthus herb

Asthma: Ten fruits are consumed in the morning for a month.

353. ***P. emblica*** **L. (Euphorbiaceae)**
Syn: *Emblica officinalis* Gaertn.

VN: Pedha usiri H: Amla S; Amalaki E: Gooseberry

Cough: Fruits are eaten.

Stem bark is crushed into paste and taken orally.

Asthma: About 2 spoonfuls of seed decoction is administered once a day in the evening for 30 days.

Decoction prepared from the fruit along with the roots of *Cyperus rotundus* and the stem of *Tinospora cordifolia* is mixed with honey and given frequently till cure.

Five g of the fruits are mixed with one tablespoon of honey and is taken every morning till cure.

354. ***P. gardnerianus*** **(Wight) Baill. (Euphorbiaceae)**
Cold: Two g of paste of aerial portion is consumed with 100 ml milk twice a day for 3 days.

355. ***P. reticulatus*** **Poir. (Euphorbiaceae)**
Syn: *Kirganelia reticulata* (Poir.) Bail.

VN: Puduguru S: Krishna-Komboji H: Kala-maimuda E: Black honey shrub

Cough, catarrh and Asthma: Decoction of the root is given orally till cure.

356. ***P. urinaria*** **L. (Euphorbiaceae)**
VN: Ettausirika H: Hazarmani S: Adhyanda

Cough: Two tablespoons of leaf extract is given twice a day.

357. ***P. virgatus*** **Forst. (Euphorbiaceae)**
Cough: Paste of whole plant is used till cure.

358. ***Physalis minima*** **L. (Solanaceae)**
VN: Lullubaba S: Tankari H: Tulatipati E: Sun berry

Cough: Two spoonfuls of whole plant paste mixed with a spoonful of honey is administered twice daily for 3 days.

359. ***Picrorhiza scrophulariiflora*** **Pennell (Scrophulariaceae)**
Cold and Cough: A decoction is prepared from about 3 g of the root in 400 ml of water and orally administered warm twice a day for 3-4 days.

360. ***Piliostigma malabaricum*** **(Roxb.) Benth. (Asclepiadaceae)**
Asthma: Crush the north bound bark and extract the juice by filtering through a cloth. Take it for one week. Avoid taking bath while using this treatment.

361. ***Piper betle*** **L. (Piperaceae)**
VN: Tamalapaku S: Tamboola H: Pan E: Betel leaf

Asthma: Leaves along with cloves, *Acorus calamus* leaves are boiled together and vapours are inhaled by covering with woollen blanket.

Leaf juice (10 ml) mixed with equal portion of honey is taken orally 5 times a day for 2-3 days.

362. ***P. cubeba*** **Rafin. (Piperaceae)**
Cough and bronchitis: Berries are powdered and taken till cure.

363. ***P. longum*** **L. (Piperaceae)**
VN: Pippallu S: Pippali H: Pipal mal E: Long pepper

Asthma: One pill of peanut size is made from long pepper and flowers of *Calotropis gigantea* in equal proportions and is given twice a day until cure.

Seeds along with honey are given in doses of 5-10 g daily till cure.

Root extract is administered orally till cure.

Whooping cough: Two to three seeds are roasted and mixed with about 3 g of the fried seeds of *Trachyspermum ammi* and a little ash of peacock feather. This preparation is eaten with a teaspoonful of honey before bed-time for 3 or more days. In course of this treatment, the patient is not allowed to take any rich food.

Cough: The mature spike is eaten raw.

Paste of powdered dry fruits of *Piper longum*, *P. nigrum* and *Zanthoxylum rhetsa* in equal quantities with a little water is made into pills of about 3 g each and 2 pills are given twice daily till cure.

Cold: Berries along with betel leaves are taken.

Fruits and seeds of *Piper nigrum* are taken in equal quantity and ground. A spoonful of paste mixed with a pinch of turmeric is administered daily twice for 3 days.

Bronchial disorders: Two spoonfuls of root decoction is administered once a day for 10 days.

364. *P. mullesua* Buch.–Ham. ex D. Don (Piperaceae)
Cold and Cough: Roasted fruits are used.

Cough: Leaves and fruits mixed with pepper and honey are taken.

365. *P. nigrum* L. (Piperaceae)
VN: Miriyalu S: Miracha H: Kali-Mirch E: Black pepper

Cough: Powdered fruit with ghee, honey and sugar is taken.

Cold: Leaf juice is given in doses of 2 spoonfuls twice a day for 5 days.

Cold and cough: A spoonful of leaf juice is administered once a day for 2 days.

Asthma and cough: Hundred ml of decoction of seeds (200 g) is administered before breakfast for 10 days.

366. *P. peepuloides* Roxb. (Piperaceae)
Cough: One to two fruits are roasted on coal fire and given to chew.

367. *P. retrofractum* Wall. (Piperaceae)
Cold and Asthma: Root juice of about 8-10 ml is used once daily till recovery.

368. *P. trichostachyon* (Miq.) DC. (Piperaceae)
Cough: Ripe fruits and leaves of *Ocimum sanctum* are ground with a cup of water. The mixture obtained is boiled with little salt and gur. This hot mixture is given 2-3 times a day.

369. *P. trioicum* Roxb. (Piperaceae)
Cough and Asthma: Paste of bark mixed with tea and sugar in hot water is given.

370. *Pistacia integerrima* Stewart (Anacardiaceae)
H: Kakra S: Chakra

Cough and throat diseases: Powder of galls and black pepper mixed in equal quantity with honey is given.

371. *Pistia stratiotes* L. (Araceae)
S: Kumbhikaa H: Jalkumbhi E: Water lettuce

Asthma and cough: Roots are made into poultice mixed with sugar is given.

Leaves are mixed with rose water and sugar and given till cure.

Leaf juice is administered in doses of 20-30 ml twice a day.

Decoction of the leaf is administered in doses of 30-40 ml twice a day.

372. *Pittosporum nepaulense* (DC.) Rehdr & Wilson (Pittosporaceae)
VN: Raka muki

Cough: Decoction of the bark is concentrated and 20 ml of it is taken orally thrice daily.

Plate 2.1

1. *Abrus precatorius;* 2. *Acorus calamus;* 3. *Adhatoda zeylanica;* 4. *Aegle marmelos;* 5. *Barringtonia acutangula;* 6. *Carissa carandas*

Plate 2.2

1. ***Clematis gouriana;*** 2. ***Clerodendrum viscosum***; 3. ***Martynia annua;*** 4. ***Solanum nigrum;*** 5. ***Terminalia arjuna;*** 6. ***Woodfordia fruticosa***

373. ***Plantago asiatica* L. (Plantaginaceae)**
Whooping Cough: Watery extract of the seed is given.

374. ***Plectranthus amboinicus* (Lour.) Spreng. (Lamiaceae)**
S: Karpuravalli H: Pashanabhedi E: Indian borage

Asthma: Leaf juice mixed with honey is given in doses of 2 spoonfuls twice a day for 5 days.

375. ***Plumeria alba* L. (Apocynaceae)**
VN: Lakshmi poolu S: Kanaka Karavisa H: Safed champa E: Pagoda tree

Cough: Ten drops of leaf latex is applied on the throat externally before bed for 2 days.

376. ***Polyalthia rufescens* Hook. f. & Thoms. (Annonaceae)**
Cough: Half a spoonful of leaf paste is consumed with water twice a day for 2 days.

377. ***Polygala arvensis* Willd. (Polygalaceae)**
Syn: *P. chinensis* L. VN: Chittodi

Cough and asthma: Root with two and a half seeds of *Piper nigrum* is made into a paste and mixed with hot cow's milk. This is prescribed twice daily for 7 days once on an empty stomach and again in the evening.

378. ***P. chinensis* L. (Polygalaceae)**
VN: Sarpakshi H : Meradu E: Common Indian milk wort

Cough: Decoction of the roots is given as an expectorant.

379. ***P. crotalarioides* Buch.-Ham. ex DC. (Polygalaceae)**
Cough: Roots are chewed to get relief thrice a day for 2 days.

380. ***P. sibirica* L. (Polygalaceae)**
E: Japanese senega

Cold, cough and bronchitis: A decoction of the root is given as an expectorant.

Asthma, chronic bronchitis and catarrhal affections: Infusion of the leaves is prescribed.

381. ***Polygonum glabrum* Willd. (Polygonaceae)**
Cough: The paste of whole plant is used.

382. ***P. plebeium* R.Br. (Polygonaceae)**
H: Raniphul S: Mastyakshi

Pneumonia: Entire plants are dried and then powdered. This powder is given internally.

383. ***Pongamia pinnata* (L.) Pierre (Fabaceae)**
Syn: *Cytisus pinnatus* L., *Pongamia glabra* Vent.

VN: Ganuga S: Karanja H: Karanj E: Indian beech tree

Whooping cough: Dry fruits are ground and mixed with pepper powder. Ten g is given once daily with pepper powder.

Ten g of dried fruit powder is given once daily with milk for 3 days.

384. *Portulaca quadrifida* L. (Portulacaceae)
VN: Paavilaku S: Waghulonika H: Chaunlayi E: Gold coast

Asthma: Dried seed powder with honey (3 : 2) is given for the treatment.

385. *Potentilla fruticosa* L. (Rosaceae)
E: Shrubby cinquefoil

Asthma and cold: Tea is prepared by boiling the roots in water till the extract becomes deep reddish and then milk and a little salt or sugar are added for taste and given.

386 *P. fulgens* Wall. ex Hook. (Rosaceae)
Cough and sore throat: Fifteen to twenty g of the fresh root is chewed in the morning and evening. The course is usually prescribed for 2 days.

387. *P. gelida* C.A. Meyer (Rosaceae)
Cold and Cough: Paste of leaves is made into tablets and given with milk.

388. *P. nepalensis* Hook. (Rosaceae)
Cold: Roots are boiled in water for 1-2 hrs till it becomes brownish red, then milk and sugar (or salt) are added according to taste. This herbal tea is taken.

389. *Pothos scandens* L. (Araceae)
Asthma: Stems are cut and smoked along with camphor.

390. *Premna herbacea* Roxb. (Verbenaceae)
VN: Adavinellikura H: Bharangi S: Bhumjambu

Tuberculosis: One glassful of root decoction is taken before bed time daily for a month.

391. *Primula macrophylla* D. Don (Primulaceae)
Cold and Cough: Extract of leaves and root stocks is taken directly.

392. *Prinsepia utilis* Royle (Rosaceae)
H: Bekkra

Cold and cough: Seed oil is warmed and smeared on the chest and throat before going to bed regularly.

393. *Prunus cerasoides* D. Don (Rosaceae)
H: Padam S: Charu E: Himalayan cherry

Asthma: Decoction of bark is taken orally.

394. *Pseudarthria viscida* (L.) Wight & Arn. (Fabaceae)
Syn: *Hedysarum viscidum* L., *Desmodium viscidum* (L.) DC.

VN: Batantu aaku S: Saliparni

Asthma: Roots are used in the form of decoction or powder till cure.

395. ***Psoralea corylifolia*** **L. (Fabaceae)**
E: Psoralea H: Babchi S: Bakuchi

Asthma: Decoction of fruits is taken internally.

396. ***Pterospermum xylocarpum*** **(Gaertn.) Sant. & Wagh (Sterculiaceae)**
Syn: *Velaga xylocarpa* Gaertn. VN: Lollugu, Tada

Asthma: Ten g decoction of the bark, 5 g of leaves and a few black peppers are given till cure.

397. ***Punica granatum*** **L. (Lythraceae)**
VN: Danimma S: Dadima H: Anar E: Pomegranate

Cough: Powdered fruit rind is taken with cold water to get relief.

398. ***Pupalia lappacea*** **(L.) Juss. (Amaranthaceae)**
Cough: Fruit is given in the form of soup.

399. ***Pyrus pashia*** **Buch.-Ham. ex D. Don (Rosaceae)**
Asthma: Four to five ripe fruits are taken twice a day for 15-30 days.

400. ***Quercus lucotrichophora*** **A. Camus (Fagaceae)**
Asthma: Twenty five to thirty ml decoction made from 50 g bark is given twice a day for 1-2 weeks.

401. ***Randia dumetorum*** **Lamk. (Rubiaceae)**
E: Common emetic nut

Cough: Root together with cloves of *Allium sativum* and salt is made into a paste and given internally.

402. ***Rauvolfia serpentina*** **(L.) Benth. ex Kurz (Apocynaceae)**
Syn: *Ophioxylon serpentinum* L.

VN: Pathala garidi S: Sarpa gandha H: Sarap-gandha E: Sarpentina root

Cold and cough: Root and whole plant of *Cissampelos pareira* are taken in equal quantity, dried and powdered. Two spoonfuls of powder mixed with a spoonful of honey is administered twice daily for 3 days.

403. ***Rhinacanthus longiflorous*** **Hook. f. (Zingiberaceae)**
Cough: About 10 ml of rhizome paste is given.

404. ***Rhodiola crenulata*** **(Hook. f. & Thoms.) Fu (Crassulaceae)**
Cold and cough: Decoction of tubers is taken.

405. ***R. tibetica*** **(Hook. f. &Thoms.) Fu (Crassulaceae)**
Cold and cough: Paste of rhizome mixed with the whole plant paste of *Polemonium caeruleum* is given.

Dry cough: Rhizome along with two to three inflorescence heads of *Spilanthes calva* is crushed and mixed in a spoon of honey and is given twice a day for 2-3 days.

Tuberculosis: Rhizome along with two spoonfuls powder of *Schrebera swietenioides* tender leaves and *Aegle marmelos* leaves is given with two spoonfuls of honey once a day till cure. Consumption of onion and potato is prohibited during the treatment.

406. *Rhus succedanea* L. (Anacardiaceae)
VN: Kakkara springi S: Karkatasrungi H: Kakkarsingi

Asthma: One to three g of powder of galls or 2-4 ml of its decoction is taken daily till cure.

Handful of leaves are mixed with seeds of *Cuminum cyminum* and made in the form of juice. One glass of juice is taken for 3 days.

407. *Rosa damacena* L. (Rosaceae)
VN: Gulabi H: Gulab S: Ati manjula E: Bussora Rose

Respiratory infections: Dry petals with sugar are fermented to produce a local preparation called *Gulkand* for relief during winter.

408. *Rotala rotundifolia* (Roxb.) Koehne (Lythraceae)
VN: Bethama S: Abhrapushpa E: Chair-bottom cane

Cough: Two tablespoons of leaf or stem extract is given twice a day.

409. *Salvadora persica* L. var. *wightiana* Verd (Oleaceae)
Vesical catarrh: Root decoction is used till cure.

410. *Salvia aegyptiaca* L. (Lamiaceae)
Asthma: Dried leaves are smoked.

411. *Sapindus emarginatus* Vahl (Sapindaceae)
Syn: *S. trifoliatus* L.

VN: Kunkudu chettu S: Arishta H: Ritha E: South Indian soapnut tree.

Asthma: Two g of fruit pulp is given orally twice a day till cure.

412. *Sauromatum venosum* (Ait.) Schott. (Araceae)
Tuberculosis: Corm powder with honey is given.

413. *Saussurea lappa* (Decne.) Sch.-Bip. (Asteraceae)
VN: Enugatippali E: Costus H: Kot S: Agada

Bronchial asthma: Powdered roots are mixed with roots of *Inula racemosa* in equal quantities and one teaspoonful of powder is given thrice a day up to one month.

414. *Schleichera ferox* L. (Sapindaceae)
Cough and cold: Fruits (50 g) are ground with 3 green peppers and a pinch of salt. Ten g of the preparation is given in the morning and evening for 3 days.

415. *S. oleosa* (Lour.) Oken (Sapindaceae)
Tuberculosis: Bark juice with seed powder of *Thespesia lampas* is prepared. One spoon is taken twice a day with hot water till cure.

416. ***Scindapsus officinalis*** **(Roxb.) Schott. (Araceae)**
Asthma: Fruit decoction is given as an expectorant.

417. ***Scoparia dulcis*** **L. (Scorphalariaceae)**
E: Sweet Broomweed H: Chinibuti

Bronchitis: Leaf infusion is administered till cure.

418. ***Semecarpus anacardium*** **L. f. (Anacardaceae)**
VN: Nalla jeedi S: Bhutamashana H: Bhilawa E: Marking nut tree

Asthma: Juice of the shell of the nut is taken in doses of 1-2 ml daily till cure.

One tablespoon of milk boiled with seeds is given orally to children to get quick relief.

419. ***Sesamum indicum*** **L. (Pedaliaceae)**
Cold and Cough: Few drops of seed oil is administered into the nostril to clear nasal congestion. It is repeated for few days to get relief from cold and cough.

420. ***Sesbania grandiflora*** **Pers. (Fabaceae)**
E: Sesben H:Basna S: Agastyah

Bronchitis: Infusion of flowers is administered till cure.

421. ***Setaria italica*** **(L.) Beauv. (Poaceae)**
Syn: *Panicum italicum* L.

VN: Korralu S: Kanguh H: Kangni E: Italian millet

Asthma: Hundred g of seeds with 3 tender leaves of *Thysanolaena maxima* are boiled in water and made into decoction. A spoonful of decoction is administered daily twice for 21 days.

422. ***Sida acuta*** **Burm. f. (Malvaceae)**
Syn: *S. carpinifolia* L. f.

VN: Gayapaku S: Bala H: Kharenti E: Hornbean leaved sida

Asthma: Bark powder is administered orally till cure.

423. ***Solanum anguivi*** **Lam. (Solanaceae)**
Syn: *S. indicum* L.

VN: Dinga bhediri S: Bakobhijri H: Birhatta E: Indian night shade

Asthma: Dried leaves and fruits are smoked through smoke pipe till cure.

One teaspoonful of root extract in water is given orally in early morning daily for two weeks.

Two teaspoons of whole plant powder is administered with honey daily once for a fortnight.

Five g of root paste is administered orally till cure.

424. *S. incanum* L. (Solanaceae)

E: Bitter apple

Asthma: Seed powder of about 500-1000 mg (according to the resistance power of the patient) is given.

425. *S. nigrum* L. (Solanaceae)

Syn: *S. americanum* Mill.

VN: Jaji kura S: Kakamachi H: Kamuni E: Black night shade

Cold and cough: Fruit is cut lengthwise, filled with purified butter and eaten.

Dry fruit and seed decoction along with black pepper is given in doses of one teaspoonful twice a day till cure.

Cough: Four to six ripe fruits are chewed and ingested as an expectorant. The course is 3-4 times each with an interval of 3-4 hours.

About 10 ripe fruits are given orally three times a day to children for a week till cure.

Lung tuberculosis: Ripe fruit paste with cow milk (5 : 3) is used for the treatment.

Whooping cough: A spoonful of root powder mixed with a spoonful of honey is administered to children once a day for 2 or 3 days.

426. *S. surattense* Burm. f. (Solanaceae)

Syn: *S. virginianum* L., *S. xanthocarpum* Schrad. & Wendl.

VN: Mulla vanga S: Katakari H: Kateli E: Dwarf lid brinjal

Asthma: One teaspoon of root powder is mixed with a pinch of pepper powder and fried ginger. This is administered with honey daily once till cure.

Forty to fifty ml decoction of whole plant is given orally twice a day till cure.

Dried seed powder is smoked till cure.

Whole plant is dried, powdered and stored. About one g of powder mixed with equal quantity of black pepper powder is given with sugar or honey till cure.

Cold, cough and catarrh: Plant decoction along with *Piper longum* root powder is taken daily on empty stomach till cure.

Cough and asthma: Flowers fried in ghee are chewed everyday for 15 days or till cure.

Decoction of root (about 5 cm long, 10 pieces of root boiled in a cup of water) is taken orally twice a day for seven days.

Forty to fifty ml decoction of whole plant is given twice a day for required period.

Cold: Five g of green fruits are crushed and boiled with 2 cups of water until the volume is reduced to half. This is strained and 5-10 ml of this decoction is given to children thrice a day for one day.

Spasmodic cough: Decoction of root is given with pepper and honey with rock salt.

Whooping cough: Fruit paste with leaf juice of *Shorea robusta* and common salt (5 : 3 : 1) is given to children.

Cold and cough: Ten fried fruits are given as cure.

Cough: About 5 g root with an equal quantity of seeds of *Piper nigrum* are boiled in 250 ml water till the volume of water gets reduced to about 50 ml. This concentrate is given twice a day.

Dried plant powder mixed with turmeric is taken orally till cure.

Unripe fruits are eaten.

427. *S. torvum* Sw. (Solanaceae)
VN: Dhungiri

Cough: Fruits are given to children.

428. *S. trilobatum* L. (Solanaceae)
S: Achuda, Alarka

Asthma and cold: Juice of leaves is taken orally for 7 days or until cure.

429. *S. tuberosum* L. (Solanaceae)
Cough: Leaves with those *of Ocimum tenuiflorum* and seeds of *Piper nigrum* are taken in equal quantity and ground. Two spoonfuls of paste is mixed with a glass of milk and administered twice daily for 3 days.

430. *Solena heterophylla* Lour. (Cucurbitaceae)
Syn: *Melothria heterophylla* (Lour.) Cogn.

VN: Adavi donda S: Amla vethasa

Throat infection: About 10 g of the fruit or root paste is taken thrice a day for 2-4 days.

431. *Sonchus arvensis* L. (Asteraceae)
VN: Nallatapata E: Corn sow-thistle

Bronchitis and asthma: Root decoction is given twice daily for 6 days.

432. *Soymida febrifuga* (Roxb.) A. Juss. (Meliaceae)
Syn: *Swietenia febrifuga* Roxb.

VN: Sami chettu H: Chandravallabha H: Rohan E: Rohan tree

Cough: Fifty ml of bark decoction is used three times daily till cure.

Cold and cough: Stem bark decoction is used for gargling.

433. *Spilanthes calva* DC. (Asteraceae)
E: Para cress

Sore throat: Two to three flower heads are chewed and swallowed 1-2 times a day regularly.

Tuberculosis: Root juice is given 2-3 times daily for at least 7 days.

Cough, cold and catarrh: One teaspoonful of the juice of the flower buds is taken orally daily in the morning till cure.

434. ***Spondias pinnata*** **(L.) Kurz (Anacardiaceae)**
VN: Adavimamidi H: Amara S: Ambarataka E: Bile tree

Whooping cough: Two to three roasted seeds are chewed 2-3 times a day for 1-2 days.

435. ***Strychnos nux-vomica*** **L. (Loganiaceae)**
VN: Mushidi S: Kupilu H: Kuchila E: Nux-vomica

Asthma: Stem bark is crushed with black pepper and this decoction is administered in doses of 2-3 spoonfuls twice a day for 45 days.

436. ***S. potatorum*** **L. f. (Loganiaceae)**
VN: Induga S: Katakah H: Nirmali E: Clearing nut tree

Asthma: Two de-seeded ripe fruits along with 1 g of turmeric and 1 g seeds of *Cuminum cyminum* are made into a paste and taken orally till cure.

437. ***Swertia chirata*** **(Wall.) C. B. Clarke (Gentianaceae)**
S:Nepalanimbah H: Chirayata E: Indian gentian

Asthma: Whole plant is taken and the infusion is made in hot water with aromatics like cloves, cinnamon and given in doses of 15-30 ml.

438. ***Syzygium aromaticum*** **(L.) Merrill & Perry (Myrtaceae)**
Asthma: A teaspoon of decoction prepared by boiling 6 cloves in 30 ml of water is given with honey thrice daily till cure.

439. ***Tacca leontopetaloides*** **(L.) O. Kuntze (Taccaceae)**
H: Diva E: Indian arrow root

Asthma and cough: About 3 teaspoons of paste obtained on mixing 10 g of dried stem bark powder with 50 ml water is given thrice a day orally for a week.

440. ***Tagetes erecta*** **L. (Asteraceae)**
E: Marigold H: Genda S: Sandu

Bronchitis: Decoction of flowers is used externally.

441. ***Tamarindus indica*** **L. (Caesalpiniaceae)**
VN: Chintha chettu S: Emlika H: Amli E: Tamarind tree

Cough: Five to ten years old fruit is given to the children.

Bark along with that of *Capparis zeylanica* is boiled in coconut oil and applied all over the body of children.

442. ***Taxus baccata*** **L. ssp. *wallichiana* (Zucc.) Pilger (Taxaceae)**
E: Yew

Cold and cough: Decoction of leaves is used as herbal tea.

443. ***Tecomella undulata*** **(Sm.) Seem. (Bignoniaceae)**
Cough: Powdered stem bark is given orally once a day for a week.

444. *Tectona grandis* L. (Verbenaceae)
VN: Teku E: Teak H: Begum S: Baka

Bronchitis: Infusion of flowers is taken internally.

445. *Tephrosia purpurea* (L.) Pers. (Fabaceae)
Syn: *Cracca purpurea* L., *Tephrosia hamiltonii* Drum.

VN: Vempali S: Sarapunka H: Sarpanka E: Wild indigo

Pulmonary diseases: Dried root bark is burned and the smoke inhaled till cure.

Asthma and cough: Pulverized roots are smoked for relief.

446. *Terminalia arjuna* (Roxb. ex DC.) Wight & Arn. (Combretaceae)
Syn: *Pentaptera arjuna* Roxb. ex DC.

VN: Tella maddi S: Arjuna H: Arjun seadada E: Arjun mytralam

Asthma and bronchitis: Bark is boiled in water and the decoction is taken orally till cure.

Bark is powdered and stored in a well-corked bottle. The patient is kept fasting on the full moon day. A dish of condensed milk and rice (*kheer*) should be prepared and placed where the full moonlight falls upon it the whole night. During the early hours, 12 g of the powdered bark is sprinkled over the *kheer* and is taken to his satisfaction. He should not sleep for at least 12 hours after consumption.

447. *T. bellirica* (Gaertn.) Roxb. (Combretaceae)
Syn: *Myrobalanus bellirica* Gaertn.

VN: Tadi S: Bibhitakah H: Badami E: Belleric myrobalan

Dry cough: A spoonful of dry unripened fruit powder with a pinch of salt mixed in a glass of hot water is administered daily thrice for 4 days.

Peel of one fruit with a little pulp is given to chew.

Sore throat and Cough: A piece of dried fruit rind is kept in mouth till cure.

Asthma: Dried fruits are ground, infusion is mixed with honey and given orally.

448. *T. chebula* Retz. (Combretaceae)
VN: Karakkai S: Harithaki H: Harra E: Black myrobalan.

Cough: Fruit powder is mixed with the water or cow's or goat's milk and taken internally twice a day for 7 days.

Asthma, bronchitis and other respiratory ailments: Equal quantity of fruits along with those of *T. bellirica* and *Phyllanthus emblica* are ground into powder. One teaspoon of the powder is mixed with half teaspoon of honey and taken orally twice a day till cure.

Tuberculosis: Fresh bark (5 g) is crushed with the seeds of *Piper nigrum* (1 g) and taken orally at noon.

Asthma: One spoonful of nut soup mixed with honey is administered daily till cure.

Fruit is coarsely powdered and smoked in a pipe.

Fruits are ground to powder with the fruits of *Terminalia chebula, Balanites aegyptiaca,* roots of *Aristolochia indica, Rauvolfia serpentina* and cloves. One spoonful of powder along with honey is given thrice a day for 30 days.

Bark along with that of *Barringtonia acutangula, Pterocarpus marsupium* are ground on a rock and is mixed in a glass of water and administered orally thrice a day.

Fruits along with those of *Terminalia chebula, Melia dubia,* flowers of *Calotropis gigantea, Datura innoxia* and leaves of *Justicia adhatoda* in equal parts are ground into paste and made into pills. These are taken twice a day till cure.

Two or three galls are crushed and taken with water till cure.

A piece of the fruit is chewed every night till cure.

A fine paste of the fruits with carrion oil effects a more rapid cure.

Cough: One roasted fruit is given to chew for instant relief.

Fruits and roots *of Piper nigrum* are taken in equal quantity and ground. Two spoonfuls of paste is administered twice daily for 2 days.

One g of fruit powder mixed in one tumbler of water is given orally.

Twenty g of bark paste is taken in a glassful of hot water once a day for 3 days.

Stem bark and dried seed powder mixed with hot milk taken orally.

Fruits are powdered and one spoonful of powder is taken daily with cold water.

Sore throat: A fruit is roasted a little on burning charcoal and chewed slowly.

Cold and cough: Fruit powder is mixed with the water or cow's or goat's milk and taken internally twice a day for 7 days to get relief.

449. *Thespesia lampas* (Cav.) Dalz. ex Dalz. & Gibs. (Malvaceae)
Syn: *Hibiscus lampas* Cav.

VN: Konda patti S: Karpasi E: Ran-bendi

Bronchitis: Dried roots and fruits are powdered, burnt and inhaled as smoke thrice daily.

Cough: Dried ripe fruit is taken.

450. *Thymus linearis* Benth. ex Wall. (Lamiaceae)
Pulmonary infections and cold: One or two teaspoonfuls of decoction of the seeds is administered once or twice a day orally to children.

451. *T. serpyllum* L. (Lamiaceae)
H: Banajwain E: Bank thyme

Cold and cough: Plant decoction is used as an effective home remedy.

452. *Thysanolaena maxima* (Roxb.) Kuntze (Poaceae)
Syn: *Agrostis maxima* Roxb.

VN: Konda chipuru gaddi S: Bushnis E: Broom grass

Tuberculosis: Pills prepared from leaf paste are taken twice daily.

453. *Tinospora cordifolia* (Willd.) Miers ex Hook. f. & Thoms. (Menispermaceae)
Syn: *Menispermum cordifolium* Willd., *Cocculus cordifolius* (Willd.) DC.

VN: Thippa theega S: Guduchi H: Amrit E: Gulancha tinospora

Cold and cough: Stem along with tuberous root of *Cyperus rotundus* and fruits of *Phyllanthus emblica* are taken in equal quantity and made into decoction. Two spoonfuls of decoction mixed with half spoon of honey is administered daily twice for 3 days.

Cough: Dried stem, dried ginger, black pepper and long pepper are powdered and taken in doses of 2 spoonfuls with honey.

Crushed root is given with milk to the children.

454. *Toddalia asiatica* (L.) Lam. (Rutaceae)
Syn: *T. aculeata* Pers.

VN: Vuppu kampa S: Thikshnaksha H: Kanj E: Forest pepper

Breathing trouble: Leaf paste is administered twice a day for getting fast relief.

455. *Trachyspermum ammi* (L.) Sprague ex Turril. (Apiaceae)
H: Ajuan E: Ajowan

Asthma: A hot and dry fomentation of the fruits is done on the chest till cure.

Half a teaspoon of the juice of the whole plant is mixed in a glass of buttermilk and taken twice daily till cure.

A hot poultice of the seeds is used for dry fomentation on the chest twice daily.

Whole plant is boiled in water and the steam is inhaled twice a day till cure.

Sore throat and catarrh of upper respiratory tract: Inhaling the smoke of burnt seeds is effective.

Cold and cough: Ten to twenty g of seed powder with hot water is given till cure.

456. *Tragia involucrata* L. (Euphorbiaceae)
VN: Chinna dulagondi H: Barhanta S: Duralabha

Whooping cough: About 5 g of dried leaves powder is given to children.

Chronic cough: Root powder is smoked.

457. *Trewia nudiflora* L. (Euphorbiaceae)
VN: Eruponaku H: Bhillaura S: Karahata

Whooping cough: Three to four ripe fruits are ground with 100 ml cow's milk. The preparation is taken once a day in the morning till relief.

458. *Trianthema portulacastrum* L. (Aizoaceae)
Syn: *T. monogyna* L., *T. obcordata* Roxb.

VN: Galijeru S: Punarnava H: Santhi E: Horse-purslane

Asthma: Decoction of the root in doses of 20-30 ml is administered orally.

459. ***Tribulus terrestris*** **L. (Zygophyllaceae)**
VN: Palleru S: Bahukantaka H: Chologokhuru E: Calthrops

Cold and cough: Dried fruits are crushed into powder form and soaked overnight in water which is taken twice or thrice a day up to one month.

460. ***Trichodesma indicum*** **(L.) R. Br. (Boraginaceae)**
VN: Guvvagutti S: Adhapushpi H: Dhadhona

Cough: Whole plant is shade dried and powdered. The powder mixed with honey is given.

461. ***Trichosanthes palmata*** **Roxb. (Cucurbitaceae)**
VN: Abuba H: Indrayan S: Mahakala

Asthma: Fruit smoke is inhaled.

462. ***Tridax procumbens*** **L. (Asteraceae)**
E: Mexican Daisy H: Pardeshi Langri

Bronchial catarrh: Leaf infusion is administered to bring out the phlegm from the bronchial tubes.

463. ***Trigonella foenum-graecum*** **L. (Fabaceae)**
VN: Menthikura E: Fenugreek H: Methi S: Methika

Bronchitis: Grounded seeds with honey are consumed.

464. ***Tylophora dalzellii*** **Hook. f. (Asclepiadaceae)**
Asthma: One or two leaves are taken in the morning and evening.

465. ***T. indica*** **(Burm. f.) Merr. (Asclepiadaceae)**
Syn: *Cynanchum indicum* Burm. f.

VN: Asthma theega S: Ajadvishta H: Antamul E: Emetic swallows wort

Bronchitis: Decoction of dried leaves is beneficial to remove mucous from bronchii.

Asthma and bronchitis: Decoction of leaves and infusion of root bark is administered.

Asthma: Leaves and roots with betel leaves, cardomom and garlic are made into tablets and used till cure.

Dried leaf powder is smoked.

One tender leaf with 3 black pepper fruits is chewed on empty stomach once a day for 30 days.

Two tender leaves are chewed daily in the early morning for a month.

Two tender leaves are chewed along with a glass of warm water early in the morning for a week.

Five ml of leaf juice or 200 mg powder is drunk in hot water daily in the morning till cure.

One cup of decoction of the whole plant is taken twice a day for 2-3 days.

Watery extract of 20 g leaves along with those of *Acalypha indica* is prepared and given twice a day.

Powder mixture is made from leaves of this plant along with *Justicia adhatoda* (leaves) 500 g; *Solanum surattense* (whole plant) 500 g; *Ziziphus mauritiana* (bark) 100 g; *Terminalia bellirica* (fruit rind) 200 g; *Datura metel* (leaves) 50 g; *Calotropis procera* (leaves) 50 g; *Clerodendrum serratum* (wood &bark) 50 g; *Piper longum* (fruits) 50 g; *Cinnamomum zeylanicum* (bark) 50 g; *Glycyrrhiza glabra* (roots) 200 g. Five g of mixture is administered four times a day. Patients are advised to drink warm water only.

466. *Urena lobata* L. (Malvaceae)
Syn: *U. sinuata* L.

VN: Menda ckikni S: Atibala H: Kunjuya

Dry and inveterate cough: Flowers are given.

467. *Urginea indica* Kunth. (Liliaceae)
VN: Adavithellagadda S: Kolakanda H:Kanda E: Indian squill

Asthma and bronchitis: Bulbs along with midrib of betel leaf are crushed and the extract is used.

468. *Vanda tessellata* (Roxb.) Hook. ex G. Don (Orchidaceae)
Syn: *V. roxburghii* R. Br., *Epidendrum tessellatum* Roxb.

VN: Rasna Gudika S: Rasana H: Rasna E: Ichneumon plant

Bronchitis: Velamen root and leaves of *Ocimum tenuiflorum* taken in equal quantities are ground with honey. A spoonful of paste is administered daily twice for 4 or 5 days.

469. *V. spathulata* Spreng. (Orchidaceae)
Tuberculosis and Asthma: Flowers are given.

470. *Verbascum thapsus* L. (Scrophulariaceae)
H: Gidartamaku E: Aaron's rod

Asthma: Leaves are smoked.

Cough: Root powder with honey is given.

Asthma and other pulmonary diseases: Leaf decoction is given till cure.

471. *Viburnum grandiflorum* Wall. (Caprifoliaceae)
Whooping cough and bronchial troubles: A teaspoonful of dried seeds is taken with lukewarm water.

472. *Viola canescens* Wall. (Violaceae)
Asthma: Flowers with cardamom and *Cinnamomum tamala* powder is prescribed.

473. *Viola hepatonicifolia* J. E. Sm. (Violaceae)
Cold and cough: Dried flowers are mixed with tea and used.

474. ***V. pilosa*** **Bl. (Violaceae)**
Syn: *V. canescens* Wall. ex Roxb.

Cold and cough: Decoction of the plant along with black pepper is given once a day in small quantity (28-30 ml).

475. ***Vitex negundo*** **L. (Verbenaceae)**
VN: Vavili S: Nirgundi H: Nirgondi E: Five leaved chaste tree

Cough: Fresh leaves are boiled with water and the vapour is inhaled twice a day.

Whooping cough: Stem decoction given 2-3 times a day till cure.

Stem decoction with paste of long pepper (3:1) is given to children.

Asthma: Dried leaf powder's smoke is inhaled till cure.

Cough, catarrh and cold: Crushed and warmed leaves are plastered on the body parts with thin cloth to get relief.

Watery sap from warmed stem is given to children.

Cold and cough: Leaf paste is applied on the forehead.

Fresh leaves are boiled with water and the vapour is inhaled twice a day

Catarrhal affections: A decoction of the leaves and the vapours are employed in bathing.

Removal of mucous from bronchial tubes: Root decoction is taken orally.

476. ***V. peduncularis*** **Wall. (Verbenaceae)**
Cough and cold: A cup of decoction of stem bark is given twice a day for 4 days.

477. ***V. trifolia*** **L. (Verbenaceae)**
VN: Chiruvavili H: Panikisanbhalu S: Indranika

Catarrh: Leaves are smoked for relief.

Decoction of the leaves and the vapours are employed in bathing.

478. ***Wedelia chinensis*** **(Osbeck) Merr. (Asteraceae)**
Cold and cough: Whole plant decoction is administered 2-3 times a day till cure.

479. ***Withania somnifera*** **(L.) Dunal (Solanaceae)**
Syn: *Physalis somnifera* L.

VN: Aswagandha S: Asswagandha H: Asgandhal E: Aswagandha

Asthma: Decoction or powder or juice of the tuber is given orally till cure.

Infusion of the bark is administered orally till cure.

Burnt root powder with butter is taken every day early morning.

Root paste of 5 g is boiled in 20 ml of mustard oil and after straining, the oil is gently massaged on chest.

480. ***Woodfordia fruticosa*** **(L.) Kurz (Lythraceae)**
Syn: *Lythrum fruticosum* L., *Woodfordia floribunda* Salisb.

VN: Jaji S: Dhataki H: Dhai E: Fire flame bush

Cough: Flower paste with honey (2:1) is administered.

481. *Wrightia tinctoria* (Roxb.) R. Br. (Apocynaceae)
Syn: *Nerium tinctorium* Roxb., *Wrightia rothii* G. Don

VN: Pala ankudu S: Krishna Kutaja H: Kala-Inderjau E: Sweet indrajao

Asthma: Three drops of leaf juice is administered into the nose as nasal drops till cure.

Latex with jaggery is taken internally in the form of pills of the bengal gram seed size twice a day for about 15 days.

Bronchitis: Latex extract is mixed with honey and taken in the form of pills of bengal gram seed size twice a day for about 15 days.

482. *Xanthium indicum* Koen. (Asteraceae)
Syn: *X. strumarium* L.

VN: Marulamatangi H: Banokra S: Arishta E: Bur-weed

Cold and cough: Decoction of the fruit in combination of *Psidium guajava* is given.

483. *Zehneria mysorensis* (Wight & Arn.) Arn. (Cucurbitaceae)
Cold and cough: Leaf powder is snuffed.

484. *Zingiber officinale* Rosc. (Zingiberaceae)
VN: Allamu S: Ardrakam H: Adrach E: Ginger

Cough and throat pain: Rhizome is crushed into paste and taken orally.

About a teaspoonful of the juice obtained from the crushed rhizome is mixed with double the amount of honey and eaten before bed time for 3-5 days.

Tuberculosis: Rhizome extract is taken.

Cough: Two to four ml of juice of rhizome with half cup of water and a pinch of common salt is used.

Four to five drops of juice of rhizome is taken with honey twice a day for 4-5 days.

Infusion of the rhizome mixed with a teaspoonful of the leaf juice of *Ocimum sanctum* and honey is given orally once a day for 3 days.

About 1 g of the common salt is put in the hollow of a piece of ginger and is placed on burning charcoal for a minute. This ginger is kept in the mouth while chewing and inhaling very slowly. This treatment is usually practiced before bedtime for 3 or more days.

About a teaspoonful of the juice from the crushed rhizome is mixed with an equal quantity of cow's butter and warmed. The warm ointment is used to massage on the chest and throat. Mostly, it is administered before bedtime for 3-4 days.

Asthma: Two to three g of stem tuber mixed with 4-5 *Ocimum sanctum* leaves is taken before sunrise on empty stomach for 5 days.

Juice of fresh ginger is given in doses of 0.5 to 2 ml daily with honey till cure.

A teaspoon of fresh juice of the rhizome mixed with a cup of fenugreek decoction and honey is taken once in the morning and evening till cure.

Cold and whooping cough: Rhizome juice alone or mixed with lemon juice and *Kaempferia galanga* rhizome is taken.

485. *Z. rubens* Rosc. (Zingiberaceae)
Cold: Paste of rhizome is applied on the forehead.

486. *Z. zerumbet* (L.) J.E. Sm. (Zingiberaceae)
Cough: About 10 ml of juice extracted from rhizomes is given thrice daily.

487. *Ziziphus mauritiana* Lam. (Rhamnaceae)
Asthma: About 10 g of root, 25 g of *Curculigo orchioides* rhizome and 5 g dried ginger are powdered with butter oil and made into pills. Two pills are given twice a day.

488. *Z. xylopyrus* (Retz.) Willd. (Rhamnaceae)
Syn: *Rhamnus xylopyrus* Retz.

VN: Gotti manu S: Ghonta H: Kat-ber

Cold: Stem bark is soaked in boiling mustard oil. Filtered oil is prescribed for massage on the body of an infant every 8 hrs.

Summary and Conclusions

The present article is an attempt to know the state of the art regarding medicinal flora useful especially against respiratory complaints like bronchitis, cold and cough, dry cough, asthma, tuberculosis, influenza, whooping cough, sore throat, throat pain, croup, catarrhal infections and some other pulmonary diseases. It revealed about 488 species that are enumerated with their recent scientific name, family, their familiar common name, wherever available in Sanskrit (S), Hindi (H) and English (E), part/s used, mode of preparation and disease to be cured.

Varied climatic and topographical conditions prevailing in India has bestowed upon it a rich and diverse flora which is responsible for the richness and uniqueness of medicinal plants. Numerous wild plants growing in Indian forest are used as folklore medicines to prevent or cure several diseases. Medicinal plants have a potential in today's era, as a number of pathogens are becoming resistant. In certain cases, plant molecules have formed excellent base for synthesis of certain life saving drugs.

Medicinal plants are important for pharmacological research and drug development, not only when plant constituents are used directly as therapeutic agents but also when they are used as basic material for synthesis of drugs or as model for pharmacologically active compounds. Ancient knowledge coupled with scientific principles can provide us with powerful remedies to eradicate the diseases. Exploration for forest based plant products for new pharmaceuticals and the demand for medicinal plants are increasing. In addition to purified plant derived drugs there is an enormous market for crude herbal medicines.

Acknowledgements

BSS is grateful to the UGC for providing fellowship under FDP scheme.

References

Agarwal, V. S. and Ghosh, B. 1985. Drug plants of India (Root Drugs). Kalyani Publishers, New Delhi.

Bhattacharjee, S. K. 2001. Hand book of medicinal plants. Pointer Publishers, Jaipur.

Bhattarai, N.K. 1993. Folk medicinal use of plants for respiratory complaints in central Nepal. Fitoterapia 64(2):

Dastur, J. F. 1962. Medicinal Plants of India and Pakistan, D.B. Taraporevala Sons & Co. Private Limited, Bombay.

Gamble, J. S. and Fischer, C.E.C.. 1916-1935. The Flora of Presidency of Madras. 3 Vols., BSI, Calcutta.

Jain, S. K. 1997. Contribution to Indian Ethnobotany, Scientific Publishers, Jodhpur, India.

Kirtikar, K. R. and Basu, B. D. 2003. Indian Medicinal Plants Vol 1-11; Oriental Enterprises, Dehradun, Uttaranchal, India.

Krishnam Nambiar, V. P. and Sesidharan, N. and Renuka, C. and Balagopalan, M. 1982. Studies on the Medicinal Plants of Kerala forests, KFRI Research Report No. 42, Kerala Forest Research Institute, Peechi, Kerala.

Pal, D. C. and Jain, S. K. 1998.. Tribal Medicine. Naya Prokash, Calcutta.

Patil, D.A. 2005. An inventory of medicinal plants useful against respiratory complaints. In: Herbal Medicine for Human Diseases, Vol-II, 244-257. Irfan Ali Khan and Atiya Khanum (Eds.) Ukaaz publications, Hyderabad, India.

Pullaiah, T. 2002. Medicinal Plants of Andhra Pradesh (India). Regency Publications, New Delhi.

Seetharami Reddi, T.V.V., Prasanthi, S. and Ramarao Naidu, B.V.A. 2005. Native Phytotherapy for Asthma. In: Herbal Medicine for Human Diseases Vol-II, 38-100. Irfan Ali Khan and Atiya Khanum (Eds.) Ukaaz publications, Hyderabad, India.

Suneetha, J., Seetharami Reddi, T.V.V. and Prasanthi, S. 2009. Herbal therapy for cold and cough from East Godavari district (Andhra Pradesh). Journal of Non-Timber Forest Products, 16: 135-138.

Thammanna and Narayana Rao, K. 1990. Medicinal Plants of Tirumala.Tirumala Tirupati Devasthanam, Tirupati.

Varier, P. S. 1994-1995. Indian Medicinal Plants Vol. 1-5; Orient Longman Limited, Hyderabad.

Yoganarasimhan, S. N. 1996. Medicinal Plants of India Vol. 1 – Karnataka.Interline Publishing Private Limited, Bangalore.

Yoganarasimhan, S.N. 2000. Medicinal Plants of India Vol. II – Tamilnadu. Interline Publishing Private Limited, Bangalore.

Chapter 3

Medicinal Plants Used Against Eye Diseases

J. Koteswara Rao, S. Prasanthi and T.V.V. Seetharami Reddi

Department of Botany, Andhra University, Visakhapatnam – 530 003, India

Introduction

Humans have used plants for medicinal purpose since time immemorial. In fact, survival of human race could, in part, be attributed to traditional uses of medicinal plants down the ages. Use of folk medicine continues to the prsent day in many societies and cultures with many plants still in use. In consequence, medicinal plants have obviously made fundamental contributions to human health. India has a rich medicinal heritage which has been utilized for thousands of years. For example, the Charaka Samhita (1000 BC) mentions the use of approximately 300 plants for medicinal purposes. Added to this, India has extremely rich and diverse natural resources base.

The importance of eye care has been stressed since man first grew into a thinking, rational being. Some of the major factors causing eye diseases today are due to heredity (*e.g.*, myopia, colour blindness), dietary deficiency (*e.g.*, cataract, jaundice), lack of basic eye care (*e.g.*, blindness due the chicken blood application to the eye), over exposure to both natural (*e.g.*, lightning) and artificial radiation (*e.g.*, television), environmental pollution (*e.g.*, exhaust fumes), injuries and accidents.

Eyes are windows to the outside world. The visual system provides us with a tremendous amount of information about our environment. Millions of people have problems with their eyes. Many of these conditions, if left untreated, can cause blindness. Some eye problems are minor and fleeting. But some lead to a permanent loss of vision.

Ophthalmic diseases/common eye problems include:

- Cataracts–clouded lenses
- Glaucoma–damage to the optic nerve from too much pressure in the eye
- Macular degeneration
- Retinopathy
- Conjunctivitis–an infection also known as pink eye
- Ophthalmia
- Red eye
- Inflammation of the eyes
- Epiphora

"*Cataract*" comes from the Latin word *cataracta* meaning "waterfall." Looking through a waterfall is similar to the vision that results from cataracts. In the normal eye, light passes through a clear lens before it reaches the retina. The lens focuses light on the retina. In an eye with a cataract, the lens becomes cloudy. This clouding distorts rays of light and prevents light from being focused on the retina. This results in distortion in one's vision.

Cataracts are a common cause of visual disturbances in the adult population. The slow, progressive opacification of the crystalline lens of the eye distorts the optical passage of light to the retina resulting in diminished visual acuity. Glare, particularly during night driving in the face of oncoming headlights, may be an early symptom of cataracts. Glare, diminished overall acuity, contrast, and color resolution are compounded by the light-scattering effect of the cataracts. The lens of the eye is composed primarily of protein. Cataracts are caused by changes in the chemical makeup of the lens. With age, the lens becomes thicker and less clear.

Risk factors are:

- *Age*: As people get older, the risk of cataracts increases. Approximately 50 per cent of all people in the US over the age of 65 years have cataracts.
- *Ultraviolet Radiation*: Excessive exposure to sunlight increases the risk of cataracts.
- *Cigarette Smoking*: Smokers have a higher chance of getting cataracts than non-smokers.
- *Eye Injuries*: Injuries to the eye increase the risk of developing cataracts.

Treatment for cataracts is surgical removal and placement of an intraocular lens.

Glaucoma refers to a group of disorders that cause increased pressure within the eyeball. In a normal eye, a liquid called the *aqueous humor* is continuously produced and drained. In glaucoma, aqueous humor builds up and increases pressure within the eye. Such increased pressure can damage the optic nerve directly or restrict blood flow, thus damaging the optic nerve indirectly. This damage may lead to blind spots in the visual field. If left untreated, glaucoma can cause permanent blindness.

Glaucoma can cause deficits in peripheral vision. The abnormal regulation of intraocular pressure can result in gradual progressive atrophy of optic nerve cells. The development of chronic elevated intraocular pressure is generally painless, and the gradual loss of peripheral visual field can progress significantly before symptoms are noticed.

Glaucoma may also affect a number of subtler visual functions, such as redirection of visual attention, night vision, and color vision. With glaucomatous damage, Snellen acuity test results may not be affected, but peripheral field test results may show deficits. Specialist examination may result in early detection and treatment before the occurrence of possibly disqualifying vision loss. Vision loss caused by glaucoma cannot be restored.

A therapeutic goal is to lower intraocular pressure to a level that preserves the existing neuronal cells and prevents further loss of the peripheral visual field deficit. Strict and ongoing compliance with prescribed ophthalmic preparations is required for successful treatment; however, antiglaucoma agents may have side effects that impact vision and interfere with safe driving.

Macular degeneration describes many ophthalmic diseases that impact the macula function and interfere with detailed, central vision. These diseases increase in prevalence with age, affecting some 30 per cent by age 70. For the majority of cases, macular degeneration is a slow process resulting in subtle visual defects; however, approximately 10 per cent of cases are a "malignant" form of the disease and cause rapid loss of central vision. Peripheral vision is generally spared in macular degeneration.

Macular degeneration causes noticeable signs and symptoms. Visual acuity drops, recovery from bright lights is lengthened, and eventually a partial or total scotoma develops in the direction of attempted gaze. Snellen-type acuity testing will detect diminishing central acuity. Telescopic lenses redirect unaffected peripheral vision to compensate for lost central acuity, resulting in a reduced peripheral field of vision.

Retinopathy is a non-inflammatory damage to the retina of the eye and has many causes. The most common cause of retinopathy is diabetes mellitus. Background retinopathy with microaneurysms and intraretinal hemorrhages is common after 5-7 years with diabetes mellitus. In many cases, the retinopathy does not progress beyond this stage; however, fluid leakage near the macula (diabetic macular edema) can create partial scotomas in central vision or cause gross hemorrhage in the eye which can obscure vision and eventually lead to retinal detachment and blindness. Subtle visual modalities such as contrast sensitivity, flicker fusion frequency and color discrimination may also be affected.

Other diseases can cause retinopathy. Carcinoma-associated retinopathy is characterized by rapid onset of blindness caused by retinal degeneration, usually of photoreceptors. Proliferative retinopathy can be a complication of sickle cell disease and sickle cell-thalassemia disease. A rare but characteristic finding of systemic lupus erythematosus is retinal exudates, usually near the disk.

Conjunctivitis is the inflammation of the conjunctiva (the membrane that lines the eyelids and covers the exposed surface of the eyeball). Conjunctivitis can be caused by allergies, bacteria, viruses, chemicals, or underlying health conditions. The eyes are susceptible to infection because they are not sterile. They rely on lysozyme (an enzyme found in the tears) to destroy bacteria. Bacteria line the surface of the eyelids (all the way down into the shaft of the eyelashes), which makes the conjunctiva predisposed to germs and conjunctivitis.

Conjunctivitis can be classified as infectious or noninfectious. Infectious conjunctivitis (pinkeye) accounts for 70 per cent of all cases and is caused by either a bacteria (usually staphylococci, pneumococci, streptococci or chlamydia trachomatis) or a virus. Bacterial conjunctivitis occurs in 50 per cent of all cases, and viral conjunctivitis occurs in 20 per cent. Infectious conjunctivitis is very contagious.

Non-infectious conjunctivitis (which accounts for the remaining 30 per cent of all cases), can be caused by allergies (such as pollen or grass), chemicals (such as air pollutants, smoke or household cleaners), or underlying diseases (such as rheumatoid arthritis, cytomegalovirus (CMV), lupus, Kawasaki's disease, ulcerative colitis or Crohn's disease.

Additionally, conjunctivitis can be caused by hemorrhaging from trauma or changes in pressure within the head (subconjunctival hemorrhage). An incompletely opened tear duct or herpetic conditions (such as herpes simplex or herpes zoster) also can cause conjunctivitis.

The symptoms of infectious conjunctivitis caused by a bacteria or viruses are:

- Eye pain
- Swelling of the eyes
- Redness in the eyes
- Yellow, green or watery discharge from the eyes which collects overnight and crusts over the eye
- A gritty feeling in the eye
- Itching of the eye
- An enlarged lymph gland in front of the ear

Red eye usually refers to hyperemia of the superficial blood vessels of the conjunctiva, sclera or episclera, and may be caused by diseases or disorders of these structures or adjacent structures that may affect them directly or indirectly. There are many causes of a red eye including conjunctivitis, blepharitis, acute glaucoma, injury, subconjunctival hemorrhage, inflamed pterygium, inflamed pinguecula, dry eye syndrome, Brooklyn eye, a burst blood vessel, tick borne illnesses like Rocky Mountain spotted fever, high stress levels and drug use including cannabis.

Night blindness can result from a nutrient deficiency of Vitamin A, which is required for the pigmented compounds of the eye to function properly. Through regular eye function, vitamin A is consumed and the body must regain it so that the brain continues to receive messages from the eyes. Without enough vitamin A in the diet night blindness as well as other disorders such as macular degeneration can occur.

Corneal opacity is a disorder of the cornea, the transparent structure on the front of the eyeball, which can cause serious vision problems. Corneal opacity occurs when the cornea becomes scarred. This stops light from passing through the cornea to the retina and may cause the cornea to appear white or clouded over. Infection, injury or inflammation of the eye are the most common causes of corneal opacity.

Symptoms are:

- Vision decrease or loss
- Pain in the eye or feeling like there is something in your eye
- Eye redness or light sensitivity
- Area on the eye that appears cloudy, milky, or is not completely transparent

Ophthalmia (also called *Ophthalmitis*) is inflammation of the eye. It is a medical sign which may be indicative of various conditions, including sympathetic ophthalmia (inflammation of both eyes following trauma to one eye), ophthalmia neonatorum (a form of bacterial conjunctivitis), and actinic conjunctivitis (inflammation resulting from prolonged exposure to ultraviolet rays).

An *inflammation of the eye* generally occurs in response to viral or bacterial infection, allergies, environmental irritants, surgery or trauma. While most cases of eye inflammation are not too serious, it is important to control the severity of the inflammation as well as the duration to be sure to avoid any scarring and permanent damage. Because the eyes are such delicate organs, even a small amount of scar tissue can cause irreversible visual impairment.

Types of Eye Inflammations

There are many different types of eye inflammations depending on what area of the eye becomes inflamed, each condition differing in its symptoms and severity:

- *Conjunctivitis*–commonly known as pinkeye, conjunctivitis is an inflammation of the conjunctiva which is the clear membrane that covers the outermost layer of the eye and the inner surface of the eyelids. Many causes are associated with this condition including bacterial and viral infections, allergies and eye irritants.
- *Episcleritis*–an inflammatory condition of the episclera which is the connective tissue between the conjunctiva and sclera. The cause of episcleritis is uncertain.
- *Blepharitis*–an inflammation of the eyelids, often as the result of poor hygiene, chronically dry eyes or oily skin.
- *Keratitis*–an inflammation of the cornea region of the eye. This is often caused by bacterial or fungal infections and is increasingly prevalent in those with poor contact lens hygiene.

- ✰ *Uveitis*–an inflammation of the eyeball which is generally considered to be one of the more serious forms of eye inflammation. There are also a number of types of Uveitis depending on what area of the eye ball is infected and these may include: Iritis, Cyclitis, Retinitis and Choroiditis.
- ✰ Scleritis–an inflammation of the sclera or white of the eye.

Epiphora is overflow of tears onto the face. A clinical sign or condition that constitutes insufficient tear film drainage from the eyes in that tears will drain down the face rather than through the nasolacrimal system. Causes of epiphora include occular irritation and inflammation (including trichiasis and entropion) or an obstructed tear outflow tract which is divided according to its anatomical location (ie. ectropion, punctal, canalicular or nasolacrimal duct obstruction). The latter is often due to aging (a spontaneous process), infection (*i.e.*, dacryocystitis), rhinitis and in neonates or infants, failure of the nasolacrimal duct to open.

The present chapter deals with the development of valuable plant based drugs to treat various eye diseases and includes the impressive works of Gamble (1916-1935), Dastur (1962), Krishnam Nambiar *et al.* (1982), Agarwal and Ghosh (1985), Thammanna and Narayana Rao (1990), Rastogi and Mehrotra (1990-1999), Varier (1994-1995), Yoganarasimhan (1996), Jain (1997), Pal and Jain (1998), Bhattacharjee (2001), Pullaiah (2002), Kirtikar and Basu (2003) and Joganathan and Balu (1997).

ENUMERATION

1. *Abrus precatorius* L. (Fabaceae)

Chemical constituents: Glycyrrhizin, pinitol, crystalline compound, flavonoids; triterpene glycosides *i.e.* abrusosides A-D; 4,5,7 – trihydroxyflavon and taxifolin – 3- glucoside.

Uses: Seeds are soaked in water for about 12 hrs and later the eyes are washed with watery extract to get relief from irritation.

Root paste is prepared from goat's milk and dropped into the eye to cure irritation.

Wash the eyes in early morning with the decoction of leaves for a month to improve eyesight.

Decoction of the leaves is used to wash the eyes early in the morning for a month to cure eyesight.

2. *Acacia arabica* (Lam.) Willd. (Mimosaceae)

Uses: Decoction of leaves mixed with honey is applied as collyrium for eye flow.

A poultice of the leaves is applied on sore eye in children.

Bark extract mixed with honey is used to relieve conjunctivitis.

3. *A. catechu* (Linn. f.) Willd. (Mimosaceae)

Chemical constituents: Kaempferol, dihydrokaempferol, taxifolin, isorhamnetin, procyanidin, quercetin.

Uses: Stem bark is made into paste with 7 long peppers and mixed with the paste of 2-3 tulisi *poka* an insect found on *Ocimum santum* in ratio of 3:1 and applied thrice a day as cure for night blindness.

4. *A. nilotica* (L.) Willd. ex Del. (Mimosaceae)

Chemical constituents: Steroids, glycosides, flavonoids, tannins, gums, amino acids, fatty acids, organic acids, polyphenols.

Pods contain garnic acid, protocatechuic acid and luecocyanidin.

Bark contains mainly catechol and pyrogallol, stearic acid, kaempferol-glucoside, isoquercitin, leucocyanidin etc.

Uses: Four leaves are crushed, ground and small pills are made and placed on the eyelids once daily for 3 days.

Eyes are washed with the stem bark decoction twice a day for a week to cure eye diseases.

Two hundred grams of leaves are boiled in one litre of water until only 20 ml of water is left, this is filtered and the filtrate is applied on the eyelids in the morning and evening for about a week for treating epiphora.

5. *A. sinuata* (Lour.) Merr. (Mimosaceae)

Chemical constituents: Bark contains saponin.

Leaves contains alkaloids-calycotomine and nicotine besides rutin, triterpenoid, saponin, tannin, amino acids, proteins and other fatty acids.

Pods contain saponins – acacinin C, D, and E besides oligosaccharides, nortiterpene acacidiol. Seeds contain saponin – acacinin A and acacinin B, concinnin,–(+) -acacialactam. Also contain machaerinic acid and its lactone, sapogenin B, lactone and acacigenin-B, β-sitosterol.

Seed oil contains coronaric acid.

Uses: Two spoonfuls of seed paste mixed in a glass of water is administered for clearing the eyes once a day for 3 days.

Two hundred g of tender leaves along with 20 g of red chilly powder and a pinch of salt are fried in palm oil and made into pickle. This pickle is eaten along with meals once a day for a week for ophthalmic diseases.

6. *Acalypha fruticosa* Forssk. (Euphorbiaceae)

Uses: Leaf juice is used in ophthalmia.

7. *Achyranthes aspera* L. (Amaranthaceae)

Chemical constituents: Alkaloids, glycosides, saponins, aminoacids, fatty acids, hormone ecdysone and oleanolic acid, aliphatic dihydroxyketone.

Uses: Roots are collected on Sunday, washed and made into a paste with the saliva of the patient and applied on the eyes to cure swelling on the eyelids.

Three drops of leaf juice is poured into the eyes for curing eye injury and conjunctivitis.

One *tola* dose of root powder is given at bedtime for night blindness.

Root powder mixed with honey is applied for eye swelling.

Root paste is applied in ophthalmic diseases and opacity of the cornea.

Leaf decoction mixed in gingelly oil is anointed on the head and applied to the body an hour before bath. One to two drops of leaf juice are poured into the eyes.

8. ***Acorus calamus*** **L. (Acoraceae)**

Chemical constituents: Roots contain tricyclic sesquiterpene – callaminone, aclamendiol and isoaclamendiol.

Roots and rhizomes contain volatile oil where asarone and β-asarone are the important constituents.

Rhizome also contains phenyl indane derivatives.

Hydrocarbons present in the oil are elemene, caryophyllene, calanene, acdalene and humulene. Rhizome oil contains phenylpropane derivatives.

Presence of small quantities of palmitic, heptylic and butyric acids, asarinaldehyde, calamol, calamone and azulene are reported.

Uses: Leaf juice is poured to cure eye infections or conjunctivitis.

Roots are crushed with garlic and heated in coconut oil and applied externally on infected eyes till recovery.

9. ***Adenanthera pavonina*** **L. (Mimosaceae)**

Chemical constituents: Roots contain triterpenoids methyl oleaolate, methyl echinocystate and stigmasterol.

Bark contains stigmasterol glucoside.

Wood contains robinetin, chalcone, butein, ampelopsin and dihydrorobinetin. Leaves contain octocosanol, dulcitol, glucosides of β-sitosterol and stigmasterol. Seeds contain flavonoids, non-protein amino acids.

Kernels contain stigmasterol, dulcitol and polysaccharide.

Uses: Wood paste is externally applied in chronic ophthalmia.

Two drops of root bark decoction is instilled in the eye daily twice for 3 days to cure ophthalmic diseases.

10. ***Aegle marmelos*** **(L.) Correa (Rutaceae)**

Chemical constituents: Essential oil, aegelenine and aegeline, tannins, phlobotannins, flavan-3-ols, leuco-anthocyanins, anthocyanins, flavonoid glycosides, skimmianine, β-sitosterol, rutin and marmecin.

Uses: Unripe fruits are warmed and then the pulp of fruits is mixed with butter milk, made into a paste and applied on the head, followed by headbath to get cooling effect.

Poultice of the leaves is used in the treatment of ophthalmia.

Leaves along with those of *Dolichandrone falcata*, a pinch of salt and turmeric are crushed with goat's milk and used as eye drops twice a day until ophthalmia is cured.

Ripened fruit pulp paste is applied on head to get cooling effect.

11. ***Agaricus campestris*** **L. (Agaricaceae)**
Uses: Whole plant is used as curry weekly twice for a month in case of ophthalmic diseases.

12. ***Ageratum conyzoides*** **L. (Asteraceae)**
Chemical constituents: 6-dimethoxyageratochromene, dimer of ageratochromene from essential oil.

Uses: A cold decoction of the root is used as a lotion in ophthalmia.

Plant juice is applied twice daily in case of conjunctivitis.

13. ***Aglaonema hookerianum*** **Schott (Araceae)**
Uses: Root infusion is taken orally for conjunctivitis.

14. ***Ajuga bracteosa*** **Wall. ex Benth. (Lamiaceae)**
Uses: Fresh leaf extract is taken orally before breakfast for curing eyeache.

15. ***Alangium salvifolium*** **(Linn. f.) Wang. (Alangiaceae)**
Chemical constituents: Alkaloid – alangamarkine, ankorine, deoxytubulosine, alangiside, sterols, organic acids and 3 triterpenoids.

Uses: Leaf juice with common salt (2:1) is applied on eyelids as cure for eye diseases like redness of eye.

One or two drops of fruit juice are poured to cure eye infections.

Half glass of fruit juice is administered daily twice for 5 days in case of ophthalmic diseases.

16. ***Albizia amara*** **(Roxb.) Boiv. (Mimosaceae)**
Chemical constituents: Echinicystic acid from seeds, β-sitosterol, α and β-unsaturated carboxylic acid methyl ester, triterpene saponin, a new flavonol glycoside from leaves.

Uses: Leaf juice is used for ophthalmia and other eye diseases.

17. ***A. lebbeck*** **(L.) Willd. (Mimosaceae)**
Chemical constituents: New flavandiols-malacacidin, lebbecacidin from heartwood; δ-catechin, leuco-cyanidin and condensed tannins in bark. Acacic acid from bark, triterpenoid sapogenin-albigenic acid from beans.

Uses: Leaf decoction is administered in the treatment of night blindness.

An ointment made of the seeds is used for eye diseases.

Leaf juice is dropped into the eyes and seed powder mixed in rice powder is made into a cake and given to eat for a fortnight to cure night blindness.

18. ***A. odoratissima*** **(L.f.) Benth. (Mimosaceae)**
Chemical constituents: Machaerinic acid, an acid genin, odoratissimin from seeds; acacic acid, glucose, rhamnose, arabinose and xylose.

Uses: Juice of the leaves is externally applied to the eyes to cure night blindness and also given internally.

Paste of the leaves is applied to treat eye diseases.

19. *Allium cepa* L. (Liliaceae)

Chemical constituents: Aminoacid – cycloallin, 2-trisaccharides-1-β-fructosyl sucrose and 6-β-fructosyl sucrose, oleanolic acid, arabinose, rhamnose, xylose and ribose.

Uses: Juice of young bulb is squeezed through a cloth and put in eyes to treat night blindness and ophthalmia.

Bulb with *Cinnamomum camphora* is taken in equal quantities and ground. Two spoonfuls of paste mixed with a spoonful of honey is administered daily once for 3 days. This paste also can be used as eyeliner for ophthalmic diseases.

20. *A. sativum* L. (Liliaceae)

Chemical constituents: A new aminoacid cycloallin, 2 trisaccharides, arabinose, rhamnose, xylose, ribose; histone from seedling roots, propenyl sulfenic acid from bulbs.

Uses: Bulb paste is used for night blindness.

Two drops of juice is used as eye drops in sunset time for night blindness.

21. *Aloe barbadensis* Mill. (Liliaceae)

Chemical constituents: Glucoside and isobarbaloin and β-barbaloin, free anthraquinone like aloe emodin, iso-emodin, free and cinnamoylate 8 methyl-7-hydroxyaloins, quercetin, rutin, chrysophanic.

Uses: The stained mixture of the leaf juice and rose water is useful as an eye wash for catarrhal and purulent ophthalmia and other eye diseases.

Pulpy portion of the leaf is put overnight as bandage over the eyes for 3 days to treat eye infection.

Leaf juice is used as eyetex for less than a month child who is unable to open the eyes.

22. *Alternanthera ficoidea* (L.) R. Br. (Amaranthaceae)

Uses: Stem bark juice is used as eye drops for eye diseases.

23. *A. sessilis* (L.) R. Br. ex DC. (Amaranthaceae)

Chemical constituents: α and β-spinasterol, lupeol from roots.

Uses: Leaf juice is mixed with cow's milk and given in morning on empty stomach to improve the eyesight.

Leaves are warmed, cooled and used as a bandage over the eyes in the case of eye sores and reddening.

Whole plant juice is mixed with an equal amount of sesame oil and the mixture is boiled and taken as oil bath for day blindness.

Fresh leaves are collected, fried with ghee and the ghee is applied externally over the eyelids during night for eye diseases.

Fresh leaves are fried with spices and taken orally for corneal diseases.

24. *A. triandra* (L.) DC. (Amaranthaceae)

Uses: Leaves are warmed, cooled and used as a bandage over the eye in case of eye sores and redness.

25. *Alysicarpus hamosus* Edgew. (Fabaceae)

Uses: Two to three drops of squashed leaf juice is dropped in the eyes before bedtime once a day for a month.

26. *Amaranthus tricolor* L. (Amaranthaceae)

Chemical constituents: Plant contains c25-35 alkanes, c24-32 aliphatic alcohols, cholesterol, campesterol, 2424-methylene cholesterol.

Inflorescence, leaves and stem contain amaranthine and isoamaranthine.

Uses: Leaf juice along with that of *Eclipa prostrata* together with milk, sesame oil are taken in quantity of a litre each. Half litre of lemon juice is added to the mixture. The mixture is boiled, reduced to one-third of the initial volume, then stored and used as hair oil for smoky vision.

27. *A. viridis* L. (Amaranthaceae)

Uses: Powder of dried seeds along with dried fruit of *Ficus carica* and sugar is taken orally with water twice a day for treating eye vision problem.

28. *Ampelocissus latifolia* (Lam.) Planch. (Vitaceae)

Uses: One to two drops of leaf juice is poured into the eyes twice a day for 3 days to treat ophthalmia.

29. *A. rugosa* (Wall.) Planch. (Vitaceae)

Uses: Three to four drops of fresh juice of stem is dropped into eyes to cure whiteness of eyeball.

30. *Anagallis arvensis* L. (Primulaceae)

Chemical constituents: Plants contain anagalligenone, quaternary alkaloids, gluco-fructoside, tannin, acetyl saponin, triterpene saponins anagalloside A, B, C and desglucanagalloside A and B, cucurbitacin glycosides arvenin I and arvenin II, cucurbitacins B, D, I, L and R, oleanane triterpene.

Roots contain two glycosidic saponin, cyclamen being one of them.

Leaves contain sterols and fatty acids.

Flowers contain stigmasterol, β-sitosterol, α-spinasterol-3-glucoside, kaempferol, quercetin and rutin.

Uses: Root juice mixed with honey is applied to remove cataract from the eyes.

31. *Andrachne aspera* Spreng. (Euphorbiaceae)

Uses: Whole plant is used for eye problems and as eye wash.

32. *Androsace rotundifolia* Hardw. (Primulaceae)

Uses: Extract of rhizome is added to a diluted solution of common salt and used as eye drops for curing a number of ophthalmic diseases such as cataract etc.

33. ***Angelica glauca*** **Edgew. (Apiaceae)**
Uses: A teaspoonful of dried powdered roots is taken with lukewarm water to cure swelling of the eyes.

34. ***Anthocephalus chinensis*** **(Lam.) A. Rich ex Walp. (Rubiaceae)**
Uses: Juice of stem bark with common salt (3:1) is used as cure for eye inflammation.

35. ***Aphanamixis polystachya*** **(Wall.) Wight & Arn. (Meliaceae)**
Chemical constituents: Plant contains aphanomixin lactone and aphanamixolide. Root contains terpenoid glycoside, 8C- methylquercetin-3-O-D-xylopyranoside. Stem bark contain β-sitosterol, stigmasterol, saponin, triterpenoid glycoside, flavone glycosides, limonoid amoorinin, polystachin, saponin- poriferasterol-3-O-α-L-rhamnopyranoside.

Bark contains tetranortriterpenoid aphanamixinin. Air dried leaf contains diterpene, aphanamixol and β-sitosterol.

Fruit contains triterpene – aphananin.

Seeds yield semidrying oil containing limonoid rohitukin.

Uses: Seed oil is used as liniment in diseases of eye.

36. ***Ardisia solanacea*** **Roxb. (Myrsinaceae)**
Chemical constituents: Leaves contain bauerenol, α-amyrin, β-amyrin.

Uses: Fruit juice is poured into eyes for eye sores.

37. ***Argemone mexicana*** **L. (Papaveraceae)**
Chemical constituents: Linoleic, linolenic, oleic, palmitic, ricinoleic, stearic, malvalic, myristic, sterculic acids in seed oil.

Uses: Latex is applied on eyes to cure white patches, conjunctivitis, ophthalmia and also cataract.

Two or three drops of latex are dropped into eyes regularly for 5 days or more to cure cataract.

One drop of latex is poured into the eye for eye infection and sores.

Three to four roots along with 1-2 flowers of neem and a pinch of foam of sea are crushed and ground together and *kajal* made from it is applied around the corners of the eye to treat eye infections.

Yellow latex is applied to the eyes for about a fortnight or the treatment is prolonged if necessary for white patches in the eyes.

Latex is used as smut/eye drops for night blindness.

38. ***Argyreia cymosa*** **(Roxb.) Sweet. (Convolvulaceae)**
Uses: Fresh root juice (1-2 drops) is used as eye drops to treat redness of eyes.

39. ***A. nervosa*** **(Burm. f.) Boj. (Convolvulaceae)**
Chemical constituents: Seeds contain ergoline alkaloids–agroclanine, chanoclanineI, II, racemic chacoclanine-II, elymoclarine, festuclarine, lysergene,

lysergol, isolysergol, molliclanine, penniclanine, stoclanine, isosestoclanine, ergine, isoergine, ergometrine, ergometrinine, lysergic acid, effeic acid and α-hydroxythylamide.

Uses: Leaf and root juice are used to treat eye infections.

40. ***Artemisia indica*** **Willd. (Asteraceae)**

Uses: Fresh leaf juice is dropped in eye to cure redness of eye but is painful.

41. ***Artocarpus heterophyllus*** **Lam. (Moraceae)**

Chemical constituents: Artocarpetin, artocarpanone, artocarpin, isoartocarpin from heartwood. Cyanomaclurin, dihydromorin, artocarpesin from wood.

Uses: Leaf juice is used to cure eye problems.

42. ***Asparagus racemosus*** **Willd. (Liliaceae)**

Chemical costituents: Sitosterol, udecanylcetanoate, glycosides and saponins.

Uses: Root powder along with milk is given orally to improve eyesight.

Dried root powder is given with ghee (clarified butter) in ratio of 5:2 to children for improving eyesight.

Twenty g of fresh root is boiled in one litre of water till it reduces to one-fourth and taken twice a day for eye infection.

43. ***Azadirachta indica*** **Juss. (Meliaceae)**

Chemical constituents: Leaves contain flavonone, meliacinsolannolide, solannin, triterpenoidnimocinol, azadirachtannin, tetranorterpenoid isoazaditolide, quercetin, nimbocinolide and β-sitosterol.

Uses: One to two drops of leaf juice is put into each eye for eye pains.

Seed oil is taken orally to cure reddening and inflammation of eyes.

Leaf extract is mixed with sugar, filtered and the filtrate is instilled as drops for 3 days for eye infection.

44. ***Azima tetracantha*** **Lam. (Salvadoraceae)**

Chemical constituents: Leaves and stem contain dimeric piperidine alkaloids – azimine, azcarpine and carpaine. Leaves contain friedelin, glutinol, lupeol and β-sitosterol.

Uses: Seed paste is applied to cure redness and other ailments of eyes.

45. ***Bambusa arundinacea*** **(Retz.) Willd. (Poaceae)**

Chemical constituents: Glycosides, flavanoids, organic acids, resin, waxes, enzymes young leafy shoots contains cyanogenetic glucoside.

Uses: Dew water accumulated on the leaves is collected and 2 to 3 drops are applied externally into the eyes in the morning once a day for 10 days for ophthalmic diseases like conjunctivitis.

46. ***Barringtonia acutangula*** **(L.) Gaertn. (Barringtoniaceae)**

Chemical constituents: Barringtogenol D from fruits, a triterpene alcohol – tanginol, barringtonic acid and barrintogenol from branch wood.

Uses: Two drops of tender leaf juice is poured into the eyes for ophthalmic infections.

Seeds are accumulated with breast milk and poured into the eyes in the form of *anjan* in the morning for 5 days to treat eye infections.

47. *Bauhinia racemosa* Lam. (Caesalpiniaceae)

Chemical constituents: Lauric, myristic, palmitic, stearic, oleic and linoleic acids from seed oil.

Uses: Flower juice is applied for eye diseases.

48. *Begonia palmata* D. Don (Begoniaceae)

Uses: Leaf extract is used for eye injury.

49. *B. trichocarpa* Dalz. (Begoniaceae)

Uses: Fresh leaf juice is dropped in eye to cure opacity.

50. *Berberis aristata* DC. (Berberidaceae)

Uses: Extract of root bark is used in eye affections.

51. *B. asiatica* Roxb. ex DC. (Berberidaceae)

Uses: Paste is made from stem wood and applied in eyes to cure conjunctivitis.

Yellow bark juice is used for conjunctivitis, as 2 drops per dose in each eye twice a day for 5-7 days.

Extract of root is used as eye drops to cure eye infections and ophthalmia.

52. *B. chitria* Lindl. (Berberidaceae)

Uses: Fruit juice is used for treatment of eye diseases.

53. *B. lyceum* Royle. (Berberidaceae)

Chemical constituents: Roots contains alkaloids berberine, berbericine hydrochloride, berberivinine hydroiodiide, berberine chloroform, palmatine choroform and oxyberberine also umbellatine, berbamine.

Uses: Filtered bark infusion, soaked for a night in water is taken before breakfast for eyeache.

Root bark extract is used externally in eye diseases.

54. *B. microcarpa* Warcz (Berberidaceae)

Uses: The filtered leaf juice is used as eye drops in the treatment of eye infection.

55. *B. wallichiana* DC. (Berberidaceae)

Uses: Diluted decoction of leaves is used for conjunctivitis.

56. *Bergenia ligulata* (Wall.) Engl. (Saxifragaceae)

Uses: Root bark paste is applied in ophthalmia.

57. *Blepharis maderaspatensis* (L.) Heyne ex Roth. (Acanthaceae)

Uses: Seed extract is used as eye drops for eye diseases.

Approximately 200 g of the whole plant is collected, cut into small pieces and tied up in a fine cloth. The plant material is then dipped into a container of 400 ml pure sesame oil. The container is made airtight and kept in the sun for 15 days, filtered and 2 drops are applied in the morning and evening for blindness.

58. *Bidens bipinnata* L. (Asteraceae)

Uses: Juice of fresh herb is used as eye drops for eye diseases.

59. *Boerhavia diffusa* L. (Nyctaginaceae)

Chemical constituents: Leaf contains crystalline acids.

Root contains alkaloid punarnavine, C-methyl flavone, rotenoid analogues – boeravinone A-F, punarnavoside, aminoacid, fatty acids, hentriacontanes, β-sitosterol.

Uses: Root grounded with snail's shell is applied to eye for night blindness.

Root rubbed with milk is applied for eye itching.

Plant extract is given with goat milk (3:2) to children as cure for night blindness.

Decoction of roots is used in the treatment of eye diseases.

Leaf juice is used as a lotion for treating ophthalmia.

60. *B. erecta* L. (Nyctaginaceae)

Uses: Leaf juice is filtered and mixed with mother's milk and 2-3 drops are applied for 5 days against eye diseases.

61. *Bombax ceiba* L. (Bombacaceae)

Chemical constituents: Lupeol, β-sitosterol, naphthaquinone, lactones, biosides.

Uses: Petal paste mixed with breast milk is applied externally for conjunctivitis.

62. *Borreria articularis* (L.f.) F.N. Williams (Rubiaceae)

Uses: Juice of leaves mixed with little salt is poured in eyes in case of conjunctivitis and other eye diseases.

One tablespoonful of leaf juice is mixed with a pinch of salt and one or two drops of this mixture are poured in eyes twice a day to cure conjunctivitis.

63. *Boswellia serrata* Roxb. ex Colebr. (Burseraceae)

Chemical constituents: Steroids, glycosides, phenols, carbohydrates, diterpene alcohol serratol, oleoresin oil containing α-thujene.

Uses: Leaf paste is applied on eyes against eye infection.

Leaf paste of young sapling is applied on the cloth and placed over infected eyes and tied around the head at night to cure conjunctivitis.

64. *Brassica campestris* L. (Brassicaceae)

Uses: Seed oil is applied on eyes before bathing for eyesight.

65. *Bridelia retusa* (L.) Spreng (Euphorbiaceae)

Uses: Leaf paste is applied on the eyes to cure conjunctivitis.

66. ***Bryophyllum calycinum*** **Sails (Crassulaceae)**
Uses: Leaves are crushed and applied on eye sores.

67. ***Butea monosperma*** **(Lam.) Taub. (Fabaceae)**
Chemical constituents: β-sitosterol, alkaloids, flavonoids, fatty acids, enzymes, acoumaranone glucoside-palsitrin from the plant and palasonin from seeds.

Uses: Fifty g of gum is dissolved in water and poured in the eye for 3 days to cure ophthalmia.

Root decoction with hen's egg (1:1) is given to children in the treatment of night blindness, cataract and other eye defects.

One or two drops of root juice are put in eyes thrice a day for 3 days for eye infection.

68. ***Cajanus cajan*** **(L.) Millsp. (Fabaceae)**
Uses: Leaf juice in doses of 2-3 drops is poured into eyes for 5 days to treat corneal opacity.

69. ***Calendula officinalis*** **L. (Asteraceae)**
Uses: A lotion of the flower is used as eye wash after surgical operations and for inflammed and sore eyes.

70. ***Calophyllum inophyllum*** **L. (Clusiaceae)**
Uses: Leaves soaked in water are used for sore eyes.

71. ***Calotropis gigantea*** **(L.) R. Br. (Asclepiadaceae)**
Chemical constituents: Alkaloids, steroids, glycosides, saponins, terpenes, enzymes, alcohol, resin, fatty acids, esters of α and β-calotropols.

Uses: Latex is used for conjunctivitis.

Leaf juice is applied to the nails of the feet to cure red eye.

72. ***Camellia sinensis*** **(L.) Ktze. (Theaceae)**
Chemical constituents: Polyphenols.

Uses: Infusion of leaves is used in conjunctivitis.

73. ***Capsella bursa-pastoris*** **Medic. (Brassicaceae)**
Uses: Juice of plant is used to cure the eye ailments.

74. ***Cardiospermum halicacabum*** **L. (Sapindaceae)**
Chemical constituents: Saponin, arachidic acid, β-sitosterol, β-D-galactoside and apigenin.

Uses: Fresh leaf juice is applied as drops for curing eye diseases like conjunctivitis.

75. ***Carissa carandas*** **L. (Apocynaceae)**
Uses: One hundred pure fresh flowers are collected, placed in 100 ml of sesame oil and kept in the sunlight until the flowers become submerged (it may take 10-13 days). The mixture is filtered and kept in a container. One or two drops is applied externally on the eyes for cataract.

Roots are rubbed with the stem juice of *Indigofera tinctoria* and applied into the eyes in the morning and evening for 5 days to treat disease of eyelids.

76. ***C. spinarum* L. (Apocynaceae)**
Uses: Two to three drops of fresh flower juice is poured into eyes once a day for 9 days to treat opacity of the cornea.

77. ***Cassia absus* L. (Caesalpiniaceae)**
Chemical constituents: β-sitosterol-β-glucoside from seed, a D-galacto-D-mannan composed of galactose and mannose.

Uses: Decoction of seeds is used in eye diseases.

Dried seed powder is applied in the eyes for curing conjunctivitis and ophthalmia.

78. ***C. auriculata* L. (Caesalpiniaceae)**
Chemical constituents: Steroids, alcohols, pyrrolizidine alkaloids, sennosides A & B, anthraquinones.

Uses: Leaf juice is poured into eyes for ophthalmic infections.

A fine powder of the seeds is blown into the affected eyes or an ointment is made with coconut oil or sesame oil and applied externally.

Two to three drops of leaf juice is poured thrice a day for 3 days in case of eye infections.

Leaf juice is mixed with fried onion and ghee and ground. Two spoonfuls of paste is administered daily twice for 40 days for night blindness.

Fresh leaf paste mixed with a pinch of turmeric is applied on the affected areas once a day till cure for night blindness.

79. ***C. fistula* L. (Caesalpiniaceae)**
Chemical constituents: Steroids, glycosides, flavonoids, tannins, quinones, disaccharides, phlobaphenes, oxyanthraquinone substances.

Uses: Powder of the stem bark is used for blindness.

One or two drops of leaf juice are put in eyes to cure cataract.

Stem bark paste is given with paste of black pepper (3:1) as cure for night blindness.

Seeds are rubbed with hot water and applied in the morning as collyrum for 3 days for night blindness.

80. ***C. occidentalis* L. (Caesalpiniaceae)**
Chemical constituents: Steroids, glycosides, flavones, dianthronic heteroside, sennosides A & B anthraquinones.

Uses: Leaf juice mixed with butter milk is applied on eyes for eye infection.

Three to four drops of leaf juice is poured once at night for 3-4 days for eye infections.

81. ***C. tora*** **L. (Caesalpiniaceae)**

Chemical constituents: Steroids, glycosides, quinones, aminoacids and fatty acids.

Uses: Paste of the leaves is applied in the eye and also on eyelids to treat eye diseases.

82. ***Cassytha filiformis*** **L. (Lauraceae)**

Chemical constituents: Alkaloids, 2 new oxoaporphine bases, cassyfiline, cassythine and cassythidine.

Uses: Juice is extracted from the whole plant and 1 or 2 drops are poured into eyes to treat redness due to infection of eyes.

One to two drops of leaf juice are instilled into eyes once a day for 3 days for eye infection and inflammation.

83. ***Catharanthus roseus*** **(L.) G. Don (Apocynaceae)**

Chemical constituents: Flowers contain diglucoside, pseudoindoxyl alkaloid – rosamine, β-carboline, bannucine and leurosinone.

Leaves contain oleoresin, volatile oil.

Roots contain ajmalacine.

Fruits and seeds contain vindoline, dehyraouin dolinine and coromaridine.

Uses: Two to three drops of leaf juice is applied daily once for a week at bedtime for conjunctivits.

84. ***Celosia argentea*** **L. (Amaranthaceae)**

Chemical constituents: Celosianin and isocelasianin.

Uses: Seeds are useful for clearing the vision and for the diseases of the eye.

Leaf juice is used as eye drops for eye diseases.

85. ***Centella asiatica*** **(L.) Urban (Apiacaeae)**

Chemical constituents: 2 terpene acids – brahmic acid, isobrahmic acid and 2 saponins – brahmoside and brahminoside.

Uses: Leaf decoction is taken for curing eye problems.

86. ***Centipeda minima*** **(L.) A. Br. & Asch. (Asteraceae)**

Uses: Infusion of herb is used in ophthalmia.

87. ***Chasalia chartacea*** **Craib (Rubiaceae)**

Uses: Decoction of root is used in the treatment of eye diseases.

88. ***Chloris variegata*** **Sw. (Poaceae)**

Uses: Watery juice of inner nodal portion of the grass is used as eye drops in conjunctivitis.

89. ***Cochlorospermum religiosum*** **(L.) Alston (Cochlorospermaceae)**

Uses: Gum is used in the treatment of eye diseases.

90. ***Chloroxylon swietenia*** **DC. (Rutaceae)**

Chemical constituents: Alkaloids, tannins, coumarins, lignans, structure of xylotenin, a new furanocoumarin.

Uses: Powdered bark is put in a thin cloth and soaked in breast milk and then applied into eyes 2-3 times for 4 days for eye infections and to heal eye injuries.

91. ***Christolea erassifolia*** **Comb. (Brassicaceae)**

Uses: Fresh leaves are used to improve eyesight.

92. ***Citrus aurantifolia*** **(Christm. & Panz.) Swingle (Rutaceae)**

Uses: Leaf is heated in rectified butter and tied over eyes to cure redness of eye.

93. ***C. limon*** **(L.) Burm. (Rutaceae)**

Chemical constituents: Eriocitrin from peel, eriodictyol, apigenin, luteolin, quercetin, isorhamnetin, limocitrin, limocitrol, sinapic acid, scorpoletin, umbelliferone from peels.

Uses: Leaf is boiled in water and then washed properly. After cooling it is placed on eyes to treat eye troubles.

94. ***C. medica*** **L. (Rutaceae)**

Uses: Two drops of mixture made from fruit juice, onion, ginger, honey and rose water are put in the eyes daily for eye problems.

95. ***Clematis heynei*** **M.A. Rau (Rananculaceae)**

Uses: Fresh leaf juice is dropped in eyes to cure opacity.

96. ***Clerodendrum serratum*** **(L.) Spreng. (Verbenaceae)**

Chemical constituents: Glucoside and D-(-) mannitol; hydrolysis of crude saponin from bark yielded oleanlic acid, queretaroic acid and serratagenic acid.

Uses: Root and leaves are used in ophthalmia and other eye diseases.

97. ***C. viscosum*** **Vent. (Verbenaceae)**

Uses: Root bark extract is instilled into eyes to treat eye infection.

98. ***Clitoria ternatea*** **L. (Fabaceae)**

Chemical constituents: γ-sitosterol from seeds, taragerol from roots.

Uses: Four to five drops of purple flower juice is poured directly to cure conjunctivitis.

99. ***Coccinia grandis*** **(L.) Voigt. (Cucurbitaceae)**

Chemical constituents: Roots contain talbutamic β-amyrin, heperol and a bitter glycoside containing cucurbitacin B from fruits. Cephalandrol, triacontane, β-sitosterol, cephalandrene B from aerial parts.

Uses: Juice of the leaves is mixed with salt and breast milk and given in case of eye diseases.

Two drops of leaf juice is poured into eyes thrice a day for about 3 days to treat eye sores.

Leaf juice is instilled into the eyes daily twice for 3 days for conjunctivitis.

100. ***Cocculus hirsutus*** **(L.) Diels (Menispermaceae)**
Chemical constituents: Trilobine, isotrilobine, syringaresinol and protoquercitol, jamtine-N-oxide, erythrina alkaloids cocculidine, isococculidine, coccoline, coccuvine, coccuvinine, cocculitine, isococculine, dibenzonine alkaloids – laurifonine, laurifine and laurifinine. D-trilobine and dicoclaurine from roots.

Uses: One teaspoon leaf powder is taken orally once a day for 8 days to cure night blindness.

101. ***Colebrookea oppositifolia*** **Sm. (Lamiaceae)**
Chemical constituents: 2 new flavones -5 – hydroxyl – 4, 6, 7, 8 – tetramethoxy-flavone isolated from aerial parts; triacontane, triacontanol, palmitic, stearic and oleic acids from roots.

Uses: Leaf juice is used to treat cataract and other eye diseases.

102. ***Coleus amboinicus*** **Lour. (Lamiaceae)**
Chemical constituents: Cirsimatrin, β-sitosterol-β-D-glucoside in leaves; oleanolic, crategolic, promolic, euscophic, tormentic, ursolic acids from leaves; salvigenin, quercetin, luteolin, apigenin, taxifolin.

Uses: Leaf juice is applied as eye drops for eye swelling.

103. ***Combretum albidum*** **G.Don (Combretaceae)**
Uses: Stem juice is used for eye diseases.

104. ***C. roxburghii*** **Spreng. (Combretaceae)**
Uses: Stem sap is put into the eyes to cure eye diseases.

105. ***Commelina benghalensis*** **L. (Commelinaceae)**
Uses: Juice of the petiole is applied externally for eye problems.

106. ***Commiphora caudata*** **(Wight & Arn.) Engl. (Burseraceae)**
Uses: A resinous gum extracted from the stem bark mixed in a glass of hot water is gargled once a day for 3 days for ophthalmic diseases.

107. ***Connarus monocarpus*** **L. (Connaraceae)**
Uses: Fruit pulp is used for eye troubles.

108. ***Cordia wallichii*** **G. Don (Boraginaceae)**
Uses: The ash of the leaves is used as a wash in burning sensation of the eyes.

109. ***Coriandrum sativum*** **L. (Apiaceae)**
Uses: Plant juice is used as eye drops for eye problems.

A dried fruit infusion is used as an eye wash in conjunctivitis.

Leaf extract is useful in conjunctivitis.

110. ***Corydalis gavaniana*** **Wall. (Fumariaceae)**
Uses: Powdered seeds are pasted externally around the eyes to cure ophthalmic diseases.

Plate 3.1

1. ***Acasia sinuata;*** 2. ***Ardisia solanacea;*** 3. ***Argyreia nervosa;*** 4. ***Butea monosperma;*** 5. ***Cassia fistula;*** 6. ***Cassytha filiformis***

Plate 3.2

1. ***Clitoria ternatea***; 2. ***Heliotropium indicum***; 3. ***Opuntia dillenii***; 4. ***Rubia cordifolia***; 5. ***Scoparia dulcis***; 6. ***Thysanolaena maxima***.

111. ***Costus speciosus*** **(Koen.) Sm. (Zingiberaceae)**
Chemical constituents: Tigogenin and diosgenin from rhizomes and stems; saponins A, B and C; β-sitosterol glucoside, α-amyrin stearate, β-amyrin and lupeol palmitates from leaves.

Uses: Rhizome paste is taken inernally and also applied to the head to treat burning sensation in the eyes.

Juice of fresh stem is applied over the scalp to cure redness of eye due to traumatic affection.

Stem is heated and juice is extracted and cooled. One drop is poured into eye to cure eye diseases.

112. ***Crotalaria prostrata*** **Rott. (Fabaceae)**
Uses: Decoction of the leaves is used as eye wash twice a day for 3 days for eye diseases.

113. ***Croton roxburghii*** **L. (Euphorbiaceae)**
Chemical constituents: β-sitosterol, α-diterpene alcohol, diterpene-oblongifoliol from seeds, root and bark.

Uses: One to two drops of juice extracted from fresh young twigs is dropped in the eyes to cure blindness and to remove cataract in each eye once daily in the evening for 3 days.

114. ***Cuminum cyminum*** **L. (Apiaceae)**
Chemical constituents: Apigenin-7-o-glucoside from fruits.

Uses: Seed powder is used as collyrium to eyes for cataract.

115. ***Curculigo capitulata*** **(Lour.) O. Kuntze (Hypoxidaceae)**
Uses: Rhizome is soaked in cold water for a few hours till a gel like substance is formed and used for eyeache or injury to relieve pain.

116. ***C. orchioides*** **Gaertn. (Hypoxidaceae)**
Chemical constituents: Roots contain methoxy ketone. Rhizome contains phenolic glycoside curculigoside, curculigine A, orcinol glycoside corchioside A, hentriacontanol, sitosterol, stogmasterol, cycloartenol, glycosides of cyclo-aratane sapogenin-curculigosaponins G-M, curculigenin B & C, glycoside yuccagenin and lycorine.

Root oil contains fatty acids.

Uses: Three spoonfuls of root decoction is mixed in a glass of water and then 3 or 4 drops of decoction is instilled into eyes daily twice for 2 days for ophthalmic diseases.

117. ***Curcuma domestica*** **Valeton. (Zingiberaceae)**
Uses: A decoction of the rhizome is said to relieve the pain of purulent ophthalmia.

118. ***C. longa*** **L. (Zingiberaceae)**
Chemical constituents: Essential oil, colouring matter, curcumin, β-termenone,

dimethoxy curcumin, bis-demethoxy curcumin, neutral polysaccharide ukonan D.

Uses: Fresh rhizome paste diluted with water is filtered 3-4 times with clean muslin cloth and applied to treat eye diseases.

119. ***Cuscuta reflexa*** **Roxb. (Cuscutaceae)**
Chemical constituents: Dulcitol, luteolin, quercetin and a glycoside of luteolin isolated from stem.

Uses: Plant juice is applied as drops to treat eye diseases.

120. ***Cyathus hookeri*** **Berk. (Nidulariaceae)**
Uses: The peridioles are ground with water, filtered through cotton and the filtrate is used as eye drops for curing disorders of eye.

121. ***C. stercoreus*** **(Schw.) de Toni (Nidulariaceae)**
Uses: Peridioles grinded, filtered and the filterate is given for eye troubles like pain, redness or conjunctivitis.

122. ***Cyclea peltata*** **(Lam.) Hook. f. & Thoms. (Menispermaceae)**
Uses: Leaf juice is used in burning sensation of the eyes.

123. ***Cynodon dactylon*** **(L.) Pers. (Poaceae)**
Chemical constituents: Alkaloids, steroids, glycosides, flavones, terpenes, organic acids, alcohols, wax, sugars.

Uses: Plant paste with root of *Cyperus rotundus* is applied externally on eyelids for swelling and redness of eyes.

124. ***Cyperus rotundus*** **L. (Cyperaceae)**
Chemical constituents: Flavonoid glycoside, cyperene 1 & 2, patchoulenone, sesquiterpene, ketone mustakone, essential oil, cyperotunndone, cyperolone, sesquiterpenoid penoid-sugetriol triacetate (from tubers of Japanese nut grass); essential oil containing copadiene, epoxyguaiene, rotundone and cyperone. Sesquiterpenic ketol contains sugeonol, cyperol and isocyperol, α-rotanol, β-rotanol, 2 norsesquiterpenoids – kobusone and isokobusone.

Uses: Tubers are grounded with goat's urine and used as a collyrium for redness and swelling of eye.

125. ***Datura innoxia*** **Mill. (Solanaceae)**
Uses: Latex is applied for eyeache.

126. ***Daucus carota*** **L. (Apiaceae)**
Chemical constituents: Vitamin A, B, $ZnSO_4$

Uses: Half cup of carrot juice in a glass of cow's milk is administered daily once for 10 days for ophthalmic diseases.

127. ***Delonix elata*** **(L.) Gamble (Caesalpiniaceae)**
Chemical constituents: (-) L – asparagine and aspartic acid isolated from trunk bark.

Uses: Fresh leaves with liver and pancreas of chicken are roasted in gingelly oil and taken internally for night blindness.

128. *Dendrocalamus strictus* (Roxb.) Nees (Bambusaceae)
Uses: Tender leaves are boiled, cooled and tied over the eye for eye sores.

129. *Dentella repens* (L.) Forst. (Rubiaceae)
Uses: Plant is taken as a vegetable to improve eyesight.

130. *Desmodium pulchellum* (Houtt.) Merr. (Fabaceae)
Uses: Leaf juice is used for eye infection.

131. *Dichrostachys cinerea* (L.) Wt. & Arn. (Mimosaceae)
Chemical constituents: Friedelin, α-amyrin from bark while heartwood yielded octacosanol; leaves yielded hentriacontanol and β-amyrin; sitosterol from bark, heartwood and leaves.

Uses: Tender shoots are crushed and 1-2 drops of juice is poured into eyes for 5 days to treat ophthalmia.

Two spoonfuls of root bark paste mixed with a spoonful of honey is administered daily twice for 3 days for ophthalmic diseases.

132. *Dioscorea deltoidea* Wall. (Dioscoreaceae)
Chemical constituents: Smilagenone, epismilagenin from yams.

Uses: An extract of the rhizome is used externally for the treatment of ophthalmic diseases.

133. *Diospyros melanoxylon* Roxb. (Ebenaceae)
Uses: Decoction of the bark is used as lotion for the eye.

Decoction of young leaves with unboiled egg and rice bran (3:2:2) is used for the treatment of night blindness.

134. *Dipteracanthus prostratus* (Poir.) Nees (Acanthaceae)
Uses: Infusion of root is applied to treat eye diseases.

135. *D. suffruticosa* Voigt (Acanthaceae)
Uses: Root paste is used in the treatment of sore eyes.

136. *Dolichos biflorus* L. (Fabaceae)
Uses: Seed paste is applied for eye diseases.

137. *D. falcatus* Klein. (Fabaceae)
Uses: Root paste is used in ophthalmia.

138. *Dysoxylum malabaricum* Bedd. ex Hiern (Meliaceae)
Uses: Oil obtained from wood is used to treat eye diseases.

139. *Eclipta prostrata* (L.) L. (Asteraceae)
Chemical constituents: Polyacetylenic thiophenes, desmethylwedelolactone-7-O-glucoside.

Uses: Root boiled in cow's milk is dried and paste is applied to get relief from burning sensation of eye.

Fresh leaf extract is mixed with equal amount of coconut oil, boiled until reduced to half the initial volume, then filtered and cooled and used as hair oil continuously for 3-6 months to improve vision.

Young fresh twig juice is given twice daily for 2-3 days in case of conjunctivitis.

Fresh leaf juice with common salt (2:1) is applied as drops in the treatment of eye troubles.

140. *Elephantopus scaber* L. (Asteraceae)
Chemical constituents: Epifriedenol, lupeol, stigmasterol and a mixture of triacontan-1-ol and dotriacontan-1-ol.

Uses: Leaf juice is used as eye drops for eye diseases.

141. *Emilia sonchifolia* (L.) DC. (Asteraceae)
Uses: Juice of the fresh leaves is used for sore eyes and night blindness.

Leaf juice of 5 ml mixed with breast milk is used as eye drops for eye injuries.

Leaf paste in doses of one spoonful is given once a day at bedtime for about 2-3 months to treat night blindness.

One or two drops of leaf extract is poured thrice a day for night blindness and ophthalmia for 5-7 days.

Two drops of leaf juice is instilled daily twice for 5 days for conjunctivitis.

142. *Eriolaena hookeriana* Wt. & Arn. (Sterculiaceae)
Chemical constituents: Lupeol, α-amyrin, β-sitosterol and β-amyrin acetate.

Uses: Green bark is powdered and mixed in castor oil and applied to the head to cure pains of eyeball.

143. *Ervatamia coronaria* Stapf (Apocynaceae)
Uses: Flowers are washed thoroughly with clean water and extract is used as eye wash.

Juice of root with lime juice is used to clear corneal opacity.

144. *E. divaricata* (L.) Burkill (Apocynaceae)
Uses: Roots are dried, thoroughly cleaned, rubbed together with ghee and applied into the eyes in the form of *ajan* to cure many kinds of corneal diseases.

145. *E. heyneana* (Wall.) Cooke (Apocynaceae)
Uses: Flowers are used to treat inflammation of cornea.

146. *Erycibe paniculata* Roxb. (Convolvulaceae)
Uses: Decoction of the young leaves is used as eyewash, twice a day for 15 days to cure night blindness.

147. *Euphorbia hirta* L. (Euphorbiaceae)
Uses: Latex is applied to remove redness of the eyes during conjunctivitis.

Whole plant juice is given for 3 days to cure white dot in eye.

Leaf juice is directly squeezed into the eye to cure eye troubles.

148. *E. ligularia* Roxb. (Euphorbiaceae)
Uses: The juice of the leaves is used in case of ophthalmia.

149. *Euphrasia officinalis* L. (Scrophulariaceae)
Uses: Fresh juice of whole plant is used to relieve eye troubles.

150. *Ficus amplissima* J. E. Sm. (Moraceae)
Uses: Latex of the root and fruit is an effective remedy in case of eye diseases.

151. *F. benghalensis* Linn. (Moraceae)
Chemical constituents: Stem bark contains leuco-anthocyanins.

Uses: One drop of fresh latex mixed with a pinch of karpur is applied externally in case of cataract till cure.

152. *Firmiana colorata* (Roxb.) R. Br. (Sterculiaceae)
Uses: Tender leaf juice with that of *Cassia tora* is poured for eye infection.

153. *Flacourtia indica* (Burm.f.) Merr. (Flacourtiaceae)
Uses: Leaf sap is dropped into eyes to cure conjunctivitis.

154. *Floscopa scandens* Lour. (Commelinaceae)
Uses: Stem juice is used in sore eyes.

155. *Foeniculum vulgare* Mill. (Apiaceae)
Uses: Leaf juice is used to improve the eye sight.

156. *Garuga pinnata* Roxb. (Burseraceae)
Uses: Stem juice is applied for conjunctivitis.

Stem bark juice is dropped into the eyes to cure opacity of cornea.

157. *Geranium wallichianum* Wall. (Geraniaceae)
Uses: Decoction of the roots is mixed with wheat flour and an adequate amount of sugar and is used for ophthalmic diseases.

158. *Globba orixensis* Roxb. (Zingiberaceae)
Uses: Flower juice is poured into eyes for sore eyes.

159. *Gossypium arboreum* L. (Malvaceae)
Uses: Two drops of leaf juice is instilled into the eyes daily twice for 3 days for conjunctivitis.

160. *Gymnema sylvestre* R. Br. (Asclepiadaceae)
Chemical constituents: Gymnamine, gymnimagenin, gymnemic acid I-VII, XV-XVIII, Dammarone saponins-gymnemiosides, gypenosides II, V, XLIII, LXXIV, gymnosaponin TN-2, quercitol, lupeol, β-amyrin, stigmasterol.

Uses: Juice of fresh leaves is applied for clean vision and cataract.

Fresh leaves are macerated and filtered and few drops are dropped in eyes thrice a day for a week to treat cataract.

Root paste prepared with breast milk is applied like eyetex for curing eye diseases.

A spoonful of leaf juice is mixed with 10 g of butter and used as eyeliner once a day to cure for cataract.

161. *Hedychium coronarium* J. Koenig (Zingiberaceae)
Uses: About 50 g of fresh flowers are pounded and made into 10 ml extract with water. Two drops of the solution is applied externally to cure conjunctivitis.

Roots are crushed, juice is extracted and warmed with water and put into eyes to cure eye troubles like red eye, swollen eye, eye pain, excessive excretions, watering etc.

162. *Hedyotis scandens* Roxb. (Rubiaceae)
Uses: Stem juice is applied as eye drops against conjunctivitis and also cleaning of dust and gum from eyes.

Leaf juice is used for eye diseases.

163. *Heliotropium indicum* L. (Boraginaceae)
Uses: Plant juice is applied for eye diseases.

Leaf juice is used for eye sores and eye infection.

Leaf juice with common salt (2:1) is applied as drops to cure conjunctivitis.

Root juice is useful against ophthalmia.

164. *H. strigosus* Willd. (Boraginaceae)
Uses: Juice of leaves is applied to sore eyes.

165. *Hemidesmus indicus* R. Br. (Asclepiadaceae)
Chemical constituents: Hexatriacontanes, lupeol, α-amyrin, β-amyrin, sitosterol, coumarinolignoids hemidesmin-1 and hemidesmin-2.

Uses: Two to three drops of leaf juice mixed in mother's milk is applied to eyes in redness and burning.

166. *Hibiscus aculeatus* Roxb. (Malvaceae)
Uses: The juice of leaves mixed with honey is used for eye diseases.

167. *H. furcatus* Roxb. (Malvaceae)
Uses: The juice of the leaves with honey is used for eye diseases.

168. *H. rosa-sinensis* L. (Malvaceae)
Chemical constituents: Alkaloids, steroids, glycosides, flavones, organic acids, vitamins, anthocyanin pigment and cyanidin diglucoside.

Uses: Root paste with honey (1:1) is applied in eye sore.

169. ***Holostemma adakodien*** **Schult. (Asclepiadaceae)**
Uses: The paste made from roots is applied in the case of ophthalmia.

170. ***H. annulare*** **(Roxb.) Schum. (Asclepiadaceae)**
Uses: Paste of root is applied to the eyes in ophthalmia.

171. ***Hyoscyamus niger*** **L. (Solanaceae)**
Uses: Powdered seeds are given to relieve eye pain.

172. ***Hypericum japonicum*** **Hook. f. & Thunb. (Hypericaceae)**
Uses: Whole plant juice is used as eye drops to treat conjunctivitis.

173. ***Hypoxis aurea*** **L. (Amaryllidaceae)**
Uses: Paste of root used to treat yellowing of eyes.

174. ***Hyptis suaveolens*** **(L.) Poit (Lamiaceae)**
Chemical constituents: Diterpenes-suaveolic acid and suaveolol.

Uses: Leaf infusion is poured in doses of 2-3 drops thrice a day till cure for ophthalmia.

175. ***Hyssopus officinalis*** **L. (Lamiaceae)**
Uses: The infusion of plant used as a wash for sore eyes.

176. ***Ipomoea aquatica*** **Forsk. (Convolvulaceae)**
Uses: Plant is used as a coolant for inflammed eyes.

Juice of fresh flowers is applied as drops for treatment of eye diseases.

177. ***Ixora coccinea*** **L. (Rubiaceae)**
Uses: Flower buds are used in the treatment of conjunctivitis.

Decoction of flowers is used in eye troubles.

178. ***Jasminum malabaricum*** **Wt. (Oleaceae)**
Uses: Leaf juice is used in the treatment of cataract.

179. ***J. sambac*** **(L.) Ait. (Oleaceae)**
Uses: A lotion made from flowers is used for washing the eyes.

One or two drops of leaf juice is instilled into eyes daily once for 3 days for ophthalmic diseases.

180. ***Jatropha curcas*** **L. (Euphorbiaceae)**
Chemical constituents: Ash of seeds contain Ca, Mg, Na, K and trace of P; oleic, linoleic, myristic, palmitic, stearic, arachidic acids and sitosterol, glucose, arabinose, xylose and rhamnose in seeds.

Uses: Leaves along with those of *J. gossypifolia* are made into paste and the eyes are covered with the paste for 2-3 days for night blindness.

181. ***Justicia adhatoda*** **L. (Acanthaceae)**
Chemical constituents: Alkaloids adhatodine, vasicine, 1-vascicinone, vasicinol, vasicinine, vasicoline, vasicolinone, adhatodine, anisotien, quinozoline

alkaloids, pyrroloquinazoline alkaloids – desmethoxy – aniflorine, 7-methoxyvasicinone besides fats, resins, sugar, mucilage, vit. C, essential oil containing limonene.

Uses: Ten to fifteen leaves are made into paste and bandaged externally on one eye for few minutes once a day till recovery for eye infection.

182. *J. procumbens* L. (Acanthaceae)
Uses: Juice of leaves is used to treat ophthalmia.

183. *Kalanchoe pinnata* (Lam.) Pers. (Crassulaceae)
Uses: Mildy heated leaf juice in doses of 2-3 drops is poured at an interval of 8 h for 3 days to treat eye sores.

184. *Kaempferia galanga* L. (Zingiberaceae)
Chemical constituents: Rhizome contains deoxypodophyllotoxin and ethyl-p-methoxy-trans-cinnamate. Tubers contain essential oil consisting of p-methoxycinnamic acid and its methyl and ethyl esters.

Uses: Rhizome juice is used for eye infection.

185. *Lagerstromia parviflora* Roxb. (Lythraceae)
Uses: Lukewarm leaves are placed on eyelids to cure redness of eyes.

186. *Lannea grandis* Engl. (Anacardiaceae)
Uses: Decoction of the bark is used as eye wash for sore eyes.

187. *Lantana camara* L. (Verbenaceae)
Chemical constituents: A new oxosteroid – lancamarone; tritepene – lantanolic acid; lantic acid.

Uses: Fresh leaf juice is used as eye drops for eye injuries.

188. *Lawsonia inermis* L. (Lythraceae)
Chemical constituents: Flowers contain essential oil.

Seeds contain fatty oil.

Plant contains xanthones laxanthones I, II & III.

Uses: Leaf paste is applied to cure conjunctivitis.

189. *Leonotis nepetiifolia* (L.) R. Br. (Lamiaceae)
Uses: Seed paste is kept in the eye for bringing out foreign particles.

190. *Leucas aspera* (Willd.) Link (Lamiaceae)
Chemical constituents: 2 sterols, 2 alkaloids, galactose, oleanolic acid, ursolic acid and β-sitosterol.

Uses: Leaf juice is used as eye drops for eye diseases.

Flower juice is administered daily once for a week for eye infection.

191. *L. indica* (L.) R. Br. (Lamiaceae)
Uses: Two to three drops of leaf or flower juice is poured into eyes for eye infections.

192. *Luffa acutangula* (L.) Roxb. (Cucurbitaceae)
Uses: Leaf juice is instilled daily twice for 2 days for conjunctivitis.

193. *L. acutangula* (L.) Roxb. var *amara* (Roxb.) C. B. Clarke (Cucurbitaceae)
Uses: Fresh leaf juice is dropped into the eyes of children in granular conjunctivitis.

194. *Madhuca indica* J. Gmel. (Sapotaceae)
Chemical constituents: Lupeol acetate, β-amyrin acetate, α-spinasterol, erythrodiol monocaprylate, betulinic acid and oleanolic acid.

Uses: Three drops of milky latex is used as eye drops for conjunctivitis.

195. *Mahonia pycnophylla* (Fedde) Deing-la-ranong Takeda (Berberidaceae)
Uses: Leaf is crushed and juice is diluted with water and used for various eye diseases.

196. *Mallotus philippensis* (Lam.) Muell.-Arg. (Euphorbiaceae)
Chemical constituents: Fruits contains cardenolides rottlerin and isorottlerin. Seeds contain corotoxigenin and its rhamnoside, coroglaucigenin and its rhamnoside.

Uses: Glandular hairs of the fruits are used in eye diseases.

197. *Manilkara hexandra* (Roxb.) Dub. (Sapotaceae)
Chemical constituents: α-spinasterol, α- and β-amyrin acetates, taraxerol and taraxeryl acetate, ursolic acid and triterpenoids, quercotol, quercetin and dihydro-quercetin from bark, fruit and seeds.

Uses: Root paste is applied on the forehead till the sight is improved.

Root paste is applied on forehead once at bedtime for 7 days for eye infections.

198. *Marsilea minuta* L. (Marsileaceae)
Uses: Two drops of plant juice is dropped in the eyes thrice a day for treatment of eye diseases.

199. *Melilotus officinalis* (L.) Pall. (Fabaceae)
Uses: One or two drops of fresh leaf juice is poured to cure conjunctivitis.

200. *Memecylon edule* Roxb. (Melastomaceae)
Chemical constituents: Plants contains umbelactone, β-amyrin, its glucoside, oleanolic and ursolic acids. Leaves contain yellow dye, yellow glycosidal substance, tartaric and malic acids, a resin.

Uses: Leaf decoction is used as eye wash.

201. *M. umbellatum* Burm. f. (Melastomaceae)
Uses: A lotion prepared from leaves is used in eye troubles.

202. *Mesua ferrea* L. (Clusiaceae)
Uses: Sap is soaked in clean water for about 2-3 hours and used for eye problems.

203. ***Michelia champaca* L. (Magnoliaceae)**

Chemical constituents: Magnaspengerine, salicifoline from bark and root cortex.

Uses: Leaf decoction is used to cure eye diseases.

Oil extracted from the flowers is locally used in ophthalmia.

204. ***Mikania micrantha* H.B.K. (Asteraceae)**

Uses: Leaves are warmed on fire and kept on the eyes repeatedly (4-5) times to cure any type of eye trouble.

205. ***Mitragyna parvifolia* (Roxb.) Korth. (Rubiaceae)**

Chemical constituents: Root bark contains isorhynchophyllin and rhynchophylline.

Uses: Root bark decoction is taken with egg (2:1) as cure for night blindness.

206. ***Mollugo nudicaulis* Lam. (Aizoaceae)**

Uses: Leaf juice is mixed with mother's milk and dropped (2-3 drops) for curing eye diseases.

207. ***Moghania chapper* (Ham. ex Benth.) O. Kuntze (Fabaceae)**

Uses: Fresh young leaf juice with common salt diluted in water (2:1) is used as eye drops for treatment for eye diseases

208. ***Momordica charantia* L. (Cucurbitaceae)**

Chemical constituents: Leaves contain mordicines I, II and III, cucurbitane, triterpenoids.

Fruits contain acylglucosyl sterols, lipids, charantin, β-sitosterol, glucoside, ascorbigen and β-O-glucoside.

Uses: Five ml of fruit juice mixed with 5 to 10 fruits of black pepper powder is given early in the morning for 3-6 weeks to treat night blindness.

Leaves and black pepper are ground together, mixed with water and the liquid is put in each eye daily for 7 days.

209. ***Moringa oleifera* Lam. (Moringaceae)**

Chemical constituents: Aldotriouronic acid from gum, aspartic acid, glutamic acid, glycine, threonine, alanine, valine, leucine, histidine, lysine, argenine, cysteine and methionine in leaves.

Uses: Leaves are cooked with green pepper (*Capsicum annuum*) and common salt for taste and is given as a vegetable to the patient for 15 days to cure night blindness.

Leaf juice mixed with an unequal amount of honey is used as collyrium for all types of eye diseases.

One to two drops of leaf juice is applied to the eye for about a week for pain and inflammation of the eye.

Flowers are collected and thoroughly cleaned, juice is extracted and applied in the eyes both in the morning and evening for 7-9 days.

Young leaf tips are eaten to prevent blindness.

Juice of leaves mixed with few drops of honey is applied to cure redness of eyes.

Ten g of cotyledons are ground with a spoonful of honey and used as eye liner once in a day for 2 days for ophthalmic diseases.

210. *Musa paradisiaca* L. (Musaceae)

Chemical constituents: Steroids, glycosides, terpenes and organic acids.

Uses: One or two drops of sap coming out of the cut leaf is dropped into the eye to cure conjunctivitis.

211. *Mussaenda belilla* Buch.-Ham. (Rubiaceae)

Uses: Stem juice is put into the eyes to cure soreness and redness of eyes.

Root juice is used to cure eye troubles.

212. *M. frondosa* L. (Rubiaceae)

Uses: Root decoction of dried shoots is used to cure eye troubles.

Leaf juice mixed with cumin seeds is used to cure eye diseases.

213. *M. glabrata* Hutch. (Rubiaceae)

Uses: Root extract is employed in eye troubles.

214. *Nardostachys grandiflora* DC. (Valerianaceae)

Uses: Whole plant is used for conjunctivitis.

215. *Nasturtium officinale* R. Br. (Brassicaceae)

Uses: Whole plant is used as a vegetable to improve eyesight.

216. *Nelumbo nucifera* Gaertn. (Nelumbonaceae)

Uses: Perianth is soaked in water in an earthen pot overnight and the eyes are cleaned with this water during conjuctivitis.

Flower nectar is applied to treat conjunctivitis.

217. *Nerium indicum* Mill. (Apocynaceae)

Uses: Watery latex from fresh twigs in diluted form is used as drops in conjunctivitis.

Leaf juice is used for inducing lachrymation in ophthalmia.

218. *Nicotiana tabacum* L. (Solanaceae)

Chemical constituents: Anatabine and (+) nornicotine from roots; quercetin – 3 – Me ether from flowers; 3 new gibberellins – nicotiana α, β and γ-along with gibberellins A and A3 from shoot apices and flower beds; new alkaloid – anatalline; solenone.

Uses: Dried leaf powder is soaked in water for 10-15 minutes and two drops of water is administered into the infected eye.

219. ***Nyctanthes arbor-tristis*** **L. (Oleaceae)**

Chemical constituents: Aborisides A, B & C, flavonol glycosides astragalin and nicotiglorin.

Uses: Stem bark is used in the treatment of pain in eyes.

220. ***Nymphaea nouchali*** **Burm.f. (Nymphaeaceae)**

Uses: Petals are boiled in water and used to wash eyes to cure eye infection.

221. ***Ocimum americanum*** **L. (Lamiaceae)**

Uses: Flower extract is applied on the eyes as drops thrice a day till cure.

222. ***O. basilicum*** **L. (Lamiaceae)**

Chemical constituents: Me cinnamate, methyl chavicol, linalool, cineole, ocimene, borneol, sambulene, safrole from essential oil.

Uses: Decoction of the leaves is used for washing sore eyes.

Leaf juice is used as eye drops for blindness.

223. ***O. gratissimum*** **L. (Lamiaceae)**

Chemical constituents: Eugenol, ocimene, cadinene, perillyl alcohol from essential oil; ocimol, gratiddimin from leaves.

Uses: Seeds are put in water overnight and in early morning the eyes are washed with this water which is believed to improve the eyesight and also to get rid of any foreign particles if present in the eyes.

224. ***O. tenuiflorum*** **L. (Lamiaceae)**

Chemical constituents: Roots contain alkaloids, glucosides, saponins and tannins. Leaves contain volatile oil, ascorbic acid. Essential oil consists of methyl clavicol, camphor, β-caryophyllene, eugenol, caryophyllin, camphene, α-pinene.

Uses: Two drops of fresh leaf juice is used as ear drops daily once before going to bed for a month for night blindness.

225. ***Opuntia dillenii*** **(Ker-Gawl.) Haworth (Cactaceae)**

Chemical constituents: Isoquercitrin, β-sitosterol, opuntiol, opuntiol acetate; glucoside – narcissin from flowers, a polysaccharide containing galactose and arabinose isolated from pods.

Uses: Leaf juice is applied to the eyes in case of ophthalmia.

Fruit pulp is used in ophthalmia.

226. ***Oryza rufipogon*** **Griff. (Poaceae)**

Uses: Water obtained in internodes is used as drops in conjunctivitis.

227. ***Ougeinia oogeinensis*** **(Roxb.) Hochr. (Fabaceae)**

Uses: Stem bark juice is dropped in the eyes for the treatment of cataract twice a day for a month.

228. *Oxalis corniculata* L. (Oxalidaceae)
Uses: An infusion of leaves is used to remove opacities of the cornea.

Two drops of the leaf juice is administered in the morning and evening daily for 5 days to treat whiteness on eyeball.

Juice of leaves is used as eye drops for removal of dust from eyes or against redness of eyes.

229. *Paederia scandens* (Lour.) Merr. (Rubiaceae)
Uses: Leaf decoction is given with unboiled egg (3:1) for treatment of night blindness.

230. *Panicum repens* L. (Poaceae)
Uses: Plant juice is used as drops in eye diseases.

231. *Pavetta indica* L. (Rubiaceae)
Uses: Fruits are eaten as preventive to red eye.

232. *Pavonia odorata* Willd. (Malvaceae)
Uses: Leaf juice is taken internally for improving vision and cooling of eyes.

233. *Phaseolus trilobus* Ait (Fabaceae)
Uses: Leaf juice in doses of 2-3 drops at an interval of 4 h is poured into eyes for eye troubles.

234. *Phyla nodiflora* (L.) Greene (Verbenaceae)
Uses: Fresh leaf extract along with black pepper and water is taken in morning for eye ache.

235. *Phyllanthus amarus* Schum. & Thonn. (Euphorbiaceae)
Uses: Whole plant is ground and applied externally over both eyelids to treat corneal diseases and conjunctivitis.

Equal amount of the leaves along with those of *Boerhavia diffusa, Alternanthera sessilis* are ground together. Five g of this preparation is taken in the morning and evening for 45 days for night blindness.

Equal amount of the whole plant juice along with the leaf juice of *Alternanthera sessilis* and sesame oil are mixed and boiled until reduced to one-third the initial volume. An oil bath is taken from the preparation once every 4 days for atleast six months for all types of eye diseases.

236. *P. emblica* L. (Euphorbiaceae)
Uses: Fruit is washed, ground and 1-2 drops of juice is administered as eye drops in conjunctivitis.

Fresh leaf juice with diluted solution of common salt (2:1) is used as eye drops for improving weak eyesight.

Dried fruit infusion is an efficacious eyewash in disease of the eye.

237. ***P. virgatus*** **Forst. (Euphorbiaceae)**
Uses: Juice of the leaves is employed in eye diseases.

238. ***Picrorhiza kurroa*** **Royle ex Benth. (Scrophulariaceae)**
Uses: Root paste is used to improve eyesight.

239. ***Piper betle*** **L. (Piperaceae)**
Chemical constituents: Leaves contain iodine, vitamin B, essential oil.

Leaf oil contains chavibetol, chavibetol acetate, caryophyllene, allypyrocatechol diacetate, comphene, chavibetol methyl ether, eugenol, α-pinene, β-pinene, α-limonene, safrole, 1,8-cineole, *p*-cymene and allypyrocatechol monoacetate.

Uses: Leaves are mixed with soamp seeds in equal proportions and made into a paste with honey and applied to the eyes daily till cure for night blindness.

240. ***P. longum*** **L. (Piperaceae)**
Uses: Fruit is placed inside mutton liver and boiled and then taken together in case of night blindness.

241. ***P. trioicum*** **Roxb. (Piperaceae)**
Uses: Fresh stem piece is dipped in water and a drop of water is poured during conjunctivitis.

242. ***Plantago asiatica*** **L. (Plantaginaceae)**
Uses: Leaf juice is used as eyewash in ophthalmia.

243. ***Plectranthus rugosus*** **Wall. ex Benth. (Lamiaceae)**
Uses: Filtered extract of fresh leaves is used in the treatment of ophthalmic diseases.

244. ***Plumeria rubra*** **L. (Apocynaceae)**
Uses: Five spoonfuls of leaf extract mixed in a glass of water is used to clean the eyes only once for red eyes.

245. ***Portulaca oleracea*** **L. (Portulacaceae)**
Uses: Whole plant is used in the treatment of eye diseases.

246. ***Pouzolzia zeylanica*** **(L.) Benn. (Urticaceae)**
Uses: Juice of aerial parts is applied twice a day in eye injury.

247. ***Pratia begonifolia*** **Lindl. (Campanulaceae)**
Uses: Decoction of the leaves is used externally in eye infection.

248. ***Priva cordifolia*** **(L.f.) Druce. (Verbenaceae)**
Uses: An infusion of the leaves is used to cure inflammations of the eyeball.

249. ***Prunus persica*** **(L.) Batsch. (Rosaceae)**
Uses: Leaves are warmed over fire and rubbed against eye pain.

250. ***Pterocarpus santalinus*** **L. f. (Fabaceae)**
Uses: Paste of the wood is a cooling application for ophthalmia and sore eyes.

251. ***Rauvolfia serpentina*** **(L.) Benth. (Apocynaceae)**
Chemical constituents: Alkaloids, steroids, fatty acids and aminoacids.

Uses: Juice of leaves is used for clearing corneal opacity.

Two or three drops of leaf juice is instilled into eyes daily once for 2 days for ophthalmic diseases.

252. ***Rhynchosia hirta*** **(Andr.) Meikle & Verdc. (Fabaceae)**
Uses: About 2-3 teaspoons of leaf extract per day is taken internally for few months and also applied around eyes as a remedy against night blindness.

253. ***Ricinus communis*** **L. (Euphorbiaceae)**
Uses: Two or three drops of extract of young leaves is poured into eyes for 2-3 days to cure redness of eyes.

Root bark is boiled in water and cooled. The sore eyes are washed with this solution to cure redness in the eyes.

Seed oil is applied for eye pain.

Leaf decoction with unboiled egg (3:2) is given to children to treat night blindness.

Latex is dropped into the eyes to treat cataract.

254. ***Rivea hypocrateriformis*** **(Desr.) Choisy (Convolvulaceae)**
Uses: Flower juice is used as drops for eye diseases.

255. ***Rosa chinensis*** **L. (Rosaceae)**
Uses: Juice of flowers is applied in the morning to reduce redness of the eye due to inflammation.

256. ***Rostellularia procumbens*** **(L.) Nees (Acanthaceae)**
Uses: Juice of the leaves is squeezed into the eyes in case of ophthalmia.

257. ***Rubia cordifolia*** **L. (Rubiaceae)**
Uses: Root infusion with brassica oil (4:1) is used for treatment of eye diseases.

Leaf extract is used to treat cataract, conjunctivitis and also to clean the eyes.

258. ***Rubus himalayense*** **Royle ex D. Don. (Rosaceae)**
Uses: Root is ground with water and the paste is applied on eyelids to cure eye infections.

259. ***Ruellia prostrata*** **Poiret (Acanthaceae)**
Uses: Two g of flowers and young fruits are collected, placed in 30 ml of rose water, left for 4 hrs, then filtered and stored. Two drops are applied externally into the eyes four times a day for 5-7 days to clear cloudy vision.

260. ***Salmalia malabarica*** **Schott. & Endl. (Bombacaceae)**
Uses: A paste of the petals mixed with milk is used as a soothing application in conjunctivitis of infants.

261. ***Salvia aegyptiaca* L. (Lamiaceae)**
Uses: A seed grain when placed in the eye forms mucilage and help in removing a foreign body from the eye.

262. ***Sapindus emarginatus* Vahl (Sapindaceae)**
Chemical constituents: A new saponin – emarginatoside from fruits.

Uses: Fine paste of seeds is gently applied over the eyelids daily at bed time for about 5 months to cure night blindness.

263. ***S. laurifolius* Vahl (Sapindaceae)**
Uses: Root is used as a collyrium in sore eye and ophthalmia.

264. ***Sarcochlamys pulcherrima* Gaud. (Urticaceae)**
Uses: Fresh leaf extract is applied to cure itching of eyes.

265. ***Scoparia dulcis* L. (Scrophulariaceae)**
Chemical constituents: Diterpenes – scopadulcic acids, scoparic acids, scopadulciol, ifflaionic acid, flavone glycoside – 8 – hydroxytricetin – 7 – glucoronide.

Uses: Mixture of the stem and leaf extract is applied externally during eye troubles.

266. ***Silene gallica* L. (Caryophyllaceae)**
Uses: Plant juice is used in the treatment of ophthalmia.

267. ***Senecio vulgaris* L. (Asteraceae)**
Uses: Powder of root is used in the treatment of eye diseases.

268. ***Sesamum orientale* L. (Pedaliaceae)**
Chemical constituents: Furofuran lignans – sesamolin and sesangolin, antioxidant – sesamolinol, sesamol and-γ-tacopherol.

Uses: Fresh flowers are collected, applied over the eyelids and tied down with fine cloth in the night hours for smoky vision.

269. ***Sesbania procumbens* (Roxb.) Wt. ex Arn. (Fabaceae)**
Uses: Two to three drops of juice of leaves and flowers are poured into eyes once at bedtime for 20-30 days for ophthalmia.

Fifty g of flowers are ground with 100 g of leaves of *Alternanthera sessilis* and a spoonful of honey and made into juice. Two drops of juice is instilled into the eyes daily twice for 21 days for early stages of cataract to disappear.

270. ***S. grandiflora* (L.) Poir. (Fabaceae)**
Chemical constituents: A saponin from leaves which yielded oleanolic acid, galactose, rhamnose and glucuronic acid.

Uses: Juice of flowers is put in the eyes to cure dimness of vision.

Leaf juice with honey (3:1) is given to children for treatment of night blindness.

A pinch of fried alum powder is dissolved in a glass of boiled, cooled water and kept for one day. The filtered water is instilled into the eyes accompanied by the application of flower juice. Two to three drops are instilled in six hours and continued till cure.

Flowers are used as curry for night blindness.

Flower juice with pure honey is poured into the eyes in doses of 2-3 drops once a day at bedtime for about 40 days to treat cataract.

271. *Sida acuta* Burm. *f.* (Malvaceae)
Uses: Root bark paste with honey (2:1) is applied for the treatment of cataract.

272. *S. cordifolia* L. (Malvaceae)
Uses: Leaf juice is applied in eyes for eye diseases.

273. *S. rhombifolia* L. (Malvaceae)
Uses: Plant is crushed, boiled and paste is applied externally on the head to treat inflammation of the eyes.

274. *Solanum americanum* Mill. (Solanaceae)
Uses: Leaves are cooked as curry and eaten daily twice for 3 days for night blindness.

275. *S. nigrum* L. (Solanaceae)
Uses: Extract of leaves (1-2 drops) is applied over eye two or three times a day to cure conjunctivitis.

Stem infusion is used for eye diseases.

276. *S. viarum* Dunal (Solanaceae)
Uses: Raw fruits are eaten once a day for 15 days to treat night blindness.

277. *S. virginianum* L. (Solanaceae)
Uses: About 20 g of roots are boiled in goat's milk and applied on eye to cure eye infection.

Fresh root juice extract mixed with equal quantity of lemon juice is applied daily for a month as a preventive measure against cataract.

278. *Solena amplexicaulis* (Lamk.) Gandhi (Cucurbitaceae)
Uses: Juice of petals of male flowers is used as drops for eye diseases.

Two to three drops of leaf juice is poured into eyes till the sore eye is cured.

279. *Sonchus brachytus* DC. (Asteraceae)
Uses: Latex is used in the treatment of eye troubles.

280. *Spatholobus parviflorus* (Roxb. ex DC.) O. Ktze. (Fabaceae)
Uses: Hundred g of leaf paste is applied on head half an hour before bath for conjunctivitis.

281. *Sphaeranthus indicus* L. (Asteraceae)
Chemical constituents: Sterol glycoside, essential oil, alkaloid sphaeranthine.

Capitula contains eudesmanolides, fatty oil, tannins and volatile oil.

Uses: Flower juice is given as eye drops for eye disease and inflammed eyes.

282. ***Spilanthus calva*** **DC. (Asteraceae)**
Uses: Decoction of the bark is used as astringent lotion for the eye.

283. ***Spondias pinnata*** **(L. f.) Kurz (Anacardiaceae)**
Uses: Unripe fruits are eaten once a day for 3 days for ophthalmic diseases.

284. ***Stachytarpheta jamaicensis*** **(L.) Vahl (Verbenaceae)**
Uses: The juice of the plant is used for treating cataract.

285. ***Streblus asper*** **Lour. (Moraceae)**
Chemical constituents: Kamaloside, asperoside, indroside, lucknoside from root bark; cannodimethoside, strophalloside, sarmethoside.

Uses: Tender twigs are worn around neck to cure inflammatory eyes.

One drop of leaf juice is poured into eyes twice a day for 3-4 days for eye infection.

286. ***Strychnos potatorum*** **L. f. (Loganiaceae)**
Chemical constituents: Steroids, glycosides, saponins, alkaloids – diaboline, mannogalactone.

Uses: Seed paste with common salt (3:1) is used in the treatment of eye diseases and conjunctivitis.

Seed powder mixed with honey and a little camphor is applied to the eyes in eye diseases such as copious watery discharge, opacity or ulceration of the cornea and purulent discharge from the conjunctiva.

Seeds are thoroughly cleaned, rubbed together with honey and applied to the eyes in the form of *anjan* in the morning for 15 days to cure congestion of eyes.

287. ***Symphorema involucratum*** **Roxb. (Verbenaceae)**
Uses: Bark of stem powder along with breast milk is used to heal eye injuries.

288. ***Symplocos racemosa*** **Roxb. (Symplocaceae)**
Uses: Infusion of stem bark is used as drops for the treatment of conjunctivitis.

Bark paste with bark of *Boswellia serrata* is applied externally on eyelids for reducing redness of the eyes thrice a day for 3 days.

289. ***Syzygium cumini*** **(L.) Skeels (Myrtaceae)**
Chemical constituents: Leaves yield essential oil including α-pinene, β-pinene, limonene, cis-ocimene and trans-ocimene.

Fruits contain anthocyanins delphinidin-3-gentiobioside, malvidine-3-laminaribioside.

Seed oil contains oleic, myristic, linoleic, stearic, vernolic, lauric, sterculic and malvalic acid.

Uses: Leaf paste is applied on forehead to cure redness of eye.

290. ***Tabernaemontana divaricata*** **(L.) R. Br. ex Roem. & Schult. (Apocynaceae)**
Uses: Root charcoal and milky juice of the leaves are used in ophthalmia.

Juice of the flowers mixed with oil is employed for relieving the burning sensation in sore eyes.

One or two drops of leaf juice is dropped twice daily to cure conjunctivitis.

Latex mixed with edible oil is applied on the head for eye pain.

Flower juice along with breast milk is used as eye drops in case of eye diseases and inflammation of cornea.

The milky juice from tender branches is poured into eye for eye infections.

Two spoonfuls of root paste mixed in a glass of goat's milk is administered daily twice for 3 days for ophthalmic diseases.

291. *Tagetes erecta* L. (Asteraceae)
Uses: Juice of the flower is used as eye drops in the treatment of ulcers in the eye.

292. *Tamarindus indica* L. (Caesalpiniaceae)
Uses: A poultice of the fresh leaves is locally applied for sore eyes.

Flowers/tender leaves are collected, ground and applied externally around the eyelids for congestion of eyes.

293. *Tectona grandis* L. f. (Verbenaceae)
Uses: Ash of the wood mixed with water is applied externally for eyelid swelling.

Stem bark ash mixed with mustard oil is used as eyeliner to make eyesight strong for ophthalmic diseases.

294. *Tephrosia purpurea* (L.) Pers. (Fabaceae)
Chemical constituents: Steroids, glycosides, flavones and fatty acids.

Uses: Fine powder of the seeds is used as collyrium to eye for cataract.

Flower juice is applied for the treatment of eye inflammation.

Root is rubbed with cow's urine and the juice is used as eye drops daily once for conjunctivitis.

295. *Terminalia bellirica* (Gaertn.) Roxb. (Combretaceae)
Chemical constituents: Hexahydroxydiphenic acid ester, β-sitosterol, gallic acid, ellagic acid, ethyl gallate, galloyl glucose, chebulagalic acid.

Fruit pulp contains cardioglycoside bellericanin.

Kernel contains non-edible oil.

Uses: Seed kernel mixed with honey is applied for eye diseases.

Seed oil is applied on the skull at night for night blindness.

Seed paste mixed with a pinch of castor oil is applied as eyeliner.

296. *T. chebula* Retz. (Combretaceae)
Uses: Infusion of dry fruit is used as eye drops in eye infection.

Fruit powder is soaked in water overnight in an earthen pot and eyes are washed with this water in the morning for eye problems.

297. ***Thalictrum foliolosum*** **DC. (Ranunculaceae)**
Uses: Root paste is used for treatment of eye diseases.

Root decoction is used for the treatment of ophthalmia and eye inflammation.

298. ***T. javanicum*** **Bl. (Ranunculaceae)**
Uses: Root paste is used in the treatment of ophthalmia.

299. ***T. reniforme*** **Wall. (Ranunculaceae)**
Uses: Root juice is used for treatment of cataract.

300. ***Thespesia populnea*** **(L.) Soland. ex Corr. (Malvaceae)**
Uses: Juice of flower petals is used as cure for eye diseases.

301. ***Thunbergia fragrans*** **Roxb. (Acanthaceae)**
Uses: Fresh root juice (1-2 drops) is used as eye drops to treat redness of eyes.

302. ***T. grandiflora*** **Roxb. (Acanthaceae)**
Uses: Water sap of stem is applied as eye drops to treat eye infections.

303. ***Thysanolaena maxima*** **O. Kuntze (Poaceae)**
Uses: Young stem juice is applied on the eye with a feather when eyes become red and dirty.

304. ***Tinospora cordifolia*** **(Willd.) Hook. f. & Thoms. (Menispermaceae)**
Chemical constituents: Furanoid diterpene 10-hydroxycoumarin, clerodane, clerodanefuranoditerpene I, II, III, IV, tinosporaside, tinosporadine, β-sitosterol.

Uses: Stem is dried thoroughly, rubbed with breast milk and applied into the eyes in the form of *anjan* for cataract.

305. ***Trachyspermum ammi*** **(L.) Sprague (Apiaceae)**
Uses: Cotton pods are soaked with seed paste and applied over eyelids to cure eye diseases.

306. ***Trianthema decandra*** **L. (Aizoaceae)**
Uses: Root is rubbed with breast milk and applied in the form of collyrium for ulcerous cornea.

Root is rubbed together with cow's milk and applied in the form of *anjan* for colour blindness.

307. ***T. portulacastrum*** **L. (Aizoaceae)**
Uses: Plant decoction with yolk (2:1) is given to children for treatment of night blindness.

Equal quantities of the roots along with those of *Amaranthus tricolor* are mixed together, rubbed with breast milk and applied to the eyes to cure the eye pain.

308. ***Tridax procumbens*** **L. (Asteraceae)**
Uses: Leaf juice is used in eye diseases.

309. ***Triumfetta rhomboidea*** **Jacq. (Tiliaceae)**
Uses: Pounded root is applied over inflammed eyelids.

310. ***Urena lobata*** **L. (Malvaceae)**
Uses: One spoon of freshly boiled rice is rolled in seven equal sized leaves. Through this roll, air is whooped into the eye for seven times and the roll is thrown backside only.

311. ***Ventilago denticulata*** **Willd. (Rhamnaceae)**
Uses: Root bark decoction is used as cure for eye disease like redness of eyes.

Stem pulp mixed with curd is given to cure eye inflammation.

312. ***Vepris bilocularis*** **(Wight & Arn.) Engl. (Rutaceae)**
Uses: A decoction of wood is boiled in oil and used for eye diseases.

313. ***Vernonia cinerea*** **(L.) Less. (Asteraceae)**
Uses: Leaf paste is used as eyetex for conjunctivitis.

Leaf juice is dropped into eyes for curing eye diseases.

314. ***V. patula*** **(Dryand.) Merr. (Asteraceae)**
Uses: Plant extract is used as eye drops to treat conjunctivitis.

315. ***Vigna trilobata*** **(L.) Verdcourt (Fabaceae)**
Uses: Fresh seed decoction with paste of ginger (3:1) is used as cure for night blindness.

316. ***Vitex negundo*** **L. (Verbenaceae)**
Chemical constituents: Flavonoids – acerosin-5-O-glucoside monoacetate isolated from bark; α-pinene, limonene, camphene, β-phellandrene, *p*-cymene, linalool, camphor, 4-terpineol, citral, terpinyl acetate, caryophyllene.

Uses: Leaf juice mixed with the seeds of *Trachyspermum ammi* is used for conjunctivitis.

Fresh leaves are crushed along with salt and the extract is taken internally once a day for 7 days for night blindness.

A drop of leaf juice is instilled into eyes only once for inflammation of eye.

317. ***Vitis vinifera*** **L. (Vitaceae)**
Uses: The sap of the young branches is used for ophthalmia.

318. ***Wattakaka volubilis*** **(L. f.) Stapf (Asclepiadaceae)**
Uses: Leaf paste prepared in salt is applied to get quick relief from exudation of eyes.

Two to three drops of leaf juice is poured into the eyes 2-3 times a day for 3 days to cure eye infections.

319. ***Wedelia chinensis*** **(Osbeck) Merr. (Asteraceae)**
Uses: Leaf decoction is administered internally to improve the eyesight.

320. *Withania somnifera* Dunal (Solanaceae)

Chemical constituents: Plant contains withains, withanolides, 5-dehydroxywithanolide-R, withasomniferin-A.

Roots contain alkaloids, glycosides, dulcitol, withanol, acylstery glycoside – sitoindoside XI, sitondoside X.

Leaves contain steroids, alkaloids- withasomine, visamine, glycoside.

Uses: A fomentation of the leaves is used for sore eyes and swelling of eyes.

321. *Xanthium strumarium* L. (Asteraceae)

Chemical constituents: Strumaroside from fruits, xanthatin, xanthinin, xanthumin, stigmasterol, lactones – xanthanol and iso-xanthanol.

Uses: Seeds are worn around the neck of children to cure chronic conjunctivitis and inflammation of eyes.

Leaf sap is dropped into eyes to cure conjunctivitis.

322. *Zingiber officinale* Rosc. (Zingiberaceae)

Chemical constituents: Rhizome contains diarylheptanoid, essential oil, ginger diol and related compounds. Predominant sesquiterpene is zingiberene; other present are: cis-sesquisabinene hydrate, zingiberneol, ar-curcumine, β-bisabolene, α-selinene, β-elemente, β-sesquiphellandrene, zingiberenol, sesqui-thujene. Monoterpene hydrocarbons presents in the oil include camphene, β-pinene, cumene, myrcene, limonene, *p*-cymene and β-phellandrene; oleoresin gingerin contains gingerol, shogaol, zingerone and paradol.

Uses: Dried rhizome is rubbed with breast milk and used as a collyrium for cloudy vision.

Summary and Conclusions

This review article depicts 322 plant species used against eye diseases along with their family. Preparation of the drug and mode of administration along with chemical constituents are also depicted. Medicinal plants are the oldest known health care products. Their importance is still growing although it varies depending on the ethnomedical, medical and historical background of each country.

Today herbal medicine is making a dramatic come back and scientists are turning to natural products for answer to various ailments. Consequently, the therapeutic properties of higher plants offer a virtually untapped reservoir of potentially useful sources of drugs that will continue to serve mankind in this century as they have done since the dawn of history.

Folklore knowledge of medicinal plants has significantly contributed in discovering many important drugs of modern system of medicine. In fact, modern pharmaceuticals still contain atleast 25 per cent drugs derived from plants. Modern medical research has finally realized the vast healing potentials hidden within the plant kingdom. Interest in medicinal plants as re-emerging health aid has been fuelled by the rising costs of prescription drugs in maintenance of personal health and well being and the bioprospecting of new plant derived drugs. Since most of the plants are procured from natural habitat, there is an urgent need to explore alternative methods to cultivate plants. In such cases, biotechnological approaches play an important role.

Acknowledgement

JKR grateful to UGC for the awarding of JRF under RFMS scheme.

References

Agarwal, V. S. and Ghosh, B. 1985. Drug plants of India (Root Drugs). Kalyani Publishers, New Delhi.

Bhattacharjee, S. K. 2001. Hand book of medicinal plants. Pointer Publishers, Jaipur.

Dastur, J. F. 1962. Medicinal Plants of India and Pakistan, D.B. Taraporevala Sons & Co. Private Limited, Bombay.

Gamble, J. S. and Fischer, C.E.C. 1916-1935. The Flora of Presidency of Madras. 3 Vols., BSI, Calcutta.

Jain, S. K. 1997. Contribution to Indian Ethnobotany, Scientific Publishers, Jodhpur, India.

Joganathan, J. N. and Balu, S. 1997.Treatment of eye diseases by the healers of Marakanam bio- region. Ancient science of life. 16: 215-221.

Kirtikar, K. R. and Basu, B. D. 2003. Indian Medicinal Plants vol 1-11; Oriental Enterprises, Dehradun, Uttaranchal, India.

Krishnam Nambiar, V. P. and Sesidharan, N. and Renuka, C. and Balagopalan, M. 1982. Studies on the Medicinal Plants of Kerala forests, KFRI research report no. 42, Kerala Forest Research Institute, Peechi, Kerala.

Pal, D. C. and Jain, S. K. 1998. Tribal Medicine, Naya Prokash, Calcutta.

Pullaiah, T. 2002. Medicinal Plants of Andhra Pradesh (India). Regency Publications, New Delhi.

Rastogi, R. P. and Mehrotra, B. N. 1990-1999. Compendium of Indian Medicinal Plants Vol. 1-6; Council of Scientific and Industrial Research, New Delhi.

Thammanna and Narayana Rao, K. 1990. Medicinal Plants of Tirumala. Tirumala Tirupati Devasthanams, Tirupati.

Varier, P. S. 1994-1995. Indian Medicinal Plants Vol. 1-5; Orient Longman Limited, Hyderabad.

Yoganarasimhan, S. N. 1996. Medicinal Plants of India Vol. 1 – Karnataka, Interline Publishing Private Limited, Bangalore.

Yoganarasimhan, S. N. 2000. Medicinal Plants of India Vol. II – Tamilnadu, Interline Publishing Private Limited, Bangalore.

Chapter 4

Economic Valuation of Ethnobotanical Species of East Godavari District (Quantitative Analysis)

J. Suneetha, S. Prasanthi and T.V.V. Seetharami Reddi

Department of Botany, Andhra University, Visakhapatnam – 530 003, India

Economic valuation of all the 600 ethnobotanical species was carried out to calculate the Total Importance Value as in the table given. Based on the present field knowledge they are put on 1-10 scale (Timber, food, medicine, fuel, fibre, dye, tannin, gum/resin, oil, latex and others) as per Belal and Springuel (1996). It is revealed that species like *Acacia leucophloea, A. nilotica, Albizia lebbeck, Artocarpus heterophyllus, Azadirachta indica, Bambusa arundinacea, Bauhinia vahlii, Buchanania lanzan, Butea monosperma, Cocos nucifera, Dendrocalamus strictus, Oryza sativa, Pongamia pinnata, Pterocarpus marsupium, Schleichera oleosa, Syzygium cumini, Tectona grandis, Terminalia bellirica, T. chebula, Triticum aestivum* and *Zizyphus maurtiana* are found to have high economic potential.

Acacia chundra, A. leucophloea, Albizia lebbeck, Anogeissus latifolia, Artocarpus heterophyllus, Azadirachta indica, Borassus flabellifer, Bridelia retusa, Chloroxylon swietenia, Dalbergia lanceolaria, D. latifolia, D. paniculata, Erythroxylum monogynum, Haldinia cordifolia, Hardwickia binata, Holoptelea integrifolia, Lagerstroemia parviflora, Mangifera indica, Mitragyna parvifolia, Morinda pubescens, Ougeinia oojeinensis, Polyalthia cerasoides,

P. longifolia, Pterocarpus marsupium, Schleichera oleosa, Syzygium cumini, Tectona grandis, Terminalia alata, T. arjuna, T. bellirica, T. chebula and *T. coriacea* are used for timber.

Acacia sinuata, Acalypha indica, Achyranthes aspera, Aegle marmelos, Agaricus campestris, Albizia lebbeck, Allium cepa, A.sativum, Allmania nodiflora, Amorphophallus paeoniifolius, Anacardium occidentale, Arachis hypogaea, Artocarpus heterophyllus, Asparagus recemosus, Bauhinia variegata, Beta vulgaris, Cajanus cajan, Carica papaya, Carissa carandas, C. spinarum, Cicer arietinum, Cocos nucifera, Coriandrum sativum, Dioscorea alata, D. bulbifera, D. hispida, D. oppositifolia, D. pentaphylla, D. tomentosa, Diospyros chloroxylon, D. melanoxylon, D. montana, Eleusine coracana, Ipomoea batatus, Manilkara hexandra, Oryza sativa, Pennisetum typhoides, Pueraria tuberosa, Sesbania grandiflora, Syzygium cumini and *Tamarindus indica* are used as food.

Acacia leucophloea, A. nilotica, Albizia lebbeck, Apluda mutica, Arachis hypogaea, Argyreia nervosa, Bauhinia purpurea, B. vahlii, Capparis zeylanica, Crotalaria juncea, Dioscorea bulbifera, D. pentaphylla, Eleusine coracana, Leucaena leucocephala, Mollugo pentaphylla, Oryza sativa, Panicum sumatrense, Pennisetum typhoides, Saccharum officinarum, S. spontaneum, Setaria italica, S. pumila, Sorghum vulgare, Syzygium cumini, Triticum aestivum, Vigna mungo, V. trilobata, Zizyphus mauritiana and *Z. oenoplia* are used as fodder.

Abrus precatorius, Achyranthes aspera, Acorus calamus, Aloe vera, Andrographis paniculata, Asparagus recemosus, Azadirachta indica, Boerhavia chinensis, Catharanthus roseus, Centella asiatica, Coleus ambonicus, Commiphora caudata, Curcuma longa, Desmodium gangeticum, Justicia adhatoda, Leucas aspera, Mentha spicata, Ocimum tenuiflorum, Phyllanthus emblica, Phyllanthus amarus, Piper nigrum, Rauvolfia serpentina, Strychnos nux-vomica, Terminalia bellirica, Tinospora cordifolia and *Withania somnifera* rank high in medicinal value.

Acacia leucophloea, A.nilotica, Balanites aegyptiaca, Bauhinia racemosa, Casuarina equisetifolia, Cipadessa baccifera, Cocos nucifera, Dalbergia spinosa, Ehretia laevis, Eichhornia crassipes, Leucaena leucocephala, Mangifera indica, Maytenus emarginata, Oryza sativa, Pongamia pinnata, Samanea saman, Sesbania sesban, Syzygium cumini, Tamarix troupii, Trema orientalis and *Xylia xylocarpa* are used as fuel.

Abelmoschus moschatus, Abutilon indicum, Acacia leucophloea, Agave cantula, Ampelocissus tomentosa, Argyreia nervosa, Arundo donax, Bauhinia recemosa, B. vahlii, B. variegata, Bombax ceiba, Borassus flabellifer, Cannabis sativa, Caryota urens, Ceiba pentandra, Cocculus hirsutus, Cochlospermum religiosum, Cocos nucifera, Crotalaria juncea, Cryptolepis buchanani, Erythrina suberosa, Eulaliopsis binata, Hibiscus cannabinus, Sansevieria roxburghiana, Sterculia foetida, S. urens, Thespesia lampas, Trema orientalis and *Wrightia arborea* are used as fibre.

Anisomeles indica, Bixa orellana, Butea monosperma, Indigofera tinctoria, Lawsonia inermis, Leonotis nepetiifolia, Mallotus philippensis, Memecylon umbellatum, Nyctanthes arbortristis, Semecarpus anacardium, Toddalia asiatica, Ventilago denticulata, Woodfordia fruticosa, Wrightia tinctoria and *Zizyphus xylocarpa* are mainly used for dyes.

Acacia chundra, A. leucophloea, A. nilotica, Albizia amara, Anacardium occidentale, Anogeissus latifolia, Bridelia montana, B. retusa, Buchanania lanzan, Butea monosperma,

Table 4.1: Total Importance Value of Ethnobotanical Plants in East Godavari District

Sl.No.	Botanical Name	Timb		Food		Med		Fuel	Fiber	Dye	Tan	G/R	Oil	Latex	Tb	Others					Total	TIV
		Maj	Mir	Hum	Vet	Hum	Vet									Bm Tg Fn	B M Mp Um	Poi-son	Bev Liq	Misc.		
1.	*Abelmoschus esculentus*	0	0	2	0	3	0	0	0	0	0	0	0	0	0	0	0	0	0	0	5	06.25
2.	*A.manihot*	0	0	0	0	2	0	0	0	0	0	0	0	0	0	0	0	0	0	0	2	02.50
3.	*A.moschatus*	0	0	0	0	4	0	0	4	0	0	0	3	0	0	0	2	0	0	0	13	16.25
4.	*Abrus precatorius*	0	0	0	0	5	0	0	0	0	0	0	0	0	0	0	0	2	0	2	9	11.25
5.	*Abutilon crispum*	0	0	0	0	2	0	0	0	0	0	0	0	0	0	0	0	0	0	0	2	02.50
6.	*A.hirtum*	0	0	0	0	2	0	0	0	0	0	0	0	0	0	0	0	0	0	0	2	02.50
7.	*A.indicum*	0	0	0	0	2	0	0	4	0	0	0	0	0	0	0	0	0	0	0	6	07.50
8.	*Acacia auriculiformis*	2	0	0	0	0	3	0	0	0	0	0	0	0	0	0	0	0	0	0	5	06.25
9.	*A.caesia*	0	0	0	0	2	0	0	0	0	0	0	0	0	0	0	0	0	0	0	2	02.50
10.	*A.catechu*	0	0	0	0	2	2	0	0	0	0	0	0	0	0	0	0	0	0	0	4	05.00
11.	*A.chundra*	4	4	0	0	3	0	0	0	0	0	3	0	0	0	0	0	0	0	0	14	17.50
12.	*A.leucophloea*	3	2	0	3	0	0	2	2	0	3	0	0	0	0	0	0	0	0	0	15	18.75
13.	*A.nilotica*	2	2	0	2	3	2	2	0	0	2	3	0	0	0	0	0	0	0	2	20	25.00
14.	*A.pennata*	0	0	0	0	2	0	0	0	0	2	0	0	0	0	0	0	0	0	0	4	05.00
15.	*A.sinuata*	0	0	3	0	2	0	0	0	0	0	0	0	0	0	0	0	0	0	2	7	08.75
16.	*A.torta*	0	0	0	0	2	0	0	0	0	0	0	0	0	0	0	0	2	0	0	4	05.00
17.	*Acalypha indica*	0	0	2	0	3	0	0	0	0	0	0	0	0	0	0	0	0	0	0	5	06.25
18.	*Achyranthes aspera*	0	0	3	0	5	0	0	0	0	0	0	0	0	2	0	0	0	0	0	10	12.50
19.	*Acorus calamus*	0	0	0	0	5	2	0	0	0	0	0	0	0	0	0	0	0	0	0	7	08.75
20.	*Adenanthera pavonina*	3	2	0	0	3	0	0	0	0	0	0	0	0	0	0	0	0	0	2	10	12.50

Contd...

Table 4.1–Contd...

Sl.No.	Botanical Name	Timb		Food		Med		Fuel	Fiber	Dye	Tan	G/R	Oil	Latex	Tb	Others					Total	TIV
		Maj	Min	Hum	Vet	Hum	Vet									Bm Tg Fn	B M Mp Um	Poi-son	Bev Liq	Misc.		
21.	*Aegle marmelos*	0	0	**4**	0	**3**	0	0	0	0	0	0	0	0	0	0	0	0	0	**2**	**9**	**11.25**
22.	*Aerva lanata*	0	0	**2**	0	**3**	0	0	0	0	0	0	0	0	0	0	0	0	0	0	**5**	**06.25**
23.	*Aeschynomene indica*	0	**2**	0	0	**2**	0	0	0	0	0	0	0	0	0	0	0	0	0	0	**4**	**05.00**
24.	*Agaricus campestris*	0	0	**2**	0	**3**	0	0	0	0	0	0	0	0	0	0	0	0	0	0	**5**	**06.25**
25.	*Agave cantula*	0	0	0	0	**2**	**2**	0	**4**	0	0	0	0	0	0	**2**	0	0	0	0	**10**	**12.50**
26.	*Ageratum conyzoides*	0	0	0	0	**2**	0	0	0	0	0	0	0	0	0	0	0	0	0	0	**2**	**02.50**
27.	*Ailanthus excelsa*	0	**4**	0	0	**4**	**2**	0	0	0	0	0	0	0	0	0	0	0	0	0	**10**	**12.50**
28.	*Alangium salvifolium*	**3**	0	**2**	0	**4**	**2**	0	0	0	0	0	0	0	0	0	**2**	0	0	0	**13**	**16.25**
29.	*Albizia amara*	**3**	**2**	**2**	0	**2**	0	0	0	0	0	**3**	0	0	0	0	0	0	0	**2**	**14**	**17.50**
30.	*A.chinensis*	**3**	**2**	0	0	**2**	0	0	0	0	0	0	0	0	0	0	0	0	0	0	**7**	**08.75**
31.	*A.lebbeck*	**4**	**4**	**4**	**3**	**2**	0	0	0	0	0	0	0	0	0	0	0	0	0	**2**	**19**	**23.75**
32.	*A.odoratissima*	0	**2**	0	0	**2**	0	0	0	0	0	0	0	0	0	0	0	0	0	**2**	**6**	**07.50**
33.	*A.procera*	**2**	**2**	0	0	**2**	0	0	0	0	0	0	0	0	0	0	0	0	0	0	**6**	**07.50**
34.	*Allium cepa*	0	**2**	**4**	0	**3**	0	0	0	0	0	0	0	0	0	0	0	0	0	0	**7**	**08.75**
35.	*A.sativum*	0	0	**4**	0	**4**	0	0	0	0	0	0	0	0	0	0	0	0	0	0	**8**	**10.00**
36.	*Allmania nodiflora*	0	0	0	0	**2**	0	0	0	0	0	0	0	0	0	0	0	0	0	0	**2**	**02.50**
37.	*Allophylus cobbe*	0	0	**2**	0	**2**	0	0	0	0	0	0	0	0	0	0	0	0	0	0	**4**	**05.00**
38.	*Aloe vera*	0	0	0	0	**5**	0	0	0	0	0	0	0	0	0	0	0	0	0	2	**7**	**08.75**
39.	*Alpinia galanga*	0	0	0	0	**2**	0	0	0	0	0	0	0	0	0	0	0	0	0	0	**2**	**02.50**
40.	*Alstonia scholaris*	0	0	**2**	0	**3**	**2**	0	0	0	0	0	0	0	0	0	0	0	0	0	**7**	**08.75**

Contd...

Table 4.1–Contd...

Sl.No.	Botanical Name	Timb		Food		Med		Fuel	Fiber	Dye	Tan	G/R	Oil	Latex	Tb	Others					Total	TIV
		Maj	Min	Hum	Vet	Hum	Vet									Bm Tg Fn	B M Mp Um	Poi-son	Bev Liq	Misc.		
41.	*A.venenata*	0	2	0	0	2	0	0	0	0	0	0	0	0	0	0	0	0	0	0	4	05.00
42.	*Alternanthera sessilis*	0	0	2	0	3	0	0	0	0	0	0	0	0	0	0	0	0	0	0	5	06.25
43.	*Amaranthus spinosus*	0	0	2	0	2	0	0	0	0	0	0	0	0	0	0	0	0	0	0	4	05.00
44.	*A.tricolor*	0	0	2	0	2	0	0	0	0	0	0	0	0	0	0	0	0	0	0	4	05.00
45.	*A.viridis*	0	0	2	0	2	0	0	0	0	0	0	0	0	0	0	0	0	0	0	4	05.00
46.	*Ammannia baccifera*	0	0	0	0	3	0	0	0	0	0	0	0	0	0	0	0	0	0	0	3	03.75
47.	*Amorphophallus paeoniifolius* var. *campanulatus*	0	0	4	0	2	2	0	0	0	0	0	0	0	0	0	0	0	0	0	8	10.00
48.	*A.sylvaticus*	0	0	0	0	2	0	0	0	0	0	0	0	0	0	0	0	0	0	0	2	02.50
49.	*Ampelocissus latifolia*	0	0	3	0	2	0	0	0	0	0	0	0	0	0	0	0	0	0	0	5	06.25
50.	*A.tomentosa*	0	0	2	0	2	0	0	3	0	0	0	0	0	0	0	0	0	0	0	7	08.75
51.	*Anacardium occidentale*	0	0	4	0	2	0	0	0	0	3	0	0	0	0	0	0	0	0	0	9	11.25
52.	*Anamirta cocculus*	0	0	0	0	3	0	0	0	0	0	0	0	0	0	0	0	0	0	0	3	03.75
53.	*Ananas comosus*	0	0	2	0	2	0	0	0	0	0	0	0	0	0	0	0	0	0	2	4	05.00
54.	*Andrographis paniculata*	0	0	0	0	5	0	0	0	0	0	0	0	0	0	0	0	0	2	2	9	11.25
55.	*Angiopteris evecta*	0	0	0	0	2	0	0	0	0	0	0	0	0	0	0	0	0	2	0	4	05.00
56.	*Anisomeles indica*	0	0	0	0	2	0	0	0	3	0	0	0	0	0	0	0	0	0	0	5	06.25
57.	*A.malabarica*	0	0	0	0	2	0	0	0	0	0	0	0	0	0	0	0	0	0	0	2	02.50
58.	*Annona reticulata*	0	0	2	0	2	3	0	0	0	0	0	0	0	0	0	0	0	0	2	9	11.25
59.	*A.squamosa*	0	0	2	0	2	4	0	0	0	0	0	0	0	0	0	0	0	2	0	10	12.50

Contd...

Table 4.1–Contd...

Sl.No.	Botanical Name	Timb		Food		Med		Fuel	Fiber	Dye	Tan	G/R	Oil	Latex	Tb	Others					Total	TIV
		Maj	Min	Hum	Vet	Hum	Vet									Bm Tg Fn	B M Mp Um	Poi-son	Bev Liq	Misc.		
60.	*Anogeissus acuminata*	0	0	0	0	**4**	0	0	0	0	0	0	0	0	0	0	0	0	0	0	**4**	**05.00**
61.	*A.latifolia*	**3**	**4**	0	0	**2**	0	0	0	0	0	**3**	0	0	0	0	0	0	0	**2**	**14**	**17.50**
62.	*Anthocephalus chinensis*	0	0	0	0	**2**	**3**	0	0	0	0	0	0	0	0	0	0	0	0	**2**	**7**	**08.75**
63.	*Antidesma ghaesembilla*	0	**2**	**2**	0	**2**	0	0	0	0	0	0	0	0	0	0	0	0	0	0	**6**	**07.50**
64.	*Antigonon leptopus*	0	0	0	0	**2**	0	0	0	0	0	0	0	0	0	0	0	0	0	0	**2**	**02.50**
65.	*Apluda mutica*	0	0	0	**3**	0	0	0	0	0	0	0	0	0	0	0	0	0	0	0	**3**	**03.75**
66.	*Arachis hypogaea*	0	0	**4**	**2**	**2**	0	0	0	0	0	0	0	0	0	0	0	0	0	**2**	**10**	**12.50**
67.	*Ardisia solanacea*	0	0	0	0	**2**	**2**	0	0	0	0	0	0	0	0	0	0	0	0	0	**4**	**05.00**
68.	*Areca catechu*	0	0	**2**	0	**2**	0	0	0	0	0	0	0	0	0	0	0	0	0	0	**4**	**05.00**
69.	*Argemone mexicana*	0	0	0	0	**3**	0	0	0	0	0	0	0	0	0	0	0	0	0	0	**3**	**03.75**
70.	*Argyreia nervosa*	0	0	0	**3**	**2**	0	0	**3**	0	0	0	0	0	0	0	0	0	0	0	**8**	**10.00**
71.	*Aristida adscensionis*	0	0	0	0	0	0	0	0	0	0	0	0	0	0	**2**	0	0	0	0	**2**	**02.50**
72.	*A.setacea*	0	**2**	0	0	**2**	**2**	0	0	0	0	0	0	0	0	**2**	0	0	0	0	**8**	**10.00**
73.	*Aristolochia bracteolata*	0	0	0	0	**3**	**4**	0	0	0	0	0	0	0	0	0	0	0	0	0	**7**	**08.75**
74.	*A.indica*	0	0	**2**	0	**2**	**2**	0	0	0	0	0	0	0	0	0	0	0	0	0	**6**	**07.50**
75.	*Artemisia vulgaris*	0	0	0	0	**2**	0	0	0	0	0	0	0	0	0	0	0	0	0	0	**2**	**02.50**
76.	*Artocarpus heterophyllus*	**2**	**3**	**4**	0	**2**	0	0	0	0	0	0	0	0	0	0	**2**	**2**	**2**	0	**17**	**21.25**
77.	*Arundo donax*	0	0	0	0	0	**2**	0	**3**	0	0	0	0	0	0	0	**2**	0	0	0	**7**	**08.75**
78.	*Asclepias curassavica*	0	0	0	0	**2**	0	0	0	0	0	0	0	0	0	0	0	0	0	0	**2**	**02.50**
79.	*Asparagus gonoclados*	0	0	0	0	**3**	0	0	0	0	0	0	0	0	0	0	0	0	0	0	**3**	**03.75**

Contd...

Table 4.1–Contd...

Sl.No.	Botanical Name	Timb		Food		Med		Fuel	Fiber	Dye	Tan	G/R	Oil	Latex	Tb	Others					Total	TIV
		Maj	Min	Hum	Vet	Hum	Vet									Bm Tg Fn	B M Mp Um	Poi-son	Bev Liq	Misc.		
80.	*A.racemosus*	0	0	4	0	5	0	0	0	0	0	0	0	0	0	0	0	0	0	0	9	11.25
81.	*Atalantia monophylla*	0	0	2	0	2	0	0	0	0	0	0	0	0	0	0	0	0	0	0	4	05.00
82.	*Atylosia scarabaeoides*	0	0	2	0	2	0	0	0	0	0	0	0	0	0	0	0	0	0	0	4	05.00
83.	*Azadirachta indica*	5	5	2	0	5	4	0	0	0	0	0	3	0	2	0	0	0	0	2	28	35.00
84.	*Bacopa monnieri*	0	0	0	0	2	0	0	0	0	0	0	0	0	0	0	0	0	0	0	2	02.50
85.	*Balanites aegyptiaca*	0	2	2	0	2	2	3	0	0	0	0	0	0	0	0	0	2	0	0	13	16.25
86.	*Baliospermum montanum*	0	0	0	0	3	0	0	0	0	0	0	0	0	0	0	0	0	0	0	3	03.75
87.	*Bambusa arundinacea*	0	5	3	0	2	0	0	0	0	0	0	0	0	2	2	2	2	0	2	20	25.00
88.	*Barleria cristata*	0	0	0	0	4	0	0	0	0	0	0	0	0	0	0	0	0	0	0	4	05.00
89.	*B. prionitis*	0	0	0	0	3	0	0	0	0	0	0	0	0	0	0	0	0	0	0	3	03.75
90.	*Barringtonia acutangula*	0	4	0	0	2	0	0	0	0	0	0	0	0	0	0	0	2	0	0	8	10.00
91.	*Basella alba*	0	0	2	0	2	0	0	0	3	0	0	0	0	0	0	0	0	0	0	7	08.75
92.	*Bauhinia purpurea*	0	3	2	3	2	0	0	0	0	0	0	0	0	0	0	0	0	0	0	10	12.50
93.	*B.racemosa*	0	0	2	0	3	0	2	3	0	0	0	0	0	0	0	0	2	0	2	14	17.50
94.	*B.vahlii*	0	3	2	3	3	2	0	3	0	0	0	0	0	0	0	2	0	0	2	20	25.00
95.	*B.variegata*	0	2	4	2	2	0	0	2	0	0	0	0	0	0	0	2	0	0	0	14	17.50
96.	*Benincasa hispida*	0	0	0	0	2	0	0	0	0	0	0	0	0	0	0	0	0	0	2	4	05.00
97.	*Benkara malabarica*	0	0	2	0	0	0	0	0	0	0	0	0	0	0	0	0	0	0	2	4	05.00
98.	*Beta vulgaris*	0	0	4	0	2	0	0	0	0	0	0	0	0	0	0	0	0	0	0	6	07.50
99.	*Biophytum nervifolium*	0	0	0	0	3	0	0	0	0	0	0	0	0	0	0	0	0	0	0	3	03.75

Contd...

Table 4.1–Contd...

Sl.No.	Botanical Name	Timb		Food		Med		Fuel	Fiber	Dye	Tan	G/R	Oil	Latex	Tb	Others					Total	TIV
		Maj	Min	Hum	Vet	Hum	Vet									Bm Tg Fn	B M Mp Um	Poi-son	Bev Liq	Misc.		
100.	*Bixa orellana*	0	0	0	0	**2**	0	0	0	**4**	0	0	0	0	0	0	0	0	0	0	**6**	**07.50**
101.	*Boehmeria platyphylla*	0	0	0	0	**3**	0	0	0	0	0	0	0	0	0	0	0	0	0	0	**3**	**03.75**
102.	*Boerhavia chinensis*	0	0	0	0	**5**	0	0	0	0	0	0	0	0	0	0	0	0	0	0	**5**	**06.25**
103.	*B.diffusa*	0	0	**2**	0	**4**	0	0	0	0	0	0	0	0	0	0	0	0	0	**2**	**8**	**10.00**
104.	*Bombax ceiba*	0	**4**	**2**	0	**2**	**2**	0	**4**	0	0	0	0	0	0	0	0	0	0	**2**	**16**	**20.00**
105.	*Borassus flabellifer*	**3**	**3**	**3**	0	**1**	0	0	**4**	0	0	0	0	0	0	**2**	**2**	0	**2**	0	**20**	**25.00**
106.	*Borreria articularis*	0	0	**2**	0	0	0	0	0	0	0	0	0	0	0	0	0	0	0	0	**2**	**02.50**
107.	*Boswellia serrata*	0	0	0	0	**2**	0	0	0	0	0	0	0	0	0	0	0	0	0	**2**	**4**	**05.00**
108.	*Bougainvillea spectabilis*	0	0	0	0	**2**	0	0	0	0	0	0	0	0	0	0	0	0	0	**2**	**4**	**05.00**
109.	*Brassica nigra*	0	0	**4**	0	**2**	0	0	0	0	0	0	**2**	0	0	0	0	0	0	0	**8**	**10.00**
110.	*Bridelia montana*	0	**4**	**2**	0	0	0	0	0	0	**2**	0	0	0	0	0	0	0	0	0	**8**	**10.00**
111.	*B.retusa*	**2**	**4**	**2**	0	0	0	0	0	0	**2**	0	0	0	**2**	0	0	0	0	0	**12**	**15.00**
112.	*Buchanania axillaris*	0	0	**3**	0	0	0	0	0	0	0	0	0	0	0	0	0	0	0	0	**3**	**03.75**
113.	*B.lanzan*	0	**2**	**3**	0	**2**	**2**	0	0	0	**2**	0	0	0	0	0	**2**	0	0	**2**	**15**	**18.75**
114.	*Butea monosperma*	0	0	0	0	**3**	0	0	0	**3**	0	**4**	0	0	0	0	**2**	0	0	**2**	**14**	**17.50**
115.	*Byttneria herbacea*	0	0	0	0	**2**	0	0	0	0	0	0	0	0	0	0	0	0	0	0	**2**	**02.50**
116.	*Caesalpinia bonduc*	0	0	0	0	**4**	0	0	0	**2**	0	0	**3**	0	0	0	0	0	0	**2**	**11**	**13.75**
117.	*C.coriaria*	0	0	0	0	0	0	0	0	0	**3**	0	0	0	0	0	0	0	0	0	**3**	**03.75**
118.	*C.pulcherrima*	0	0	0	0	**3**	0	0	0	0	0	0	0	0	0	0	0	0	0	**2**	**5**	**06.25**
119.	*Cajanus cajan*	0	0	**4**	0	**3**	0	0	0	0	0	0	0	0	0	0	0	0	0	0	**7**	**08.75**

Contd...

Table 4.1–Contd...

Sl.No.	Botanical Name	Timb		Food		Med		Fuel	Fiber	Dye	Tan	G/R	Oil	Latex	Tb	Others					Total	TIV
		Maj	Min	Hum	Vet	Hum	Vet									Bm Tg Fn	B M Mp Um	Poi-son	Bev Liq	Misc.		
120.	*Calamus rotang*	0	0	2	0	2	0	0	0	0	0	0	0	0	0	0	2	0	0	0	6	07.50
121.	*C.viminalis*	0	3	0	0	0	0	0	0	0	0	0	0	0	0	0	2	2	0	0	7	08.75
122.	*Calophyllum inophyllum*	0	0	0	0	2	0	0	0	0	0	0	0	0	0	0	0	0	0	0	2	02.50
123.	*Calotropis gigantea*	0	0	0	0	4	0	0	3	0	0	0	0	0	2	0	0	0	0	0	9	11.25
124.	*C.procera*	0	0	0	0	3	0	0	0	0	0	0	0	0	0	0	0	0	0	0	3	03.75
125.	*Canavalia gladiata*	0	0	2	0	2	0	0	0	0	0	0	0	0	0	0	0	0	0	2	6	07.50
126.	*C.virosa*	0	0	2	0	2	0	0	0	0	0	0	0	0	0	0	0	0	0	0	4	05.00
127.	*Canna coccinea*	0	0	0	0	2	0	0	0	0	0	0	0	0	0	0	0	0	0	0	2	02.50
128.	*C.indica*	0	0	3	0	3	0	0	0	0	0	0	0	0	0	0	0	0	0	0	6	07.50
129.	*Cannabis sativa*	0	0	0	0	2	0	0	3	0	0	0	0	0	0	0	2	0	0	0	7	08.75
130.	*Canthium parviflorum*	0	0	3	0	1	0	0	0	0	0	0	0	0	0	0	0	0	0	0	4	05.00
131.	*Capparis zeylanica*	0	0	2	2	3	0	0	0	0	0	0	0	0	0	0	0	0	0	0	7	08.75
132.	*Capsicum annuum*	0	0	4	0	2	0	0	0	0	0	0	0	0	0	0	0	0	0	0	6	07.50
133.	*C.frutescens*	0	0	4	0	2	0	0	0	0	0	0	0	0	0	0	0	0	0	0	6	07.50
134.	*Caralluma adscendens*	0	0	2	0	1	0	0	0	0	0	0	0	0	0	0	0	0	0	0	3	03.75
135.	*Cardiospermum halicacabum*	0	0	2	0	2	0	0	0	0	0	0	0	0	0	0	0	0	0	0	4	05.00
136.	*Careya arborea*	0	0	0	0	2	2	0	0	0	0	0	0	0	0	0	2	0	0	0	6	07.50
137.	*Carica papaya*	0	0	4	0	3	0	0	0	0	0	0	0	0	0	0	0	0	0	0	7	08.75
138.	*Carissa carandas*	0	0	4	0	3	0	0	0	0	0	0	0	0	0	0	0	0	0	0	7	08.75

Contd...

Table 4.1–Contd...

Sl.No.	Botanical Name	Timb		Food		Med		Fuel	Fiber	Dye	Tan	G/R	Oil	Latex	Tb	Others					Total	TIV
		Maj	Min	Hum	Vet	Hum	Vet									Bm Tg Fn	B M Mp Um	Poi-son	Bev Liq	Misc.		
139.	C.spinarum	0	0	3	0	2	4	0	0	0	0	0	0	0	0	0	0	0	0	0	9	11.25
140.	Caryota urens	0	2	0	0	2	0	0	3	0	0	0	0	0	0	0	2	2	2	2	15	18.75
141.	Cascabela thevetia	0	0	0	0	1	0	0	0	0	0	0	2	0	0	0	0	0	0	0	3	03.75
142.	Casearia elliptica	0	0	0	0	0	0	0	0	0	0	0	0	0	0	0	0	2	2	0	4	05.00
143.	C.graveolens	0	0	0	0	2	0	0	0	0	0	0	0	0	0	0	0	0	0	0	2	02.50
144.	Cassia absus	0	0	0	0	2	0	0	0	0	0	0	0	0	0	0	0	0	0	0	2	02.50
145.	C.alata	0	0	0	0	2	0	0	0	0	3	0	0	0	0	0	0	0	0	0	5	06.25
146.	C.auriculata	0	0	0	0	4	0	0	0	0	0	0	0	0	0	0	0	0	0	0	4	05.00
147.	C.fistula	0	0	0	0	4	0	0	0	0	0	0	0	0	2	0	0	0	0	2	6	07.50
148.	C.occidentalis	0	0	2	0	3	0	0	0	0	2	0	0	0	0	0	0	0	0	0	9	11.25
149.	C.siamea	0	2	0	0	1	0	0	0	0	0	0	0	0	0	0	0	0	0	0	3	03.75
150.	C.sophera	0	0	2	0	2	0	0	0	0	0	0	0	0	0	0	0	0	2	0	6	07.50
151.	C.tora	0	0	2	0	4	0	0	0	0	0	0	0	0	0	0	0	0	0	0	6	07.50
152.	Cassine glauca	0	0	0	0	0	0	0	0	0	0	0	0	0	0	0	0	0	0	2	6	07.50
153.	Cassytha filiformis	0	0	0	0	2	0	0	0	0	0	0	0	0	0	0	0	0	0	0	2	02.50
154.	Casuarina equisetifolia	3	0	0	0	2	0	4	0	0	0	0	0	0	0	0	0	0	0	2	11	13.75
155.	Catharanthus pusillus	0	0	0	0	2	0	0	0	0	0	0	0	0	0	0	0	0	0	0	2	02.50
156.	C.roseus	0	0	0	0	5	0	0	0	0	0	0	0	0	0	0	0	0	0	0	5	06.25
157.	Cayratia pedata	0	0	0	0	2	0	0	0	0	0	0	0	0	0	0	0	0	0	0	2	02.50
158.	Ceiba pentandra	0	0	0	0	2	0	0	4	0	0	0	0	0	0	0	0	0	0	2	8	10.00

Contd...

Table 4.1–Contd...

Sl.No.	Botanical Name	Timb		Food		Med		Fuel	Fiber	Dye	Tan	G/R	Oil	Latex	Tb	Others					Total	TIV
		Maj	Min	Hum	Vet	Hum	Vet									Bm Tg Fn	B M Mp Um	Poi-son	Bev Liq	Misc.		
159.	*Celastrus paniculatus*	0	0	0	0	**4**	0	0	0	**3**	0	0	**3**	0	0	0	0	0	0	0	**10**	**12.50**
160.	*Celosia argentea*	0	0	**2**	0	**2**	0	0	0	0	0	0	0	0	0	0	0	0	0	0	**4**	**05.00**
161.	*Centella asiatica*	0	0	**2**	0	**5**	0	0	0	0	0	0	0	0	0	0	0	0	0	0	**7**	**08.75**
162.	*Cereus pterogonus*	0	0	0	0	**2**	0	0	0	0	0	0	0	0	0	0	0	0	0	0	**2**	**02.50**
163.	*Ceriops decandra*	0	**4**	0	0	0	0	0	0	**2**	**2**	0	0	0	0	0	0	0	0	0	**8**	**10.00**
164.	*Ceriscoides turgida*	0	0	0	0	**2**	0	0	0	0	0	0	0	0	0	0	0	0	0	0	**2**	**02.50**
165.	*Chenopodium album*	0	0	**2**	**2**	0	0	0	0	0	0	0	0	0	0	0	0	0	0	0	**4**	**05.00**
166.	*Chlorophytum arundinaceum*	0	0	0	0	**3**	0	0	0	0	0	0	0	0	0	0	0	0	0	0	**5**	**06.25**
167.	*Chloroxylon swietenia*	**2**	**4**	0	0	**2**	0	0	0	0	0	0	0	0	0	0	0	0	0	0	**8**	**10.00**
168.	*Chrysopogon aciculatus*	0	0	0	**2**	0	0	0	0	0	0	0	0	0	0	0	0	0	0	0	**2**	**02.50**
169.	*Cicer arietinum*	0	0	**4**	**2**	**2**	0	0	0	0	0	0	0	0	0	0	0	0	0	0	**8**	**10.00**
170.	*Cinnamomum zeylanicum*	0	0	**2**	0	**2**	0	0	0	0	0	0	0	0	0	0	0	0	0	0	**4**	**05.00**
171.	*Cipadessa baccifera*	0	0	0	0	**2**	0	**2**	0	0	0	0	0	0	0	0	0	0	**2**	0	**6**	**07.50**
172.	*Cissampelos pareira*	0	0	0	0	**3**	0	0	**3**	0	0	0	0	0	0	0	0	0	0	0	**6**	**07.50**
173.	*Cissus quadrangularis*	0	0	**2**	0	0	**2**	0	0	0	0	0	0	0	0	0	0	0	0	0	**4**	**05.00**
174.	*C.repanda*	0	0	0	0	**3**	0	0	0	0	0	0	0	0	0	0	0	0	0	0	**3**	**03.75**
175.	*Citrullus colocynthis*	0	0	0	0	**2**	0	0	0	0	0	0	0	0	0	0	0	0	0	0	**2**	**02.50**
176.	*Citrus aurantifolia*	0	0	**2**	0	**2**	0	0	0	0	0	0	0	0	0	0	0	0	0	0	**4**	**05.00**
177.	*C.limon*	0	0	**2**	0	**3**	0	0	0	0	0	0	0	0	0	0	0	0	0	0	**5**	**06.25**

Contd...

Table 4.1–Contd...

Sl.No.	Botanical Name	Timb		Food		Med		Fuel	Fiber	Dye	Tan	G/R	Oil	Latex	Tb	Others					Total	TIV
		Maj	Min	Hum	Vet	Hum	Vet									Bm Tg Fn	B M Mp Um	Poi-son	Bev Liq	Misc.		
178.	*C.medica*	0	0	**2**	0	**2**	0	0	0	0	0	0	0	0	0	0	0	0	0	0	**4**	**05.00**
179.	*C.sinensis*	0	0	**2**	0	**1**	0	0	0	0	0	0	0	0	0	0	0	0	0	0	**3**	**03.75**
180.	*Cleistanthus collinus*	**3**	0	0	0	**2**	**4**	0	0	0	0	0	0	0	0	**2**	0	**2**	0	0	**13**	**16.25**
181.	*Cleome aspera*	0	0	0	0	**2**	0	0	0	0	0	0	0	0	0	0	0	0	0	0	**2**	**02.50**
182.	*C.gynandra*	0	0	**3**	0	**2**	0	0	0	0	0	0	0	0	0	0	0	0	0	0	**5**	**06.25**
183.	*C.monophylla*	0	0	**2**	0	**1**	0	0	0	0	0	0	0	0	0	0	0	0	0	0	**3**	**03.75**
184.	*C.viscosa*	0	0	**2**	0	**3**	0	0	0	0	0	0	0	0	0	0	0	0	0	0	**5**	**06.25**
185.	*Clerodendrum inerme*	0	0	0	**2**	**2**	0	**3**	0	0	0	0	0	0	0	0	0	0	0	0	**7**	**08.75**
186.	*C.phlomidis*	0	0	0	0	**2**	0	0	0	0	0	0	0	0	0	0	0	0	0	0	**2**	**02.50**
187.	*C.serratum*	0	0	**2**	0	**3**	0	0	0	0	0	0	0	0	0	0	0	0	0	0	**5**	**06.25**
188.	*Clitoria ternatea*	0	0	0	0	**3**	0	0	0	0	0	0	0	0	0	0	0	0	0	0	**3**	**03.75**
189.	*Coccinia grandis*	0	0	**2**	0	**2**	0	0	0	0	0	0	0	0	0	0	0	0	0	0	**4**	**05.00**
190.	*Cocculus hirsutus*	0	0	**2**	0	**2**	0	0	**2**	**2**	0	0	0	0	0	**2**	0	0	0	**2**	**12**	**15.00**
191.	*Cochlospermum religiosum*	0	0	0	0	**2**	0	0	**4**	0	0	**2**	0	0	0	0	0	0	0	**2**	**10**	**12.50**
192.	*Cocos nucifera*	**3**	**5**	**4**	0	**2**	0	**3**	**4**	0	0	0	**4**	0	**2**	**2**	**2**	0	**2**	**2**	**35**	**43.75**
193.	*Coffea arabica*	0	0	0	0	**2**	0	0	0	0	0	0	0	0	0	0	0	0	**2**	0	**4**	**05.00**
194.	*Coldenia procumbens*	0	0	0	0	**3**	0	0	0	0	0	0	0	0	0	0	0	0	0	0	**3**	**03.75**
195.	*Coleus ambonicus*	0	0	0	0	**5**	**2**	0	0	0	0	0	0	0	0	0	0	0	0	0	**7**	**08.75**
196.	*Colocasia esculenta*	0	0	**3**	0	**2**	0	0	0	0	0	0	0	0	0	0	0	0	0	0	**5**	**06.25**
197.	*Combretum albidum*	0	0	0	0	0	0	0	**2**	0	0	0	0	0	0	0	0	0	0	0	**2**	**02.50**

Contd...

Table 4.1–Contd...

Sl.No.	Botanical Name	Timb		Food		Med		Fuel	Fiber	Dye	Tan	G/R	Oil	Latex	Tb	Others					Total	TIV
		Maj	Min	Hum	Vet	Hum	Vet									Bm Tg Fn	B M Mp Um	Poi-son	Bev Liq	Misc.		
198.	*Commelina benghalensis*	0	0	**2**	0	**1**	0	0	0	0	0	0	0	0	0	0	0	0	0	0	**3**	**03.75**
199.	*Commiphora caudata*	0	0	**3**	0	**3**	0	0	0	0	0	0	0	0	0	0	0	0	0	**2**	**8**	**10.00**
200.	*Corallocarpus epigaeus*	0	0	0	0	0	**2**	0	0	0	0	0	0	0	0	0	0	0	0	0	**2**	**02.50**
201.	*Corchorus aestuans*	0	0	0	0	**2**	0	0	0	0	0	0	0	0	0	0	0	0	0	0	**2**	**02.50**
202.	*C.capsularis*	0	0	0	0	**2**	0	0	**3**	0	0	0	0	0	0	0	0	0	0	0	**5**	**06.25**
203.	*C.olitorius*	0	0	0	0	**2**	0	0	**3**	0	0	0	0	0	0	0	0	0	0	0	**5**	**06.25**
204.	*Cordia dichotoma*	0	0	**3**	0	**2**	0	0	0	0	0	0	0	0	0	0	0	0	0	0	**5**	**06.25**
205.	*C.monoica*	0	0	**2**	0	**2**	0	0	0	0	0	0	0	0	0	0	0	0	0	0	**4**	**05.00**
206.	*Coriandrum sativum*	0	0	**4**	0	**3**	0	0	0	0	0	0	0	0	0	0	0	0	0	0	**7**	**08.75**
207.	*Costus speciosus*	0	0	**3**	0	**3**	0	0	0	0	0	0	0	0	0	0	0	0	0	0	**6**	**07.50**
208.	*Crataeva magna*	0	0	0	0	**2**	0	0	0	0	0	0	0	0	0	0	0	0	0	**2**	**4**	**05.00**
209.	*Crinum asiaticum*	0	0	0	0	**3**	0	0	0	0	0	0	0	0	0	0	0	0	0	0	**3**	**03.75**
210.	*Crotalaria juncea*	0	0	**2**	**3**	**2**	0	0	**4**	0	0	0	0	0	0	0	0	0	0	**2**	**13**	**16.25**
211.	*C.laburnifolia*	0	0	0	0	**2**	0	0	0	0	0	0	0	0	0	0	0	0	0	0	**2**	**02.50**
212.	*C.verrucosa*	0	0	0	0	**2**	0	0	0	0	0	0	0	0	0	0	0	0	0	0	**2**	**02.50**
213.	*Croton bonplandianum*	0	0	0	0	**2**	0	0	0	0	0	0	0	0	0	0	0	0	0	0	**2**	**02.50**
214.	*Cryptolepis buchanani*	0	0	0	0	**3**	0	0	**3**	0	0	0	0	0	0	0	0	0	0	0	**6**	**07.50**
215.	*Cucumis melo*	0	0	**3**	0	**2**	0	0	0	0	0	0	0	0	0	0	0	0	0	0	**5**	**06.25**
216.	*C.sativus*	0	0	**3**	0	0	0	0	0	0	0	0	0	0	0	0	0	0	0	0	**3**	**03.75**

Contd...

Table 4.1–Contd...

Sl.No.	Botanical Name	Timb		Food		Med		Fuel	Fiber	Dye	Tan	G/R	Oil	Latex	Tb	Others					Total	TIV
		Maj	Min	Hum	Vet	Hum	Vet									Bm Tg Fn	B M Mp Um	Poi-son	Bev Liq	Misc.		
217.	*Cucurbita maxima*	0	0	**3**	0	**2**	**2**	0	0	0	0	0	0	0	0	0	0	0	0	0	**7**	**08.75**
218.	*Curculigo orchioides*	0	0	**2**	0	**3**	**2**	0	0	0	0	0	0	0	0	0	0	0	0	0	**7**	**08.75**
219.	*Curcuma amada*	0	0	0	0	**3**	0	0	0	0	0	0	0	0	0	0	0	0	0	0	**3**	**03.75**
220.	*C.aromatica*	0	0	0	0	**2**	0	0	0	0	0	0	0	0	0	0	0	0	0	0	**2**	**02.50**
221.	*C.longa*	0	0	**2**	0	**5**	0	0	0	0	0	0	0	0	0	0	0	0	0	**2**	**9**	**11.25**
222.	*C.pseudomontana*	0	0	0	0	**2**	0	0	0	0	0	0	0	0	0	0	**2**	0	0	0	**4**	**05.00**
223.	*Cyamopsis tetragonaloba*	0	0	**2**	0	**2**	0	0	0	0	0	0	0	0	0	0	0	0	0	0	**4**	**05.00**
224.	*Cycas circinalis*	0	0	0	0	**2**	0	0	0	0	0	0	0	0	0	0	0	0	0	0	**2**	**02.50**
225.	*Cymbopogon coloratus*	0	0	0	0	0	0	0	0	0	0	0	0	0	0	**2**	0	0	0	0	**2**	**02.50**
226.	*C.flexuosus*	0	0	0	0	**2**	0	0	0	0	0	0	**3**	0	0	0	0	0	0	**2**	**7**	**08.75**
227.	*C.martinii*	0	0	0	0	**2**	0	0	0	0	0	0	**3**	0	0	0	0	0	0	0	**5**	**06.25**
228.	*Cynodon dactylon*	0	0	0	**2**	**3**	0	0	0	0	0	0	0	0	0	0	0	0	0	0	**5**	**06.25**
229.	*Cyperus javanicus*	0	0	0	0	0	0	0	0	0	0	0	0	0	0	**2**	0	0	0	0	**2**	**02.50**
230.	*C.rotundus*	0	0	0	0	**4**	0	0	0	0	0	0	0	0	0	0	0	0	0	0	**4**	**05.00**
231.	*Dalbergia lanceolaria*	**3**	**2**	0	0	**2**	0	0	0	0	0	0	0	0	0	0	0	0	0	0	**7**	**08.75**
232.	*D.latifolia*	**2**	**3**	0	**2**	**2**	**2**	0	0	0	0	0	0	0	0	0	0	0	0	0	**11**	**13.75**
233.	*D.paniculata*	**3**	**2**	0	0	**2**	0	0	0	0	0	0	0	0	0	0	0	0	0	0	**7**	**08.75**
234.	*D.sissoo*	0	**2**	0	0	**3**	0	0	0	0	0	0	0	0	0	0	0	0	0	0	**5**	**06.25**
235.	*D.spinosa.*	0	**2**	0	**2**	0	0	**3**	0	0	0	0	0	0	0	0	0	0	0	0	**7**	**08.75**
236.	*Datura innoxia*	0	0	0	0	**2**	0	0	0	0	0	0	0	0	0	0	0	0	0	0	**2**	**02.50**

Contd...

Table 4.1–Contd...

Sl.No.	Botanical Name	Timb		Food		Med		Fuel	Fiber	Dye	Tan	G/R	Oil	Latex	Tb	Others					Total	TIV
		Maj	Min	Hum	Vet	Hum	Vet									Bm Tg Fn	B M Mp Um	Poi-son	Bev Liq	Misc.		
237.	*D.metel*	0	0	0	0	**3**	0	0	0	0	0	0	0	0	0	0	0	0	0	0	**3**	**03.75**
238.	*Daucus carota*	0	0	**4**	0	**2**	0	0	0	0	0	0	0	0	0	0	0	0	0	0	**6**	**07.50**
239.	*Debregeasia longifolia*	0	0	**2**	0	0	0	0	**3**	0	0	0	0	0	0	0	0	0	0	0	**5**	**06.25**
240.	*Decalepis hamiltonii*	0	0	**2**	0	**2**	0	0	0	0	0	0	0	0	0	0	0	0	0	0	**4**	**05.00**
241.	*Delonix alata*	0	0	0	0	**2**	0	0	0	0	0	0	0	0	0	0	0	0	0	0	**2**	**02.50**
242.	*D.regia*	0	0	**2**	0	**1**	0	0	0	0	0	0	0	0	0	0	0	0	0	**2**	**5**	**06.25**
243.	*Dendrocalamus strictus*	**3**	**4**	**3**	**2**	**2**	**3**	0	0	0	0	0	0	0	**2**	**2**	**2**	**2**	0	**2**	**27**	**33.75**
244.	*Dendrophthoe falcata*	0	0	0	0	**2**	**3**	0	0	0	0	0	0	0	0	0	0	0	0	0	**5**	**06.25**
245.	*Derris scandens*	0	0	**2**	0	**2**	**2**	0	0	0	0	0	0	0	0	0	0	**2**	0	0	**8**	**10.00**
246.	*Desmodium gangeticum*	0	0	0	0	**3**	0	0	0	0	0	0	0	0	0	0	0	0	0	0	**3**	**03.75**
247.	*D.triflorum*	0	0	0	0	**2**	0	0	0	0	0	0	0	0	0	0	0	0	0	0	**2**	**02.50**
248.	*Dichrostachys cinerea*	0	0	0	0	**3**	0	0	0	0	0	0	0	0	0	0	0	0	0	0	**3**	**03.75**
249.	*Digera muricata*	0	0	**2**	0	**2**	0	0	0	0	0	0	0	0	0	0	0	0	0	0	**4**	**05.00**
250.	*Dillenia indica*	0	0	**2**	0	**2**	0	0	0	0	0	0	0	0	0	0	0	0	0	**2**	**6**	**07.50**
251.	*D.pentagyna*	0	0	**2**	0	**2**	0	0	0	0	0	0	0	0	0	0	0	0	0	**2**	**6**	**07.50**
252.	*Dioscorea alata*	0	0	**2**	0	0	0	0	0	0	0	0	0	0	0	0	0	0	0	**2**	**4**	**05.00**
253.	*D.bulbifera*	0	0	**2**	**2**	**2**	0	0	0	0	0	0	0	0	0	0	0	0	0	0	**6**	**07.50**
254.	*D.hispida*	0	0	**2**	0	**1**	0	0	0	0	0	0	0	0	0	0	0	0	0	0	**3**	**03.75**
255.	*D.oppositifolia*	0	0	**2**	0	**2**	0	0	0	0	0	0	0	0	0	0	0	0	0	0	**4**	**05.00**
256.	*D.pentaphylla*	0	0	**2**	**2**	**3**	0	0	0	0	0	0	0	0	0	0	0	0	0	0	**7**	**08.75**

Contd...

Table 4.1–Contd...

Sl.No.	Botanical Name	Timb		Food		Med		Fuel	Fiber	Dye	Tan	G/R	Oil	Latex	Tb	Others					Total	TIV
		Maj	Min	Hum	Vet	Hum	Vet									Bm Tg Fn	B M Mp Um	Poi-son	Bev Liq	Misc.		
257.	*D.tomentosa*	0	0	**2**	0	**2**	0	0	0	0	0	0	0	0	0	0	0	0	0	0	**4**	**05.00**
258.	*Diospyros chloroxylon*	0	**2**	**3**	0	0	0	0	0	0	0	0	0	0	0	0	0	0	0	**2**	**7**	**08.75**
259.	*D.melanoxylon*	0	**3**	**2**	0	0	0	0	0	0	0	0	0	0	0	0	0	0	0	**2**	**7**	**08.75**
260.	*D.montana*	0	**3**	**2**	0	0	0	0	0	0	0	0	0	0	0	0	0	**2**	0	0	**7**	**08.75**
261.	*Diplocyclos palmatus*	0	0	0	0	**2**	0	0	0	0	0	0	0	0	0	0	0	0	0	0	**2**	**02.50**
262.	*Dipteracanthus prostratus*	0	0	0	0	**2**	0	0	0	0	0	0	0	0	0	0	0	0	0	0	**2**	**02.50**
263.	*Dodonaea viscosa*	0	**2**	0	0	**2**	0	0	0	0	0	0	0	0	0	0	0	0	0	0	**4**	**05.00**
264.	*Dolichandrone falcata*	0	0	0	0	**2**	0	0	0	0	0	0	0	0	0	0	0	0	0	0	**2**	**02.50**
265.	*Dolichos trilobus*	0	0	**3**	0	0	0	0	0	0	0	0	0	0	0	0	0	0	0	0	**3**	**03.75**
266.	*Duranta repens*	0	0	0	0	**1**	0	0	0	0	0	0	0	0	0	**2**	0	0	0	0	**3**	**03.75**
267.	*Ecbolium viride*	0	0	0	0	**2**	0	0	0	0	0	0	0	0	0	0	0	0	0	0	**2**	**02.50**
268.	*Echinochloa frumentacea*	0	0	**2**	**3**	**2**	0	0	0	0	0	0	0	0	0	0	0	0	0	0	**7**	**08.75**
269.	*Eclipta prostrata*	0	0	0	0	**5**	**2**	0	0	0	0	0	0	0	0	0	0	0	0	0	**7**	**08.75**
270.	*Ehretia laevis* .	0	**2**	0	**2**	0	**2**	**3**	0	0	0	0	0	0	0	0	0	0	**2**	0	**11**	**13.75**
271.	*Eichhornia crassipes*	0	0	0	**2**	**2**	0	**2**	0	0	0	0	0	0	0	0	0	0	0	0	**6**	**07.50**
272.	*Elephantopus scaber*	0	0	0	**2**	**2**	**2**	0	0	0	0	0	0	0	0	0	0	0	0	0	**6**	**07.50**
273.	*Elettaria cardamomum*	0	0	**2**	0	**2**	0	0	0	0	0	0	**3**	0	0	0	0	0	0	0	**7**	**08.75**
274.	*Eleusine coracana*	0	0	**3**	**3**	**2**	0	0	0	0	0	0	0	0	0	0	0	0	**2**	0	**10**	**12.50**
275.	*Elytraria acaulis*	0	0	0	0	**2**	0	0	0	0	0	0	0	0	0	0	0	0	0	0	**2**	**02.50**
276.	*Emilia sonchifolia*	0	0	0	0	**2**	0	0	0	0	0	0	0	0	0	0	0	0	0	0	**2**	**02.50**

Contd...

Table 4.1–Contd...

Sl.No.	Botanical Name	Timb		Food		Med		Fuel	Fiber	Dye	Tan	G/R	Oil	Latex	Tb	Others					Total	TIV
		Maj	Min	Hum	Vet	Hum	Vet									Bm Tg Fn	B M Mp Um	Poi-son	Bev Liq	Misc.		
277.	*Entada rheedii*	0	0	0	0	**2**	**3**	0	0	0	0	0	0	0	0	0	0	2	0	0	**7**	**08.75**
278.	*Erythrina suberosa*	0	**2**	0	0	0	0	0	**3**	0	0	0	0	0	0	0	0	0	0	0	**5**	**06.25**
279.	*E.variegata*	**3**	0	0	0	**2**	0	0	0	0	0	0	0	0	0	0	0	0	0	0	**5**	**06.25**
280.	*Erythroxylum monogynum*	**3**	**3**	**3**	0	0	0	0	0	0	0	0	0	0	0	2	0	0	0	0	**11**	**13.75**
281.	*Eucalyptus globulus*	0	0	0	0	**4**	0	0	0	0	0	0	**3**	0	0	0	0	0	0	0	**7**	**08.75**
282.	*Eulaliopsis binata*	0	0	0	0	0	0	0	**3**	0	0	0	0	0	0	**2**	0	0	0	0	**5**	**06.25**
283.	*Euphorbia nivulia*	0	0	0	0	**2**	0	0	0	0	0	0	0	0	0	0	0	0	0	0	**2**	**02.50**
284.	*E.heterophylla*	0	0	0	0	**2**	0	0	0	0	0	0	0	**2**	0	0	0	0	0	0	**4**	**05.00**
285.	*E.hirta*	0	0	**2**	0	**2**	0	0	0	0	0	0	0	**2**	0	0	0	0	0	0	**6**	**07.50**
286.	*E.ligularia*	0	0	0	0	**3**	0	0	0	0	0	0	0	**2**	0	0	0	0	0	0	**5**	**06.25**
287.	*E.thymifolia*	0	0	0	0	**2**	0	0	0	0	0	0	0	**2**	0	0	0	0	0	0	**4**	**05.00**
288.	*E.tirucalli*	0	0	0	0	**2**	0	0	0	0	0	0	0	**3**	0	0	0	0	0	0	**5**	**06.25**
289.	*Evolvulus alsinoides*	0	0	0	0	**2**	0	0	0	0	0	0	0	0	0	0	0	0	0	0	**2**	**02.50**
290.	*E.nummularius*	0	0	0	0	**2**	0	0	0	0	0	0	0	0	0	0	0	0	0	0	**2**	**02.50**
291.	*Excoecaria agallocha*	0	0	0	0	**2**	0	0	0	0	0	0	0	0	0	0	0	**2**	0	0	**4**	**05.00**
292.	*Ferula asafoetida*	0	0	**2**	0	**3**	0	0	0	0	0	0	0	0	0	0	0	0	0	0	**5**	**06.25**
293.	*Ficus benghalensis*	0	0	**2**	0	**3**	0	0	**2**	0	0	0	0	0	0	0	**2**	0	0	**2**	**11**	**13.75**
294.	*F.carica*	0	0	**2**	0	**2**	0	0	0	0	0	0	0	0	0	0	0	0	0	0	**4**	**05.00**
295.	*F.hispida*	0	0	**2**	0	**3**	0	0	0	0	0	0	0	0	0	0	0	0	0	0	**5**	**06.25**
296.	*F.microcarpa*	**2**	**3**	0	0	**2**	0	0	0	0	0	0	0	0	0	0	0	0	0	0	**7**	**08.75**

Contd...

Table 4.1–Contd...

Sl.No.	Botanical Name	Timb		Food		Med		Fuel	Fiber	Dye	Tan	G/R	Oil	Latex	Tb	Others					Total	TIV
		Maj	Min	Hum	Vet	Hum	Vet									Bm Tg Fn	B M Mp Um	Poi-son	Bev Liq	Misc.		
297.	*F.racemosa*	0	0	**3**	0	**2**	**2**	0	0	0	0	0	0	0	0	0	0	0	**2**	0	**9**	**11.25**
298.	*F.religiosa*	0	0	**2**	0	**3**	0	0	0	0	0	0	0	0	0	0	0	0	0	**2**	**7**	**08.75**
299.	*Flacourtia indica*	0	0	0	0	**3**	0	0	0	0	0	0	0	0	0	0	0	0	0	0	**3**	**03.75**
300.	*Flemingia stricta*	0	0	0	0	**2**	0	0	0	0	0	0	0	0	0	0	0	0	0	0	**2**	**02.50**
301.	*Foeniculum vulgare*	0	0	0	0	**3**	0	0	0	0	0	0	0	0	0	0	0	0	0	0	**3**	**03.75**
302.	*Gardenia gummifera*	0	0	**2**	0	**2**	0	0	0	0	0	**3**	0	0	0	0	0	0	0	0	**7**	**08.75**
303.	*G.latifolia*	0	0	**2**	0	**2**	0	0	0	0	0	0	0	0	0	0	0	0	0	0	**4**	**05.00**
304.	*G.resinifera*	0	**2**	0	0	0	0	0	0	0	0	**3**	0	0	0	0	0	0	0	0	**5**	**06.25**
305.	*Garuga pinnata*	0	0	**3**	0	0	0	0	0	0	0	0	0	0	0	0	0	0	0	0	**3**	**03.75**
306.	*Gloriosa superba*	0	0	0	0	**3**	0	0	0	0	0	0	0	0	0	0	0	**2**	0	**2**	**7**	**08.75**
307.	*Glycosmis pentaphylla*	0	0	**2**	0	**3**	0	0	0	0	0	0	0	0	0	0	0	0	0	0	**5**	**06.25**
308.	*Glycyrhiza glabra*	0	0	0	0	**2**	0	0	0	0	0	0	0	0	0	0	0	0	0	2	**4**	**05.00**
309.	*Gmelina arborea*	0	**4**	0	0	**2**	0	0	0	0	0	0	0	0	0	0	0	0	0	0	**6**	**07.50**
310.	*G.asiatica*	0	**3**	0	0	**3**	0	0	0	0	0	0	0	0	0	0	0	0	0	0	**6**	**07.50**
311.	*Gnetum ula*	0	0	**2**	0	0	0	0	0	0	0	0	0	0	0	0	0	0	0	**2**	**4**	**05.00**
312.	*Gomphrena serrata*	0	0	0	0	0	**3**	0	0	0	0	0	0	0	0	0	0	0	0	0	**3**	**03.75**
313.	*Gossypium arboreum*	0	0	0	0	**2**	0	0	**4**	0	0	0	0	0	0	0	0	0	0	0	**6**	**07.50**
314.	*G.barbadense*	0	0	0	**3**	0	0	0	**4**	0	0	0	0	0	0	0	0	0	0	0	**7**	**08.75**
315.	*G.herbaceum*	0	0	0	**2**	**2**	0	0	**3**	0	0	0	0	0	0	0	0	0	0	0	**7**	**08.75**
316.	*Grewia hirsuta*	0	0	**3**	0	0	0	0	0	0	0	0	0	0	0	0	0	0	0	0	**3**	**03.75**

Contd...

Table 4.1–Contd...

Sl.No.	Botanical Name	Timb		Food		Med		Fuel	Fiber	Dye	Tan	G/R	Oil	Latex	Tb	Others					Total	TIV
		Maj	Min	Hum	Vet	Hum	Vet									Bm Tg Fn	B M Mp Um	Poi-son	Bev Liq	Misc.		
317.	*G.rhamnifolia*	0	0	2	3	2	0	0	0	0	0	0	0	0	0	0	0	0	0	0	7	08.75
318.	*G.tiliaefolia*	2	4	0	2	0	0	0	3	0	0	0	0	0	0	0	0	0	0	2	13	16.25
319.	*Guazuma ulmifolia*	0	2	3	2	2	0	0	0	0	0	0	0	0	0	0	0	0	0	0	9	11.25
320.	*Guizotia abyssinica*	0	0	0	0	2	0	0	0	0	0	0	3	0	0	0	0	0	0	0	5	06.25
321.	*Gymnema sylvestre*	0	0	0	0	3	0	0	0	0	0	0	0	0	0	0	0	0	0	0	3	03.75
322.	*Gyrocarpus asiaticus*	0	3	0	0	0	0	0	0	0	0	0	0	0	0	0	0	0	0	2	5	06.25
323.	*Haldinia cordifolia*	3	3	0	0	3	0	0	0	0	0	0	0	0	0	0	0	0	0	0	9	11.25
324.	*Hardwickia binata*	3	2	0	0	0	0	0	3	0	0	3	0	0	0	0	0	0	0	2	13	16.25
325.	*Helianthus annuus*	0	0	2	3	2	0	3	0	0	0	0	4	0	0	0	0	0	0	0	14	17.50
326.	*Helicteres isora*	0	0	0	0	2	0	0	3	0	0	0	0	0	0	0	0	0	0	0	5	06.25
327.	*Hemidesmus indicus*	0	0	0	0	4	0	0	0	0	0	0	0	0	0	0	0	0	2	0	6	07.50
328.	*H.indicus* var.*pubescens*	0	0	0	0	3	0	0	0	0	0	0	0	0	0	0	0	0	2	0	5	06.25
329.	*Heteropogon contortus*	0	0	0	3	0	0	0	0	0	0	0	0	0	0	0	0	0	0	0	3	03.75
330.	*Hibiscus cannabinus*	0	0	4	0	2	0	0	4	0	0	0	0	0	0	0	2	0	0	2	14	17.50
331.	*H.ovalifolius*	0	0	0	0	2	0	0	0	0	0	0	0	0	0	0	0	0	0	2	4	05.00
332.	*H.rosa-sinensis*	0	0	0	0	3	0	0	0	0	0	0	0	0	0	0	0	0	0	2	5	06.25
333.	*H.sabdariffa*	0	0	4	0	2	0	0	3	0	0	0	0	0	0	0	0	0	0	0	9	11.25
334.	*H.surattensis*	0	0	2	0	0	0	0	0	0	0	0	0	0	0	0	0	0	0	0	2	02.50
335.	*H.vitifolius*	0	0	0	0	0	0	0	3	0	0	0	0	0	0	0	0	0	0	0	3	03.75
336.	*Holarrhena pubescens*	0	0	0	0	3	0	0	0	0	0	0	0	0	0	0	0	0	0	0	3	03.75

Contd...

Table 4.1–Contd...

Sl.No.	Botanical Name	Timb		Food		Med		Fuel	Fiber	Dye	Tan	G/R	Oil	Latex	Tb	Others					Total	TIV
		Maj	Min	Hum	Vet	Hum	Vet									Bm Tg Fn	B M Mp Um	Poi-son	Bev Liq	Misc.		
337.	*Holoptelea integrifolia*	**2**	**4**	**2**	0	**3**	0	0	0	0	0	0	0	0	0	0	0	0	0	0	**11**	**13.75**
338.	*Holostemma ada-kodien*	0	0	0	0	**2**	0	0	0	0	0	0	0	0	0	0	0	0	0	0	**2**	**02.50**
339.	*Hybanthus enneaspermus*	0	0	0	0	**3**	0	0	0	0	0	0	0	0	0	0	0	0	0	0	**3**	**03.75**
340.	*Hyptis suaveolens*	0	0	0	0	**2**	0	0	0	**3**	0	0	0	0	0	0	0	0	0	0	**5**	**06.25**
341.	*Ichnocarpus frutescens*	0	0	0	0	**2**	0	0	0	0	0	0	0	0	0	0	0	0	0	0	**2**	**02.50**
342.	*Imperata cylindrica*	0	0	0	0	**2**	0	0	**3**	0	0	0	0	0	0	0	**2**	0	0	0	**7**	**08.75**
343.	*Indigofera tinctoria*	0	0	0	0	**3**	0	0	0	**3**	0	0	0	0	0	0	0	0	0	0	**6**	**07.50**
344.	*Indoneesiella echioides*	0	0	0	0	**2**	0	0	0	0	0	0	0	0	0	0	0	0	0	0	**2**	**02.50**
345.	*Ipomoea aquatica*	0	0	**2**	0	**2**	0	0	0	0	0	0	0	0	0	0	0	0	0	0	**4**	**05.00**
346.	*I.batatas*	0	0	**4**	0	**2**	0	0	0	0	0	0	0	0	0	0	0	0	0	0	**6**	**07.50**
347.	*I.carnea*	0	0	0	0	**2**	0	0	0	0	0	0	0	0	0	0	0	0	0	0	**2**	**02.50**
348.	*I.muricata*	0	0	**2**	0	0	0	0	0	0	0	0	0	0	0	0	0	0	0	0	**2**	**02.50**
349.	*I.obscura*	0	0	0	0	**2**	0	0	0	0	0	0	0	0	0	0	0	0	0	0	**2**	**02.50**
350.	*I.pes-tigridis*	0	0	0	0	**3**	0	0	0	0	0	0	0	0	0	0	0	0	0	0	**3**	**03.75**
351.	*Ixora coccinea* .	0	0	0	0	**2**	0	0	0	0	0	0	0	0	0	0	0	0	0	0	**2**	**02.50**
352.	*Jasminum angustifolium*	0	0	0	0	**3**	0	0	0	0	0	0	0	0	0	0	0	0	0	**2**	**5**	**06.25**
353.	*J.auriculatum*	0	0	0	0	**2**	0	0	0	0	0	0	0	0	0	0	0	0	0	**2**	**4**	**05.00**
354.	*J.grandiflorum*	0	0	0	0	**3**	0	0	0	0	0	0	0	0	0	0	0	0	0	**2**	**5**	**06.25**
355.	*J.sambac*	0	0	0	0	**2**	0	0	0	0	0	0	0	0	0	0	0	0	0	**2**	**4**	**05.00**
356.	*Jatropha curcas*	0	0	0	0	**3**	0	0	0	**3**	0	0	**4**	0	**2**	**2**	0	0	0	0	**14**	**17.50**

Contd...

Table 4.1–Contd...

Sl.No.	Botanical Name	Timb		Food		Med		Fuel	Fiber	Dye	Tan	G/R	Oil	Latex	Tb	Others					Total	TIV
		Maj	Min	Hum	Vet	Hum	Vet									Bm Tg Fn	B M Mp Um	Poi-son	Bev Liq	Misc.		
357.	*J.glandulifera*	0	0	0	0	2	0	0	0	0	0	0	3	0	0	0	0	0	0	0	5	06.25
358.	*J.gossypifolia*	0	0	0	0	2	0	0	0	0	0	0	0	0	2	0	0	0	0	0	4	05.00
359.	*Justicia adhatoda*	0	0	0	0	3	0	0	0	2	0	0	0	0	0	0	0	0	0	0	5	06.25
360.	*J.glauca*	0	0	2	0	2	0	0	0	0	0	0	0	0	0	0	0	0	0	0	4	05.00
361.	*Kalanchoe pinnata*	0	0	0	0	2	0	0	0	0	0	0	0	0	0	0	0	0	0	0	2	02.50
362.	*Knoxia sumatrensis*	0	0	0	0	2	0	0	0	0	0	0	0	0	0	0	0	0	0	0	2	02.50
363.	*Kydia calycina*	0	0	0	0	2	0	0	3	0	0	0	0	0	0	0	0	0	0	2	7	08.75
364.	*Lablab purpureus*	0	0	3	0	2	0	0	0	0	0	0	0	0	0	0	0	0	0	0	5	06.25
365.	*Lagenaria siceraria*	0	0	3	0	2	0	0	0	0	0	0	0	0	0	0	0	0	0	2	7	08.75
366.	*Lagerstroemia parviflora*	2	4	0	0	2	0	0	0	0	2	0	0	0	0	0	0	0	0	0	10	12.50
367.	*Lannea coromandelica*	0	4	0	3	0	0	0	0	0	0	3	0	0	0	0	0	0	0	0	10	12.50
368.	*Lantana camara*	0	0	2	0	2	0	0	0	0	0	0	0	0	2	0	2	0	0	0	8	10.00
369.	*Lawsonia inermis*	0	0	0	0	2	0	0	0	4	0	0	0	0	0	0	0	0	0	2	8	10.00
370.	*Leonotis nepetifolia*	0	0	0	0	2	0	0	0	2	0	0	0	0	0	0	0	0	0	2	6	07.50
371.	*Leptadenia reticulata*	0	0	0	0	2	0	0	0	0	0	0	0	0	0	0	0	0	0	0	2	02.50
372.	*Leucaena leucocephala*	0	0	0	2	1	0	3	0	0	0	0	0	0	0	0	0	0	0	0	6	07.50
373.	*Leucas aspera*	0	0	2	0	3	0	0	0	0	0	0	0	0	0	0	0	0	0	2	7	08.75
374.	*L.cephalotes*	0	0	0	0	2	0	0	0	0	0	0	0	0	0	0	0	0	0	0	2	02.50
375.	*Limonia acidissima*	0	0	3	0	2	0	0	0	0	0	3	0	0	0	0	0	0	0	2	10	12.50
376.	*Litsea glutinosa*	0	0	0	0	2	0	0	0	0	0	0	0	0	0	0	0	0	0	0	2	02.50

Contd...

Table 4.1–Contd...

Sl.No.	Botanical Name	Timb		Food		Med		Fuel	Fiber	Dye	Tan	G/R	Oil	Latex	Tb	Others					Total	TIV
		Maj	Min	Hum	Vet	Hum	Vet									Bm Tg Fn	B M Mp Um	Poi-son	Bev Liq	Misc.		
377.	*Lutla acutangula*	0	0	3	0	2	0	0	0	0	0	0	0	0	0	0	0	0	0	0	5	06.25
378.	*L.cylindrica*	0	0	3	0	2	0	0	0	0	0	0	0	0	0	0	0	0	0	0	5	06.25
379.	*Lycopersicon esculentum*	0	0	4	0	0	0	0	0	0	0	0	0	0	0	0	0	0	0	0	4	05.00
380.	*Macaranga peltata*	0	0	0	0	2	0	0	0	0	0	0	0	0	0	0	2	0	0	0	4	05.00
381.	*Macrotyloma uniflorum*	0	0	3	0	0	3	0	0	0	0	0	0	0	0	0	0	0	0	0	6	07.50
382.	*Madhuca indica*	0	0	2	0	4	0	0	0	0	0	0	2	0	0	0	0	0	2	2	12	15.00
383.	*Maerua oblongifolia*	0	0	0	0	2	0	0	0	0	0	0	0	0	0	0	0	0	0	0	2	02.50
384.	*Mallotus philippensis*	0	0	0	0	2	3	2	0	4	0	0	0	0	0	0	0	0	0	2	13	16.25
385.	*Malvastrum coromandelianum*	0	0	0	0	2	0	0	0	0	0	0	0	0	0	0	0	0	0	0	2	02.50
386.	*Mangifera indica*	3	4	3	0	2	2	2	0	0	0	0	0	0	0	0	0	0	2	2	20	25.00
387.	*Manihot esculenta*	0	0	2	0	2	0	0	0	0	0	0	0	0	0	0	0	0	0	2	6	07.50
388.	*Manilkara hexandra*	0	4	3	2	2	0	0	0	0	0	0	0	0	0	0	0	0	0	2	13	16.25
389.	*M.zapota*	0	0	2	0	2	0	0	0	0	0	0	0	0	0	0	0	0	0	2	6	07.50
390.	*Marsilea quadrifolia*	0	0	2	0	3	0	0	0	0	0	0	0	0	0	0	0	0	0	0	5	06.25
391.	*Martynia annua*	0	0	0	0	3	0	0	0	0	0	0	2	0	0	0	0	0	0	0	5	06.25
392.	*Maytenus emarginata*	0	0	0	3	2	2	2	0	0	0	0	0	0	0	2	0	0	0	0	11	13.75
393.	*Melastoma malabathricum*	0	0	0	0	3	0	0	0	0	0	0	0	0	0	0	0	0	0	0	3	03.75
394.	*Melia azedarach*	0	4	0	0	2	2	0	0	0	0	0	0	0	0	0	0	0	0	0	8	10.00
395.	*Memecylon edule*	0	2	2	0	0	0	0	0	0	0	0	0	0	0	0	0	0	0	0	4	05.00

Contd...

Table 4.1–Contd...

Sl.No.	Botanical Name	Timb		Food		Med		Fuel	Fiber	Dye	Tan	G/R	Oil	Latex	Tb	Others					Total	TIV
		Maj	Min	Hum	Vet	Hum	Vet									Bm Tg Fn	B M Mp Um	Poi-son	Bev Liq	Misc.		
396.	*M.umbellatum*	0	0	**2**	0	0	0	0	0	**3**	0	0	0	0	0	0	0	0	0	0	**5**	**06.25**
397.	*Mentha spicata*	0	0	**3**	0	**4**	0	0	0	0	0	0	0	0	0	0	0	0	0	**2**	**9**	**11.25**
398.	*Merremia tridentata*	0	0	0	0	**2**	0	0	0	0	0	0	0	0	0	0	0	0	0	0	**2**	**02.50**
399.	*M.vitifolia*	0	0	**3**	0	0	0	0	0	0	0	0	0	0	0	0	0	0	0	0	**3**	**03.75**
400.	*Michelia champaca*	0	0	0	0	**2**	0	0	0	0	0	0	**2**	0	0	0	0	0	0	**2**	**6**	**07.50**
401.	*Miliusa tomentosa*	0	0	0	0	**3**	0	0	0	0	0	0	0	0	0	0	0	0	0	0	**3**	**03.75**
402.	*Mimosa intsia*	0	0	0	0	**2**	0	0	0	0	0	0	0	0	0	0	0	0	0	0	**2**	**02.50**
403.	*M.pudica*	0	0	0	0	**3**	0	0	0	0	0	0	0	0	0	0	0	0	0	0	**3**	**03.75**
404.	*Mimusops elengi*	0	**2**	**2**	0	**2**	0	0	0	0	0	0	0	0	2	0	0	0	0	**2**	**10**	**12.50**
405.	*Mirabilis jalapa*	0	0	0	0	**4**	0	0	0	0	0	0	0	0	0	0	0	0	0	0	**4**	**05.00**
406.	*Mitragyna parvifolia*	**2**	**3**	0	0	**2**	0	0	**2**	0	0	0	0	0	0	0	0	0	0	0	**9**	**11.25**
407.	*Mollugo pentaphylla*	0	0	0	**4**	**2**	0	0	0	0	0	0	0	0	0	0	0	0	0	0	**6**	**07.50**
408.	*Momordica charantia*	0	0	**2**	0	**3**	**2**	0	0	0	0	0	0	0	0	0	0	0	0	0	**7**	**08.75**
409.	*M.dioica*	0	0	**2**	0	**2**	0	0	0	0	0	0	0	0	0	0	0	0	0	0	**4**	**05.00**
410.	*Morinda pubescens*	**2**	**3**	**2**	0	**2**	0	0	0	0	0	0	0	0	0	0	0	0	0	**2**	**11**	**13.75**
411.	*Moringa oleifera*	0	**2**	**2**	0	**2**	**2**	0	0	0	0	0	0	0	0	0	0	0	0	0	**8**	**10.00**
412.	*Mucuna monosperma*	0	0	0	0	**2**	0	0	0	0	0	0	0	0	0	0	0	0	0	0	**2**	**02.50**
413.	*M.pruriens*	0	0	0	0	**2**	0	0	0	0	0	0	0	0	0	0	0	0	0	0	**2**	**02.50**
414.	*Murraya koenigii*	0	0	**3**	0	**2**	0	0	0	0	0	0	0	0	0	0	0	0	0	**2**	**7**	**08.75**

Contd...

Table 4.1–Contd...

Sl.No.	Botanical Name	Timb		Food		Med		Fuel	Fiber	Dye	Tan	G/R	Oil	Latex	Tb	Others					Total	TIV
		Maj	Min	Hum	Vet	Hum	Vet									Bm Tg Fn	B M Mp Um	Poi-son	Bev Liq	Misc.		
415.	*Musa paradisiaca*	0	0	**3**	0	**2**	0	0	**2**	0	0	0	0	0	0	0	**2**	0	0	**2**	**11**	**13.75**
416.	*M.rosacea*	0	0	**2**	0	0	**2**	0	0	0	0	0	0	0	0	0	**2**	0	0	0	**6**	**07.50**
417.	*Naravelia zeylanica*	0	0	0	0	**2**	0	0	0	0	0	0	0	0	**2**	0	0	0	0	0	**4**	**05.00**
418.	*Naringi crenulata*	0	**3**	0	0	0	0	0	0	0	0	0	0	0	0	0	0	0	0	0	**3**	**03.75**
419.	*Nelumbo nucifera*	0	0	**2**	0	**2**	0	0	0	0	0	0	0	0	0	0	**2**	0	0	**2**	**8**	**10.00**
420.	*Nerium indicum*	0	0	0	0	**2**	0	0	0	0	0	0	0	0	0	0	0	0	0	0	**2**	**02.50**
421.	*Nicotiana tabacum*	0	0	0	0	**2**	0	0	0	0	0	0	0	0	0	0	0	0	0	**2**	**4**	**05.00**
422.	*Nyctanthes arbor-tristis*	0	0	0	0	**2**	0	0	0	**2**	0	0	0	0	0	0	0	0	0	0	**4**	**05.00**
423.	*Nymphaea nouchali*	0	0	**2**	0	**2**	0	0	0	0	0	0	0	0	0	0	0	0	0	**2**	**6**	**07.50**
424.	*Ochana obtusata*	0	**3**	0	0	0	0	0	0	0	0	0	0	0	0	0	0	0	0	0	**3**	**03.75**
425.	*Ocimum americanum*	0	0	**2**	0	**3**	0	0	0	0	0	0	0	0	0	0	0	0	0	0	**5**	**06.25**
426.	*O.basilicum*	0	0	**2**	0	**4**	**2**	0	0	0	0	0	0	0	0	0	0	0	0	**2**	**10**	**12.50**
427.	*O.gratissimum*	0	0	0	0	**2**	0	0	0	0	0	0	0	0	0	0	0	0	0	0	**2**	**02.50**
428.	*O.tenuiflorum*	0	0	0	0	**4**	0	0	0	0	0	0	**3**	0	0	0	0	0	0	**2**	**9**	**11.25**
429.	*Olax scandens*	0	0	**2**	0	0	0	0	0	0	0	0	0	0	0	0	0	0	0	0	**2**	**02.50**
430.	*Operculina turpethum*	0	0	0	0	**2**	0	0	0	0	0	0	0	0	0	0	0	0	0	0	**2**	**02.50**
431.	*Opuntia dillenii*	0	0	**2**	0	**2**	0	0	0	0	0	0	0	0	0	0	0	0	0	0	**4**	**05.00**
432.	*Oroxylum indicum*	0	0	0	0	**2**	0	0	0	0	0	0	0	0	0	0	0	0	0	0	**2**	**02.50**
433.	*Oryza sativa*	0	0	**5**	**5**	**2**	0	**3**	0	0	0	0	0	0	0	0	0	0	0	0	**15**	**18.75**
434.	*Ougeinia oojeinensis*	**2**	**2**	0	0	**2**	0	0	0	0	0	0	0	0	0	0	0	0	0	0	**6**	**07.50**

Contd...

Table 4.1–Contd...

Sl.No.	Botanical Name	Timb		Food		Med		Fuel	Fiber	Dye	Tan	G/R	Oil	Latex	Tb	Others					Total	TIV
		Maj	Min	Hum	Vet	Hum	Vet									Bm Tg Fn	B M Mp Um	Poison	Bev Liq	Misc.		
435.	*Oxalis corniculata*	0	0	**2**	0	**3**	0	0	0	0	0	0	0	0	0	0	0	0	0	0	**5**	**06.25**
436.	*Pandanus odoratissimus*	0	0	**2**	0	**2**	0	0	0	0	0	0	**3**	0	0	0	0	0	0	**2**	**9**	**11.25**
437.	*Panicum sumatrense*	0	0	**5**	**3**	**2**	0	0	0	0	0	0	0	0	0	0	0	0	0	0	**10**	**12.50**
438.	*Pavetta indica*	0	0	0	0	**2**	0	0	0	0	0	0	0	0	0	0	0	0	0	0	**2**	**02.50**
439.	*Pavonia zeylanica*	0	0	0	0	**2**	0	0	0	0	0	0	0	0	0	0	0	0	0	0	**2**	**02.50**
440.	*Pedalium murex*	0	0	**2**	0	**2**	0	0	0	0	0	0	0	0	0	0	0	0	0	0	**4**	**05.00**
441.	*Peltophorum pterocarpum*	0	0	0	0	**2**	0	0	0	0	0	0	0	0	0	0	0	0	0	0	**2**	**02.50**
442.	*Pennisetum typhoides*	0	0	**4**	**3**	**1**	0	0	0	0	0	0	0	0	0	0	0	0	0	0	**8**	**10.00**
443.	*Pergularia daemia*	0	0	**2**	0	**2**	0	0	**3**	0	0	0	0	0	0	0	0	0	0	0	**7**	**08.75**
444.	*Perilla frutescens*	0	0	0	0	**2**	0	0	0	0	0	0	0	0	0	0	0	0	0	**2**	**4**	**05.00**
445.	*Phoenix loureirii*	0	0	**2**	0	**1**	0	0	0	0	0	0	0	0	0	**2**	**2**	0	**2**	0	**9**	**11.25**
446.	*P.pusilla*	0	0	**2**	0	0	0	0	0	0	0	0	0	0	0	**2**	**2**	0	0	0	**6**	**07.50**
447.	*P.sylvestris*	0	0	**2**	0	**2**	0	0	0	0	0	0	0	0	0	**2**	**2**	0	**2**	0	**10**	**12.50**
448.	*Phyla nodiflora*	0	0	0	0	**2**	0	0	0	0	0	0	0	0	0	0	0	0	0	0	**2**	**02.50**
449.	*Phyllanthus reticulatus*	0	0	**2**	0	**2**	0	0	0	0	0	0	0	0	0	0	**2**	0	0	0	**6**	**07.50**
450.	*P.amarus*	0	0	0	0	**5**	0	0	0	0	0	0	0	0	0	0	0	0	0		**5**	**06.25**
451.	*P.debilis*	0	0	0	0	**2**	0	0	0	0	0	0	0	0	0	0	0	0	0	0	**2**	**02.50**
452.	*P.emblica*	0	0	**3**	0	**5**	**2**	0	0	0	0	0	0	0	0	0	0	0	0	0	**10**	**12.50**
453.	*P.maderaspatensis*	0	0	0	0	**2**	0	0	0	0	0	0	0	0	0	0	0	0	0	0	**2**	**02.50**
454.	*Physalis minima*	0	0	**2**	0	**2**	0	0	0	0	0	0	0	0	0	0	0	0	0	0	**4**	**05.00**

Contd...

Table 4.1–Contd...

Sl.No.	Botanical Name	Timb		Food		Med		Fuel	Fiber	Dye	Tan	G/R	Oil	Latex	Tb	Others					Total	TIV
		Maj	Min	Hum	Vet	Hum	Vet									Bm Tg Fn	B M Mp Um	Poi-son	Bev Liq	Misc.		
455.	*Pinus roxburghii*	0	0	0	0	**2**	0	0	0	0	0	0	0	0	0	0	0	0	0	0	**2**	**02.50**
456.	*Piper betle*	0	0	**2**	0	**2**	0	0	0	0	0	0	0	0	0	0	0	0	0	0	**4**	**05.00**
457.	*P.longum*	0	0	**2**	0	**4**	0	0	0	0	0	0	0	0	0	0	0	0	0	0	**6**	**07.50**
458.	*P.nigrum*	0	0	**2**	0	**5**	0	0	0	0	0	0	0	0	0	0	0	0	0	0	**7**	**08.75**
459.	*Pithecolobium dulce*	0	**3**	**2**	**2**	**2**	0	0	0	0	0	0	0	0	0	0	0	0	0	0	**9**	**11.25**
460.	*Pleurostylia opposita*	0	0	0	0	0	0	0	0	0	0	0	0	0	0	0	0	0	0	**2**	**2**	**02.50**
461.	*Plumbago indica*	0	0	0	0	**2**	0	0	0	0	0	0	0	0	0	0	0	0	0	0	**2**	**02.50**
462.	*P.zeylanica*	0	0	0	0	**3**	0	0	0	0	0	0	0	0	0	0	0	0	0	0	**3**	**03.75**
463.	*Plumeria alba*	0	0	0	0	**3**	0	0	0	0	0	0	0	0	0	0	0	0	0	**2**	**5**	**06.25**
464.	*P.rubra*	0	0	0	0	**2**	0	0	0	0	0	0	0	0	0	0	0	0	0	**2**	**4**	**05.00**
465.	*Polyalthia cerasoides*	**2**	**2**	**2**	0	0	0	0	**2**	0	0	0	0	0	0	0	0	0	0	0	**8**	**10.00**
466.	*P.longifolia*	**2**	**2**	**2**	0	**3**	**3**	0	**2**	0	0	0	0	0	0	0	0	0	0	**2**	**16**	**20.00**
467.	*Polygonum chinense*	0	0	**2**	0	**2**	0	0	0	0	0	0	0	0	0	0	0	0	0	0	**4**	**05.00**
468.	*Pongamia pinnata*	0	0	**2**	0	**3**	0	**2**	0	0	0	0	**3**	0	**2**	0	0	0	0	**2**	**14**	**17.50**
469.	*Portulaca oleracea*	0	0	**2**	0	**3**	0	0	0	0	0	0	0	0	0	0	0	0	0	0	**5**	**06.25**
470.	*Pouzolzia zeylanica*	0	0	0	0	**2**	0	0	0	0	0	0	0	0	0	0	0	0	0	0	**2**	**02.50**
471.	*Premma tomentosa*	0	**3**	**2**	0	0	0	0	0	0	0	0	0	0	0	0	0	0	0	0	**5**	**06.25**
472.	*Prosopis chilensis*	0	0	0	0	**2**	0	0	0	0	0	0	0	0	0	0	0	0	0	0	**2**	**02.50**
473.	*P.cineraria*	0	0	0	0	**2**	0	0	0	0	0	0	0	0	0	0	0	0	0	0	**2**	**02.50**
474.	*Psidium guajava*	0	0	**2**	0	**3**	0	0	0	0	0	0	0	0	0	0	0	0	0	0	**5**	**06.25**

Contd...

Table 4.1–Contd...

Sl.No.	Botanical Name	Timb		Food		Med		Fuel	Fiber	Dye	Tan	G/R	Oil	Latex	Tb	Others					Total	TIV
		Maj	Min	Hum	Vet	Hum	Vet									Bm Tg Fn	B M Mp Um	Poi-son	Bev Liq	Misc.		
475.	*Psoralea corylifolia*	0	0	2	0	3	0	0	0	0	0	0	0	0	0	0	0	0	0	0	5	06.25
476.	*Pteris vittata*	0	0	0	2	2	0	0	0	0	0	0	0	0	0	2	0	0	0	0	6	06.25
477.	*Pterocarpus marsupium*	3	4	2	0	3	0	0	0	0	0	3	0	0	0	0	0	0	0	2	17	21.25
478.	*Pterolobium hexapetalum*	0	0	0	0	0	3	0	0	0	0	0	0	0	0	0	0	0	0	0	3	03.75
479.	*Pterospermum xylocarpum*	0	3	0	0	2	0	0	0	0	0	0	0	0	0	0	2	0	0	0	7	08.75
480.	*Pueraria tuberosa*	0	0	2	0	3	0	0	0	0	0	0	0	0	0	0	0	0	2	0	7	08.75
481.	*Punica granatum*	0	0	2	0	2	0	0	0	0	0	0	0	0	0	0	0	0	0	0	4	05.00
482.	*Pupalia lappacea*	0	0	0	0	2	0	0	0	0	0	0	0	0	0	0	0	0	0	0	2	02.50
483.	*Quisqualis indica*	0	0	0	0	3	0	0	0	0	0	0	0	0	0	0	0	0	0	0	3	03.75
484.	*Rauvolfia serpentina*	0	0	0	0	5	0	0	0	0	0	0	0	0	0	0	0	0	0	2	7	08.75
485.	*R.tetraphylla*	0	0	0	0	2	0	0	0	0	0	0	0	0	0	0	0	0	0	0	2	02.50
486.	*Rhynchosia minima*	0	0	2	0	2	0	0	0	0	0	0	0	0	0	0	0	0	0	0	4	05.00
487.	*Ricinus communis*	0	0	2	0	3	0	0	0	0	0	0	4	0	0	0	0	0	0	2	11	13.75
488.	*Ruellia tuberosa*	0	0	0	0	2	0	0	0	0	0	0	0	0	0	0	0	0	2	2	6	07.50
489.	*Saccharum officinarum*	0	0	3	2	2	0	0	0	0	0	0	0	0	0	0	0	0	0	0	7	08.75
490.	*S.spontaneum*	0	0	2	2	2	2	0	2	0	0	0	0	0	0	2	0	0	0	0	12	15.00
491.	*Samanea saman*	0	0	0	2	2	0	3	0	0	0	2	0	0	0	0	0	0	0	0	9	11.25
492.	*Sansevieria roxburghiana*	0	0	0	0	2	0	0	3	0	0	0	0	0	0	0	0	0	0	2	7	08.75
493.	*Santalum album*	0	2	3	0	0	0	0	0	0	0	0	0	0	0	0	0	0	0	2	7	08.75
494.	*Sapindus emarginatus*	0	0	2	0	2	0	0	0	0	0	0	0	0	0	0	0	0	0	2	6	07.50

Contd...

Table 4.1–Contd...

Sl.No.	Botanical Name	Timb		Food		Med		Fuel	Fiber	Dye	Tan	G/R	Oil	Latex	Tb	Others					Total	TIV
		Maj	Min	Hum	Vet	Hum	Vet									Bm Tg Fn	B M Mp Um	Poi-son	Bev Liq	Misc.		
495.	*Saraca asoca*	0	0	0	0	**2**	0	0	0	0	0	0	0	0	0	0	0	0	0	0	**2**	**02.50**
496.	*Sarcolobus carinatus*	0	0	**2**	0	0	0	0	0	0	0	0	0	0	0	0	0	0	0	0	**2**	**02.50**
497.	*Sarcostemma secamone*	0	0	0	0	**2**	0	0	0	0	0	0	0	0	0	0	0	0	0	0	**2**	**02.50**
498.	*Sauropus quadrangularis*	0	0	**2**	0	**2**	0	0	0	0	0	0	0	0	0	0	0	0	0	0	**4**	**05.00**
499.	*Schleichera oleosa*	**2**	**3**	**2**	**3**	**2**	**2**	0	0	0	0	0	**3**	0	0	0	0	0	0	0	**17**	**21.25**
500.	*Schrebera swietenioides*	0	**2**	0	0	**2**	0	0	0	0	0	0	0	0	0	0	0	0	0	**2**	**6**	**07.50**
501.	*Scoparia dulcis*	0	0	0	0	**2**	0	0	0	0	0	0	0	0	0	0	0	0	0	0	**2**	**02.50**
502.	*Securinega virosa*	0	0	0	0	**2**	0	0	0	0	0	0	0	0	0	0	0	0	0	0	**2**	**02.50**
503.	*Selaginella repanda*	0	0	0	0	**2**	0	0	0	0	0	0	0	0	0	0	0	0	0	0	**2**	**02.50**
504.	*Semecarpus anacardium*	0	0	**2**	0	**2**	0	0	0	**3**	0	**2**	**2**	0	0	0	0	0	**2**	0	**13**	**16.25**
505.	*Sesamum indicum.*	0	0	**2**	0	**2**	0	0	0	0	0	0	**3**	0	0	0	0	0	0	**2**	**9**	**11.25**
506.	*Sesbania grandiflora*	0	**2**	**2**	0	**2**	**2**	0	0	0	0	0	0	0	0	0	0	0	0	0	**8**	**10.00**
507.	*S.sesban*	0	0	0	0	**2**	0	**2**	0	0	0	0	0	0	0	0	0	0	0	**2**	**6**	**07.50**
508.	*Setaria italica*	0	0	**3**	**3**	**2**	0	0	0	0	0	0	0	0	0	0	0	0	0	0	**8**	**10.00**
509.	*S.pumila*	0	0	0	**2**	**2**	**2**	0	0	0	0	0	0	0	0	0	0	0	0	0	**6**	**07.50**
510.	*Sida acuta*	0	0	0	0	**2**	0	0	**3**	0	0	0	0	0	0	**2**	0	0	0	0	**7**	**08.75**
511.	*S.cordata*	0	0	0	0	**2**	0	0	0	0	0	0	0	0	0	0	0	0	0	0	**2**	**02.50**
512.	*S.cordifolia*	0	0	0	0	**2**	0	0	0	0	0	0	0	0	0	0	0	0	0	0	**2**	**02.50**
513.	*S.rhombifolia*	0	0	0	0	**2**	0	0	0	0	0	0	0	0	0	2	0	0	0	0	**4**	**05.00**
514.	*Smilax zeylanica*	0	0	**2**	0	**3**	0	0	**2**	0	0	0	0	0	2	0	0	0	0	0	**9**	**11.25**

Contd...

Table 4.1–Contd...

Sl.No.	Botanical Name	Timb		Food		Med		Fuel	Fiber	Dye	Tan	G/R	Oil	Latex	Tb	Others					Total	TIV
		Maj	Min	Hum	Vet	Hum	Vet									Bm Tg Fn	B M Mp Um	Poi-son	Bev Liq	Misc.		
515.	*Solanum americanum*	0	0	**2**	0	**2**	0	0	0	0	0	0	0	0	0	0	0	0	0	0	**4**	**05.00**
516.	*S.anguivi*	0	0	**2**	0	**2**	0	0	0	0	0	0	0	0	0	0	0	0	0	0	**4**	**05.00**
517.	*S.melongena*	0	0	**3**	0	**2**	0	0	0	0	0	0	0	0	0	0	0	0	0	0	**5**	**06.25**
518.	*S.surattense*	0	0	**2**	0	**3**	**2**	0	0	0	0	0	0	0	0	0	0	0	0	0	**7**	**08.75**
519.	*S.torvum*	0	0	**2**	0	**2**	0	0	0	0	0	0	0	0	0	0	0	0	0	0	**4**	**05.00**
520.	*S.tuberosum*	0	0	**3**	0	**2**	0	0	0	0	0	0	0	0	0	0	0	0	0	0	**5**	**06.25**
521.	*Solena heterophylla*	0	0	**2**	0	**2**	0	0	0	0	0	0	0	0	0	0	0	0	0	0	**4**	**05.00**
522.	*Sorghum vulgare*	0	0	**3**	**2**	**2**	0	0	0	0	0	0	0	0	0	0	0	0	0	0	**7**	**08.75**
523.	*Soymida febrifuga*	0	0	0	0	**3**	0	0	**3**	0	0	0	0	0	0	0	0	0	0	0	**6**	**07.50**
524.	*Sphaeranthus indicus*	0	0	**2**	0	**2**	0	0	0	0	0	0	0	0	0	0	0	0	0	0	**4**	**05.00**
525.	*Spondias pinnata*	0	0	**3**	0	**2**	0	0	0	0	0	0	0	0	0	0	0	0	0	0	**5**	**06.25**
526.	*Stachytarpheta urticaefolia*	0	0	**2**	0	**2**	0	0	0	0	0	0	0	0	0	0	0	0	0	0	**4**	**05.00**
527.	*Sterculia foetida*	0	0	**2**	0	**2**	0	0	**3**	0	0	0	0	0	0	0	0	0	0	0	**7**	**08.75**
528.	*S.urens*	0	0	0	0	**2**	0	0	**3**	0	0	**3**	0	0	0	0	0	0	0	**2**	**10**	**12.50**
529.	*S.villosa*	0	0	**2**	0	**2**	0	0	**2**	0	0	**2**	0	0	0	0	0	0	0	0	**8**	**10.00**
530.	*Streblus asper*	0	0	0	0	**4**	0	0	0	0	0	0	0	0	0	0	0	0	0	0	**4**	**05.00**
531.	*Strychnos nux-vomica*	0	0	0	0	**5**	**2**	0	0	0	0	0	0	0	0	0	0	**2**	0	0	**9**	**11.25**
532.	*S.potatorum*	0	0	0	0	**4**	**2**	0	0	0	0	0	0	0	0	0	0	**2**	0	**2**	**10**	**12.50**
533.	*Syzygium cumini*	**5**	**4**	**2**	**2**	**4**	**2**	**2**	0	0	0	0	0	0	0	0	0	**2**	0	**2**	**25**	**31.25**
534.	*Tabernaemontana divaricata*	0	0	0	0	**2**	0	0	0	0	0	0	0	0	0	0	0	0	0	0	**2**	**02.50**

Contd...

Table 4.1–Contd...

Sl.No.	Botanical Name	Timb		Food		Med		Fuel	Fiber	Dye	Tan	G/R	Oil	Latex	Tb	Others					Total	TIV
		Maj	Min	Hum	Vet	Hum	Vet									Bm Tg Fn	B M Mp Um	Poi-son	Bev Liq	Misc.		
535.	*Tagetes erecta*	0	0	0	0	**2**	0	0	0	0	0	0	0	0	0	0	0	0	0	**2**	**4**	**05.00**
536.	*Talinum portulacifolium*	0	0	**2**	0	0	0	0	0	0	0	0	0	0	0	0	0	0	0	0	**2**	**02.50**
537.	*Tamarindus indica*	**2**	**3**	**3**	0	**2**	0	0	0	0	0	0	0	0	0	0	0	0	0	**2**	**12**	**15.00**
538.	*Tamarix troupii*	0	0	0	0	0	0	**3**	0	0	0	0	0	0	0	0	0	0	0	0	**3**	**03.75**
539.	*Tecoma stans*	0	0	0	0	0	0	0	0	0	0	0	0	0	0	0	0	0	0	**2**	**2**	**02.50**
540.	*Tectona grandis*	**5**	**5**	0	0	**2**	**2**	0	0	0	0	0	0	0	0	0	0	0	0	0	**14**	**17.50**
541.	*Tephrosia purpurea*	0	0	0	0	**2**	**2**	0	0	0	0	0	0	0	0	0	0	**2**	0	**2**	**8**	**09.75**
542.	*Terminalia alata*	**2**	**2**	0	**2**	**2**	0	0	0	0	**2**	0	0	0	0	0	0	0	0	0	**10**	**12.50**
543.	*T.arjuna*	**4**	**3**	0	0	**2**	0	0	0	0	**2**	0	0	0	0	0	0	0	0	**2**	**13**	**16.25**
544.	*T.bellirica*	**2**	**3**	**2**	**2**	**5**	0	0	0	0	0	**2**	**2**	0	0	0	0	0	0	**2**	**20**	**25.00**
545.	*T.catappa*	0	0	**2**	0	**2**	0	0	0	0	0	0	0	0	0	0	0	0	0	0	**4**	**05.00**
546.	*T.chebula*	**2**	**3**	0	**2**	**5**	**2**	0	0	0	**2**	0	0	0	0	0	0	0	0	0	**16**	**20.00**
547.	*T.coriacea*	**2**	**2**	0	0	**2**	0	0	0	0	0	0	0	0	0	0	0	0	0	**2**	**8**	**10.00**
548.	*T.paniculata*	0	0	0	0	**2**	0	0	0	0	0	0	0	0	0	0	0	0	0	0	**2**	**02.50**
549.	*Themeda cymbarica*	0	0	0	**2**	**2**	0	0	0	0	0	0	0	0	0	**2**	0	0	0	0	**6**	**07.50**
550.	*Thespesia lampas*	0	0	0	0	**2**	0	0	**3**	0	0	0	0	0	0	0	0	0	0	0	**5**	**06.25**
551.	*T.populnea*	0	**3**	0	0	**2**	0	0	0	0	0	0	0	0	0	0	0	0	0	0	**5**	**06.25**
552.	*Thunbergia fragrans*	0	0	0	0	**3**	0	0	0	0	0	0	0	0	0	0	0	0	0	0	**3**	**03.75**

Contd...

Table 4.1–Contd...

Sl.No.	Botanical Name	Timb		Food		Med		Fuel	Fiber	Dye	Tan	G/R	Oil	Latex	Tb	Others					Total	TIV
		Maj	Min	Hum	Vet	Hum	Vet									Bm Tg Fn	B M Mp Um	Poi-son	Bev Liq	Misc.		
553.	*Thysanolaena maxima*	0	0	0	**3**	0	0	0	0	0	0	0	0	0	0	**3**	0	0	0	0	**6**	**07.50**
554.	*Tiliacora acuminata*	0	0	0	0	**2**	0	0	0	0	0	0	0	0	0	0	**2**	0	0	0	**4**	**05.00**
555.	*Tinospora corditolia*	0	0	0	0	**5**	0	0	0	0	0	0	0	0	0	0	0	0	**2**	0	**7**	**08.75**
556.	*Toddalia asiatica*	0	0	0	0	**2**	0	0	0	**3**	0	0	0	0	0	0	0	0	0	0	**5**	**06.25**
557.	*Toona ciliata*	0	**3**	0	0	0	0	0	0	0	0	0	0	0	0	0	0	0	0	0	**3**	**03.75**
558.	*Trema orientalis*	0	0	**2**	0	**2**	0	**2**	**3**	0	0	0	0	0	0	0	0	0	0	0	**9**	**11.25**
559.	*Trianthema decandra*	0	0	0	0	**2**	0	0	0	0	0	0	0	0	0	0	0	0	0	0	**2**	**02.50**
560.	*T.portulacastrum*	0	0	**2**	0	**2**	0	0	0	0	0	0	0	0	0	0	0	0	0	0	**4**	**05.00**
561.	*Tribulus terrestris*	0	0	**2**	0	**2**	0	0	0	0	0	0	0	0	0	0	0	0	0	0	**4**	**05.00**
562.	*Trichosanthes anguina*	0	0	**2**	0	**2**	0	0	0	0	0	0	0	0	0	0	0	0	0	0	**4**	**05.00**
563.	*Tridax procumbens*	0	0	**2**	0	**3**	**2**	0	0	0	0	0	0	0	0	0	0	0	0	0	**7**	**08.75**
564.	*Trigonella toenum-graecum*	0	0	**3**	0	**2**	0	0	0	0	0	0	0	0	0	0	0	0	0	**2**	**7**	**08.75**
565.	*Triticum aestivum*	0	0	**4**	**3**	**2**	**2**	0	0	0	0	0	0	0	0	0	**2**	0	0	**2**	**15**	**18.75**
566.	*Triumtetta rhomboidea*	0	0	0	0	**2**	0	0	0	0	0	0	0	0	0	0	0	0	0	0	**2**	**02.50**
567.	*Tylophora indica*	0	0	0	0	**3**	0	0	0	0	0	0	0	0	0	0	0	0	0	0	**3**	**03.75**
568.	*Typha angustata*	0	0	0	**3**	**2**	0	0	0	0	0	0	0	0	0	0	**2**	0	0	0	**7**	**08.75**
569.	*Uraria picta.*	0	0	0	0	**2**	0	0	0	0	0	0	0	0	0	0	0	0	0	0	**2**	**02.50**
570.	*Urena lobata* subsp. *sinuata*	0	0	0	0	**2**	0	0	**3**	0	0	0	0	0	0	0	0	0	0	0	**5**	**06.25**
571.	*Urginea indica*	0	0	**2**	0	**2**	0	0	0	0	0	0	0	0	0	0	0	0	0	0	**4**	**05.00**
572.	*Vanda tessellata*	0	0	0	0	**2**	0	0	0	0	0	0	0	0	0	0	0	0	0	0	**2**	**02.50**

Contd...

Table 4.1–Contd...

Sl.No.	Botanical Name	Timb		Food		Med		Fuel	Fiber	Dye	Tan	G/R	Oil	Latex	Tb	Others					Total	TIV
		Maj	Min	Hum	Vet	Hum	Vet									Bm Tg Fn	B M Mp Um	Poi-son	Bev Liq	Misc.		
573.	*Ventilago denticulata*	0	0	0	0	0	0	0	0	**3**	0	0	0	0	0	0	0	0	0	0	**3**	**03.75**
574.	*Vernonia cinerea*	0	0	**2**	0	**2**	0	0	0	0	0	0	0	0	0	0	0	0	0	0	**4**	**05.00**
575.	*Vetiveria zizanioides*	0	0	0	0	**3**	0	0	**3**	0	0	0	**2**	0	0	0	**2**	0	0	**2**	**12**	**15.00**
576.	*Vigna mungo*	0	0	**3**	**3**	**2**	0	0	0	0	0	0	0	0	0	0	0	0	0	0	**8**	**10.00**
577.	*V.radiata*	0	0	**3**	0	0	0	0	0	0	0	0	0	0	0	0	0	0	0	0	**3**	**03.75**
578.	*V.trilobata*	0	0	**2**	**3**	**2**	0	0	0	0	0	0	0	0	0	0	0	0	0	**2**	**9**	**11.25**
579.	*V.unguiculata*	0	0	0	0	**2**	0	0	0	0	0	0	0	0	0	0	0	0	0	0	**2**	**02.50**
580.	*Viscum orientale*	0	0	0	0	**2**	**2**	0	0	0	0	0	0	0	0	0	0	0	0	0	**4**	**05.00**
581.	*Vitex altissima*	0	**3**	0	0	**2**	0	0	0	0	0	0	0	0	0	0	0	0	0	0	**5**	**06.25**
582.	*V.negundo*	0	0	0	0	**3**	**2**	0	0	0	0	0	0	0	0	0	**2**	0	0	**2**	**9**	**11.25**
583.	*Walsura trifolia*	0	**2**	0	0	0	0	0	0	0	0	0	0	0	0	0	0	**2**	0	0	**4**	**05.00**
584.	*Waltheria indica*	0	0	0	0	**2**	0	0	0	0	0	0	0	0	0	0	0	0	0	0	**2**	**02.50**
585.	*Wattakaka volubilis*	0	0	0	0	**2**	**2**	0	0	0	0	0	0	0	0	0	0	0	0	0	**4**	**05.00**
586.	*Withania somnifera*	0	0	0	0	**5**	**2**	0	0	0	0	0	0	0	0	0	0	0	0	0	**7**	**08.75**
587.	*Woodfordia fruticosa*	0	0	0	0	**2**	**2**	0	0	**2**	0	0	0	0	0	0	0	0	**2**	0	**8**	**10.00**
588.	*Wrightia arborea*	0	**2**	0	0	0	**2**	0	**2**	**3**	0	0	0	0	0	0	0	0	0	0	**9**	**11.25**
589.	*W.tinctoria*	0	**3**	**2**	0	**2**	0	0	**2**	0	0	0	0	0	0	0	0	0	0	0	**9**	**11.25**
590.	*Xanthium pungens*	0	0	**2**	0	**3**	0	0	0	0	0	0	0	0	0	0	0	0	0	0	**5**	**06.25**
591.	*Ximenia americana*	0	0	**2**	0	0	**4**	0	0	0	0	0	0	0	0	0	0	0	0	0	**6**	**07.50**
592.	*Xylia xylocarpa*	0	**3**	**2**	0	0	0	**3**	0	0	0	0	0	0	0	0	0	**2**	0	0	**10**	**12.50**

Contd...

Table 4.1–Contd...

Sl.No.	Botanical Name	Timb		Food		Med		Fuel	Fiber	Dye	Tan	G/R	Oil	Latex	Tb	Others					Total	TIV
		Maj	Min	Hum	Vet	Hum	Vet									Bm Tg Fn	B M Mp Um	Poi- son	Bev Liq	Misc.		
593.	*Zea mays*	0	0	3	0	2	0	0	0	0	0	0	0	0	0	0	0	0	2	0	7	08.75
594.	*Zingiber officinale*	0	0	2	0	4	0	0	0	0	0	0	0	0	0	0	0	0	0	0	6	07.50
595.	*Zizyphus mauritiana*	0	2	2	3	2	2	0	0	0	3	0	0	0	2	0	0	0	0	2	18	22.50
596.	*Z.nummularifolia*	0	0	2	0	2	0	0	0	0	0	0	0	0	0	0	0	0	0	0	4	05.00
597.	*Z.oenoplia*	0	0	2	3	2	2	0	0	0	0	0	0	0	0	0	0	0	0	0	9	11.25
598.	*Z.rugosa*	0	0	2	0	2	0	0	0	0	0	0	0	0	0	0	0	0	0	0	4	05.00
599.	*Z.xylopyrus*	0	0	2	0	3	0	0	0	3	3	0	0	0	0	0	0	0	0	0	11	13.75
600.	*Zornia gibbosa*	0	0	0	0	2	0	0	0	0	0	0	0	0	0	0	0	0	0	0	2	02.50

Timb: Timber; **Med**: Medicine; **Maj**: Major; **Min**: Minor; **Hum**: Human; **Vet**: Veterinary; **Tan**: Tannin; **G/R**: Gum,Resin; **Tb**: Tooth brush; **Br**: Broom; **Tg**: Thatching; **Fn**: Fencing; **B**: Baskets; **M**: Mats; **Mp**: Meal plate; **Um**: Umbrella; **Poison**: Fish/Arrow Poison; **Bev**: Beverage; **Liq**: Liquor; **Misc**: Miscellaneous.

Plate 4.1

1. ***Abrus precatorius***; 2. ***Aristolochia indica***; 3. ***Asparagus gonoclados***; 4. ***Buchanania lanzan***; 5. ***Careya arborea***; 6. ***Corallocarpus epigaeus***

Plate 4.2

1. *Gmelina asiatica*; 2. *Mucuna pruriens*; 3. *Psoralea corylifolia*; 4. *Rauvolfia tetraphylla*; 5. *Strychnos nux-vomica*

Caesalpinia coriaria, Cassia alata, C. occidentalis, Ceriops decandra, Cochlospermum religiosum, Gardenia gummifera, G. resinifera, Hardwickia binata, Limonia acidissima, Pterocarpus marsupium, Samanea saman, Semecarpus anacardium, Sterculia urens, S. villosa, Terminalia alata, T. arjuna, T. bellirica, T. chebula, Zizyphus mauritiana and *Z. xylopyrus* are used for tannins, gums and resins.

Azadirachta indica, Abelmoschus moschatus, Brassica nigra, Caesalpinia bonduc, Cascabela thevetia, Celastrus paniculatus, Cocos nucifera, Cymbopogon coloratus, C. flexuosus, Elettaria cardamomum, Eucalyptus globulus, Guizotia abyssinica, Helianthus annuus, Jatropha curcas, J. glandulifera, Madhuca indica, Martynia annua, Michelia champaca, Ocimum tenuiflorum, Pandanus odoratissimus, Pongamia pinnata, Ricinus communis, Schleichera oleosa, Semecarpus anacardium, Sesamum indicum, Terminalia bellirica and *Vetiveria zizanioides* are used to extract oil.

Achyranthes aspera, Azadirachta indica, Bambusa arundinacea, Bridelia retusa, Calotropis gigantea, Cassia occidentalis, Cocos nucifera, Dendrocalamus strictus, Jatropha curcas, J. gossypifolia, Lantana camara, Mimusops elengi, Naravelia zeylanica, Pongamia pinnata, Smilax zeylanica and *Zizyphus mauritiana,* the whole plant or the tender stems are used as tooth brushes.

Abrus precatorius, Adenanthera pavonina, Aeschynomene indica, Andrographis paniculata, Artocarpus heterophyllus, Bambusa arundinacea, Bauhinia vahlii, Borassus flabellifer, Caryota urens, Cocos nucifera, Bombax ceiba, Buchanania lanzan, Butea monosperma, Calamus viminalis, Casearia elliptica, Dendrocalamus strictus, Diospyros melanoxylon, Ocimum basilicum, O. tenuiflorum, Phoenix loureirii, P. pusilla, P. sylvestris, Pongamia pinnata, Aristida adscensionis, Aristida setacea, Sida acuta, S. rhombifolia, Thysanolaena maxima, Sapindus emarginatus, Xylia xylocarpa and *Zizyphus mauritiana* are used in the preparation of baskets, cages, mats, meal plates, bamboo sticks and umbrellas.

Of six hundred ethnobotanical plant species evaluated, 35.14 per cent are useful in ethnomedicine for human and 4.72 per cent for veterinary, 17.76 per cent as food for human beings, 4.40 per cent for fodder to animals, 11.3 per cent as timber yielding, 4.91 per cent as fibre yielding, 2.16 per cent oil producing, 1.81 per cent dye yielding, 1.7 per cent fuel yielding, 1.31 per cent gum and resin yielding, 1.1 per cent tannin, 0.89 per cent tooth brush, 0.3 per cent latex producing and 12.5 per cent are used for other purposes/miscellaneous like basket making, mats, meal plates, umbrellas, brooms, thatching, fencing, tattooing, toddy, liquors, beverages etc.

Method of Calculation

The Total Importance Value (TIV) of plants of ethnic use from East Godavari district was calculated.

$$\text{TIV per cent} = \frac{\text{U1+ U2+ U3.............Un}}{\text{Number of uses} \times \text{Maximum value}} \times 100$$

where,

TIV: Total Importance Value

U: Importance for each particular use (*e.g.*, Timber, food, medicine etc.).

References

Arunee Kumar and Nisteswar, K. 1990. Medicinal plants of Kakinada, East Godavari district, Andhra Pradesh. Indian Medicine 2: 229-232.

Belal, A. and Springuel, I. 1996. Economic value of plant diversity in arid environment. Nature and Resources 32: 33-39.

Krishna Prasad, V., Rajagopal, T., Yogesh Kant and Badarinath, K.V.S. 1999. Food plants of Konda Reddis of Rampa Agency, East Godavari district, Andhra Prades-A case study. Ethnobotany 11: 92-96.

Nisteswar, K. and Kumar, K.A. 1980. Utilitarian values of medical-lore of Rampa Agency (Andhra Pradesh) in primary health care. Sachitra Ayurved 33: 210-212.

Sudhakar, S. and Rao, R.S. 1985. Medicinal plants of East Godavari district, Andhra Pradesh. J. Econ. Taxon. Bot. 7: 399-406.

Chapter 5

The Natural Source of Medicinal Wealth: Orchids

M.C. Gayatri and Ananya Sarkar***

Plant Biotechnology Unit, Annexe-Department of Botany, Bangalore University, Bangalore – 560 056, India

ABSTRACT

According to the Holy Mother and Sri Aurobindo, Orchid flowers relate to the Divine as it wraps itself around the Divine and finds all its support in Him. In India, orchids are also employed for a healing touch as it possesses a variety of therapeutic uses in different systems of medicine, which are attracting the attention of many modern workers. Traditional therapeutic uses of number of orchids are mentioned and their phyto-constituents including flavonoids, terpenes, carbohydrates and glycosides are reviewed in the past. The main idea of this chapter is to review the recent advances in plant drug research and developments in orchid study, in an attempt to provide useful references for both plant drug studies and tissue culture studies. It has been confirmed by WHO that herbal medicines serve the health needs of about 80 per cent of the world's population; especially for millions of people in the vast rural belt of the developing countries. Simultaneously, the patients of developed countries are also facing disillusionment with modern healthcare and are ready for any kind of alternatives. This has led to the recent resurgence of green remedies based on the following major criteria: (1) the effectiveness of plant medicine; (2) the affordability for masses; (3) the vast wealth of green medicine; (4) the side-effect of most modern drugs; and (5) the development of science and technology.

E-mail: *gayatrimc@hotmail.com; **sarkarananya2@gmail.com.

In this context, the conservation of therapeutically important orchids and their *ex-situ* preservation on a large scale with the application of micropropagation to meet the increased demands of the drug industry is to be emphasized.

***Keywords**: Medicinal orchids, Orchids of Western Ghat, Orchids of eastern India.*

Introduction

The pages from human history indicate that, for over thousands of years in every continent, man has had a direct contact with this unique group of plants with a wide range of variability. There are innumerable stories of people being obsessed with the beauty of orchid flowers, attracted by their heavy perfume, or intrigued by their extractable essences. The wild blooming orchids are no fewer majestic in beauty than many other cultivated flowers of commercial importance. They are ultimate in rarity, beauty and allure. They are quite popular among professionals and amateur orchid lovers for collection either as a hobby or an item of cut flower trade. But apart from their ornamental value, orchids are successful in invoking a special interest because of their distinctive uses in traditional medicines.

Like roses, lilies and anthuriums, Orchids are also a particular group of flowering plants belonging to the family Orchidaceae. The family Orchidaceae is one of the most highly evolved of the entire plant kingdom. It comprises over 30,000 species in approximately 750 genera. In addition to this, new discoveries are constantly being made and described every year. It is one of the most widely spread plant with terrestial, saprophytic and epiphytic habits.

There are some features common to all orchids which enable them to get included in the Orchidaceae. As of instance, the orchid flower has the feminine and masculine organ of reproduction fusing in only one body called column or gynostemium. Another characteristic feature is the irregular form of the flower segment with 3 sepals and 3 petals, the medium petal is modified and is called as labellum or lip which is intensively colored and is the most interesting part of the flower from where the scent is exhaled in order to attract the agents of pollination. It can have the appearance of the female bee (like *Ophyrs*), female wasp like *Drakea* or *Cryptostylis*, a bucket like *Coryanthes*, having a very sweet or disagreeable, strong or light perfume. Everything is designed to attract the pollinator to guarantee the pollination of the species.

The word "Orchid" originated from the Greek word "Orchis", which literally means "Testicle". It was Theophrastus who first used the Greek word "Orchis" to indicate the particular group of plants whose roots were chopped, dried and used in the traditional pharmacopoeia of Greece and neighboring Asia Minor as antidepressants and stimulants. Even today, in some rural areas of these countries, a nutritional drink called "Salep" is prepared from the dried tubers of certain orchids. It was not until the mid 18th century that the name was applied to the whole family of orchids in Linnaeus's, *Species Plantarum*. This marked the beginning of modern plant taxonomy, where Linnaeus described orchids as two semi-globular underground structures tubers, placed side-by-side resembling human testicles.

Besides Theophrastus and Linnaeus, many other people were interested in orchids, principally as medicinal plants. This is testified by references in the medical books and herbals of men such as Dioscorides (1st century A.D.), Otto Brunfels of Strasbourg (C.1488-1534), Leonhard Fuchs (1501-66) and John Gerard (1542-1612). In China, some 2,500 years ago, Confucius commended orchids for their wondrous beauty and scent in the first book on the cultivation of orchids. It contains descriptions of species and varieties and was probably written in Chinese in around the year A.D. 1000.

In ancient times, early Aztec people used to flavor a traditional drink made from cocoa beans and the fruit of the tropical climbing orchid genus *Vanilla*.

In the beginning of the early 19th century, the sudden surge of interest in the collection of tropical orchids led to the plants frequently been featured in literature, fiction and as well as in science.

Medicinal Orchids in the Past: 0001-1579 A.D.

- ☆ 0040 AD: A Greek physician commented on the usefulness of orchids as a determinant of the sex of the offspring. Dioscorides – (0040-0090) –his "De Materia Medica"; possibly the 1st written evidence of orchids outside the Orient.
- ☆ 1368/1644: Many references are available to indicate the use of orchids as medicinal herbs: *Cymbidium* used for rheumatism, neuralgia and veneral diseases, *Gomsea* for stomach upsets, *Crispa* for diarrhea, fever, boils, abcess. Vanilla was used for flavouring. Most of these data were derived from the Ming Dynasty.
- ☆ 1427: The fruits of *Vanilla planifolia* were used to flavour "Chocolat" a Mexican drink made from cacao and honey.
- ☆ 1488-1534: Orchids were referred to as herbal remedies by Otto Brunfels of Strasbourg.
- ☆ 1501-66: The usage of the orchids as medicinal plants was demonstrated by a renowned physician–Leonard Fuchs.
- ☆ 1542-1612: The orchid extracts were added in medicine by the famous physician John Gerard.
- ☆ 1552: The Aztec herbal had the illustrations of *Vanilla* – "The Badianus MS".
- ☆ 1552: The first American book on medicinal plants "Badiano Codex" by Martin de la Cruz – Aztec physician, referred to *Vanilla planifolia* as a useful herb for the treatment of hysteria, fevers, impotence, rheumatism, and to increase the energy of muscular systems; *Encyclia citrina* used by natives on infected wounds, coughs, sunstroke, and dysentery. Some other medicinal orchids reported were *Laelia autumnalis, Stanhopea hernandezii, Arpophyllum spicatum, Bletia catenulate, Cranichis speciosa, Epidendrum pastoris* and *Ophrys apifera.*
- ☆ 1579:– The power of the species *Orchis odorata* was described by Willium Langham in his "Garden of Health".

Distribution

The geographical distribution is not that regular around the world. There are a few genera which are represented in almost every continent but, in general, they have a restricted distribution.

As popular belief, orchids are not fed by the host, they only use the host to fix themselves well. In general, they are epiphytic, living on the trees or on the rocks (in tropical zones) or rock soil as lithophytes or terrestrial (temperate zones). Few orchids also grow as myco-heterotropic which are achlorophyllous and live on ectomycorrhizal symbiosis totally depending on soil fungi. The maximum density of orchids is seen in tropical and sub-tropical regions of the world.

Extremely diversified plants, orchids are found all around the world except in the deserts so it is always possible to find one species which thrive in the conditions provided to it.

Although they can be found at the sea-level to 4,000m high, they are more numerous between 500 and 2,000m above the sea.

Some orchid species are among the world's smallest flowering plant like *Eurystyles* or many *Pleurothallidinae* plants, others can reach 4m height like *Selenipedium* (growing in Tropical America) or *Grammatophylum papuanum* (Papua- New Guinea).

The total number of orchid genera is calculated about 750 and the number of species to be 30,000 whereas the number of orchid hybrids already registered is found to be approximately 1,20,000 all over the world.

Bulbophylum is the richest genus in number of species and it is supposed to have 2,000 spcies under it. The second richest is *Pleurothallis* with more than 1,500 species attributed to this genus and many species yet to be described. The list is then followed by *Dendrobium* with 1,400 and finally *Epidendrum* with more than 800 species. Nowadays, many species of *Bulbophylum* are described as *Cirrhopetalum* and that of *Epidendrum* as *Encyclia* or *Anacheilium*.

The generalized and restricted orchid distribution can be summarized as follows:

- *Bulbophylum*: North, South and Central America, Africa, Asia and Australia.
- *Cattleya, Epidendrum, Lellia, Masdevallia, Maxilaria, Odontoglossum, Oncidium, Pleurothallis* : North (Mexico), South and Central America.
- *Coelogyne*: Asia.
- *Cymbidium*: Africa, Asia and Australia.
- *Cypripedium*: North America, Europe, Asia and Australia.
- *Dendrobium*: Tropical and sub-tropical Asia and Australia.
- *Habenaria*: North, South and Central America, Europe, Africa, Asia and Australia.
- *Paphiopedilium*: Asia and Australia.
- *Phalaenopsis, Vanda, Renanthera*: Asia.

- *Vanilla:* North, South and Central America, Equatorial Africa and Southeast Asia.
- In tropical America, there are more than 8,000 species and in tropical Asia more than 7,000.
- In Brazil, there are more than 2,500 species and 1/3 of them are found in Rio de Janeiro state.
- In Columbia, more than 3,000 species.
- In Venezuela, more than 1,500 species.
- In Mexico, more than 660 species.

Presently, most of the species are widely spread all over the world since they are so well adapted and they can thrive everywhere as a native genus. *Angreacum, Cymbidium, Dendrobium, Paphiopedilium, Phaius, Phalaenopsis, Vanda, Renanthera* and their relatives, *Cattleya* and *Oncidium* and their relatives can be cultivated in every continent based on the weather and the green house conditions of the place of cultivation.

Orchids of India

In Indian subcontinent the orchid wealth is distributed in two main hot spots such as the North-Eastern Himalayas and the Western Ghats. About 1220 species are reported from India, while 750 species are found in North East India and about 550 species (45 per cent of the country) are reported from Arunachal Pradesh alone. The Western Ghats comprise of 245 species of orchids distributed in 75 genera. Owing to the fact that the landmass of Western Ghats is far older than the Himalayan landmass, the biodiversity in Western Ghats is less compared to the Himalayan region and such biodiversity is also restricted to certain specialized niches of Western Ghats. In India, orchids are employed for a variety of therapeutic use in different systems of medicines are attracting attention of many modern workers. Traditional therapeutic uses of numerous orchids are mentioned and their phytochemical constituents including alkaloids, flavonoids, terpenes, carbohydrates and glycosides are reviewed in the past. There have been different programs to help promoting orchid cultivation, making it a 'People's' movement, partly conserving our rich national heritage and partly, improving the economy by exploiting the potential of our native orchids.

Different orchids are used in the treatment of boils and pimples, disorders of bladder and kidneys, gonorrhea, burns, diarrhea and dysentery, stones in urinary tract, muscle wasting and blood disorders.

The several medicinal qualities of the orchids of North-Eastern Himalayas and the Western Ghats are as follows:

Medicinal Orchids of North-Eastern Himalayas

1. *Acampe papillosa*–Root is employed under the name Rasna, used for treating rheumatism, sciatica, neuralgia, syphilis and uterine diseases.

2. *Aerides odorata*–The ground fruit used for healing wounds. Juice of leaves to heal boils in ears and nose.
3. *Anoectochilus regalis*–Stem and leaves are one of the ingredients in certain medicinal oils.
4. *Calanthe sylvatica*–Flowers used to stop nose bleeding.
5. *Calanthe triplicata*–Roots are ingredient of local medicine to treat swollen hands; with other ingredients, roots chewed for diarrhea, flowers as a painkiller in caries, pseudo bulbs as a masticatory, gastrointestinal disorder.
6. *Cephalanthera longifolia*–Roots and rhizome as a tonic.
7. *Corymborkis veratrifolia*–Fresh leaf juice as an emetic to cure fever in children.
8. *Crepedium acuminatum*–Bulbous stem used in drug for strength, enhances sperm formation and prevents diseases borne by vaata, pitta and kapha.
9. *Cremastra appendiculata*–Roots paste used for toothache and as emollient. Tuber used for abscesses, scrofula, and freckles and as an antidote to snakebite.
10. *Cymbidium aloifolium*–Pounding the plant with ginger and extracting the mixture with water is used to reduce vomiting and diarrhoea, to cure chronic illness, weakness of the eyes, vertigo and paralysis.Used as an ingredient of oil to cure benign and malignant tumors.
11. *Dendrobium nobile*- Fresh and dried stems used in preparation of Chinese drug Shih-hu for longevity and as an aphrodisiac, stomachic and analgesic.
12. *Dendrobium jenkinsii*- Fresh and dried stems used in preparation of Chinese drug Shih-hu.
13. *Dienia muscifera*–Decoction of tuber used as tonic to strengthen kidneys.
14. *Echioglossum williamsoni*–Leaf juice applied by Monpa tribe to cure swellings of hands and legs and also to cure bone fractures.
15. *Eria pannea*–Roots and leaves decoction used in bathing in case of ague.
16. *Goodyera repens*–Roots and leaves for female disorders, stomach and bladder diseases, chewed leaves applied to reptile bites. Mashed leaves prevent rash in infants.
17. *Goodyera schlectandaliana*–Tincture of the plant in rice wine is used as a tonic for internal injuries and to improve circulation.
18. *Gymnadenia orchidis*- Salep made of tubers used as aphrodisiac, tonic.
19. *Phaius tancarvilleae*- Paste of pseudo bulbs is used to heal swellings of hands and legs, to soothe pain in abscess.
20. *Pholidota chinensis*–Aqueous extract of pseudo bulbs is taken for scrofula, feverish stomach and tooth ache. Tincture is used for internal bleeding, hemorrhage, asthmatic cough, tuberculosis and dysentery.
21. *Pholidota imbricata*–Pseudo bulbs finely macerated in mustard oil and applied on joints for rheumatic pains.
22. *Pleione maculata*- Pseudo bulbs used in liver complaints and stomach ache.

23. *Ponerorchis chusua*–Tubers used for diarrhea, dysentery and chronic fever.
24. *Rhynchostylis retusa*- Roots (Rasna) for rheumatism. Plant used against asthma, tuberculosis, nervous twitchings, cramps, infantile epilepsy, vertigo, palpitation, kidney stone and menstrual disorders.
25. *Satyrium nepalense*- Tubers eaten by Monpa tribe for malariam dysentery also as aphrodisiac.
26. *Spathoglottis plicata*–Decoction of the boiled plant used for rheumatism and used in hot as foment.
27. *Tropidia curculioides*–Decoction of the plant given for cold stage of malaria, decoction of boiled roots given for diarrhoea.
28. *Vanda corulea*–Leaf juice for diarrhoea, dysentery, external application for skin diseases.
29. *Arundina graminifolia*- Scrapped bulbous stem applied on heels to treat cracks.
30. *Dendrobium densiflorum*–Leaves crushed to paste with salt and applied on fractured area to set bone.
31. *Cymbidium hookerianum*–Seeds applied on cuts and injuries as hemostatic.

Medicinal Orchids of Western Ghats

1. *Aerides crispum*–Whole plants are powdered, boiled in neem oil and filtered for use in case of ear ache.
2. *Acampae praemorsa*–Roots used for treating rheumatism.
3. *Cymbidium aloifolium*–Used as emetic and purgative by locals in the form of salep.
4. *Coelogyne ovalis*–Aptly called as Jeevanti and whole plant is used for cough, urinary infections and eye disorders.
5. *Limodorum sapthulatum*–Flowers used to cure asthma.
6. *Habenaria* species–Used in Siddha and Vriddhi tonics to treat unconsciousness and as blood purifiers.
7. *Dendrobium ovatum*–Juice of hand crushed stems are used for treating constipation and stomach ache.
8. *Eulophia nuda*–Used as an appetizer and also for treating tumours and bronchitis.
9. *Flickingeria nodosa*–Used to treat asthma, bronchitis and throat infections.
10. *Malaxis rheedi*- Used as tonic and is also a constituent of the famous Ayurvedic drug called Ashtavarga.
11. *Zeuxine strateaumatica*–Tubers are used as tonic in combination with the roots of *Cymbidium aloifolium*.
12. *Nervilia aragoana*–Helps to cure eye infections.
13. *Luisia zeylanica*–Stems are used as emollient for boil, abscess and burns.

14. *Vanda testacea*–Roots used in dyspepsia, bronchitis, inflammation and hiccough. Under Unani system, roots are being used as tonic to the liver and brain and to lessen the inflammation and heal minor fractures.
15. *Vanda tessellata*–Roots used in dyspepsia, bronchitis, inflammation and hiccough. Under Unani system, roots are being used as tonic to the liver and brain and to lessen the inflammation and heal minor fractures.

Orchids are quite rich in their medicinal value and hence have been harvested indiscriminately. The need of the hour is to create awareness and a conservation strategy to save these wonder plants. The orchids of the North Eastern Himalayas and Western Ghats are having tremendous potential to uplift the economy of the entire country and sustainable development. On the contrary, losing out to human greed with wanton destruction of forests, the orchids of India are struggling to survive. Degradation of the environment over the years has resulted in disappearance of many species. The orchids also serve as ecological indicators and consequently their disappearance indicates a change in the quality of soil and in the climate which is a matter of concern for the entire nation.

Conclusion

Today, more and more people take plant medicine as an alternative therapy. The resurgence in Western herbal remedies mainly results from the lack of side effects, its holistic emphasis, respect for the individual and emphasis on self help. In addition to these factors, the economic advantages also contribute to this resurgence. The role of plants in medicine is expanding beyond their traditional and continuing role as a pharmacopoeia. The basic similarity of all life chemistry has inspired the use of plants as manufacturers for tumor-attacking monoclonal antibodies and other biopharmaceuticals. Genes coding for clinically important proteins are engineered and inserted into the plant cells, which can be programmed to take up and express DNA. Large amounts of these therapeutically significant proteins can be harvested and extracted from the different parts of the adult plants. Investigations into traditional plant medicine conducted with modern theories and techniques can enrich Western medicine by absorbing new ideas and concepts from traditional plant medicine from all over the world. Traditional plant medicine will become an area of ever- increasing importance in the health- care system in the future.

The medicinal and aesthetic attributes of the Orchids have resulted in indiscriminate and unscrupulous harvesting of these Divine species. Moreover, the constant deterioration of the climatic condition over the years, can also be blamed for untimely disappearance of many species. The orchids also serve as ecological indicators. Their disappearance indicates a change in the quality of soil and in the climate. The need of the hour is to create awareness and a conservation strategy to save these wonder plants.

To conclude, one must say that nature has been extremely kind in endowing this world with colorful and spectacular orchids. These natural jewels are commercially important not only for their ornamental value, but also for their wide and extensive medicinal uses. Realizing the precarious situation in all the orchid habitats within the country and in the entire world, efforts are being carried out for creating sufficient awareness amongst native people. At last the time has arrived when the economically backward section of the society are taking up orchid cultivation as an alternative source of their income.

References

Anon, Datamonitor (Trade source), UK, 1996.

Cragg G.M., Newmann D.J. and Snader K.M. 1997. Natural products in drug discovery and development. J Nat Prod. 60: 52-60.

Dhananjaya S.P. 2009. Conservation and *in vitro* multiplication of important orchids of Western Ghats. Thesis, Bangalore University, Bangalore. Education, Research and Training (NCERT), Bangalore.

Hooker J.D. 1894. The Flora of British India.Ashford:L.Reeve & Co., Ltd., 6:1-793.

Joelle Q.L., Ameenah G.F. and Anwar H.S. 2006. Screening of anti-microbial and medicinal importance. Bioresource Technology 98(16): 3037-3043.

Kong J.M, Goh N.K., Chia L.S. and Chia T.F. 2003. Recent advances in traditional plant drugs and orchids. Acta Pharmacol Sin. 24(1): 7-21.

Oumadevi R., Guy R., Francisco E.R., Kiban C., Suzanne U.R., Rao A.T and Sridhar S. 2007. Wild orchids in Karnataka- A Pictorial.

Rao N.A. 2004. Indian Medicinal Plants of conservation concern. News Letter of Envis Node on Indian Medicinal Plant. 1(2).

Chapter 6

Diversity in Green Medicine of Tirupati–Kadapa-Nallamalai Hotspot

N. Savithramma[1], P. Ganesh[1], P. Venkateswarlu[1] and S.K.M. Basha[2]

[1]*Department of Botany, Sri Venkateswara University, Tirupati – 517 502*
[2]*N.B.K.R. Research Centre, Vidhya Nagar – 524 413*

Eastern and Western Ghats are identified as the richest natural reserves in India. The Seshachalam hill range is the southern part of Eastern Ghats located in Kadapa and Chittoor Districts of Andhra Pradesh. The hills are with deep and shallow ravines and massive peaks, which support perennial and temporary water falls. Majority of the hilly area is composed of granite complex. The climatic factors are favourable for the distribution of unique endemic plants. The environment is very much suitable for growth and development of important medicinal plants. Medicinal plants have been widely used in traditional forms of medicine in India several generations. Medicine is an important fundamental branch of science and being as an art in the treatment of disease. (Savithramma *et al.*, 2002). Nagaraju and Rao (1990) and Savithramma and Sulochana (1998) made an attempt to enlist locally rare, endangered and endemic medicinal plants of Tirumala hills.

India is bestowed with a variety of agroclimatic conditions ranging from tropical – subtropical to subtemperate – temperate. Ten biogeographic zones of India enthral botanists with the diversity and endemism of flora (Rana and Ranade 2009). Consequently, our country is one of the 12 mega biodiversity rich centres of the world besides housing two of the 25 recognized biodiversity hotspots (Myers *et al.* 2000).

Boswellia ovalifoliolata

Bridelia retusa

Capparis sepiaria

Carica papaya

Cochlospermum religiosum

Commiphora caudata

Deccania pubescens

Datura metel

Gardenia gummifera

Gelonium angustifolium

Gmelina arborea

Holarrhena antidysenterica

Madhuca indica

Melhania incana

Memecylon umbellatum

Oncoba spinosa

Ruellia prostrata

Schefflera stellata

Scutia myrtina

Sesamum indicum

Shorea talura

Sida acuta

Sida cardifolia

Shorea tumbuggaia

Talinum cuneifolium

There are 47000 plant species in India including 18000 angiosperms which form a major part of the flora (Karthikeyan and Mudgal 1995). Diversity is unmatched due to the presence of 16 different agroclimatic zones, 10 vegetative zones and 15 biotic provinces. About 1500 plants with medicinal uses are mentioned in ancient texts and around 800 plants have been used in traditional medicine (Kamboj 2000).

Nevertheless, 30 per cent of the angiosperms and 18 per cent of the total flora are endemic to India (Nayar 1977). According to the IUCN report (2007), 45 tree species from India have been included in the *Red list of Threatened Plant Species,* and declared as "Critically endangered" facing high risk of extinction while 247 tree species have been declared 'threatened' (Anon 2007). Around one third of flowering plants are endemic and nearly 1000 species are endangered (Pullaiah, 1998). Alongside the widespread use of medicinal plants in non – commercial, semi – commercial and commercial ways, it is estimated that "wild populations" of around 1000 species of India's medicinal plants are under threat due to habitat loss and in specific cases due to over exploitation. Medicinal Plants ought to be given the status of a "national resource" because their continued availability is essential to sustain one of the world's oldest medical traditions, a priceless legacy of the India people.

Table 6.1

Sl.No.	*Scientific Name, Family and Telugu Name*	*Part Used*	*Used for Curing Disease*
1.	*Boswellia ovalifoliolata* Bal. & Henry (Burseraceae) Konda sambrani.	Stem Leaf Gum	Stomach ulcers, diabetes. Mouth and throat ulcers. Dysentery, inflammations, joint pains, ulcers, arthritis, hydrocele, diabetes.
2.	*Bridelia retusa* (L) Spr. (Euphorbiaceae) Verri karaka.	Root & leaf Root & bark Fruit	Anti fungal. Hepatoprotective. Sciatica, lumbago.
3.	*Capparis sepiaria* L. (Capparidaceae) Nalla uppi, China uppili.	Whole plant Bark Leaf	Febrifuge, tonic, skin troubles. Rheumatic pains, herpes virus infections. Gonorrhoea.
4.	*Carica papaya* L. (Caricaceae) Boppaya, Boppasa.	Root Ripe fruit Unripe fruit Seed	Yaws, piles, generative tonic. Stomachic, digestive, carminative, diuretic. Anthelmintic, freckles. Vermifuge, thirst.
5.	*Cochlospermum religiosum* (L) Alston. (Cochlospermaceae) Konda gogu.	Gum Young leaves	Sweetish, cooling, sedative, cough, demulcent, astringent. Cooling wash for the hairs.
6.	*Commiphora caudata* Engl. (Wt. & Arn.) (Burseraceae) Konda mamidi, Vatapatri.	Wood Resin Fruit	Body pains. Fissures of feet to relieve pain and swelling. Tannins, gums used as pickle.
7.	*Datura metel* L. (Solananceae) Nalla ummetta.	Whole plant	Anti-spasmodic, wounds, narcotic, scrofula, skin diseases, psychosis.
8.	*Deccania pubescens* (Roth) Tirven. (Rubiaceae) Pedda manga, Yerri bikki.	Whole plant	Sores.

Contd...

Table 6.1–Contd...

Sl.No.	Scientific Name, Family and Telugu Name	Part Used	Used for Curing Disease
9.	*Gardenia gummifera* L. f. (Rubiaceae) Bikki, Manchi bikki, Tella manga.	Resin	Anti spasmodic, expectorant, carminative, diaphoretic, anthelmintic, skin diseases, liver disorders.
10.	*Gelonium angustifolium* Baill ex Muell. - Arg. (Euphorbiaceae) Suragada.	Leaf Seed	Ankle pains. Purgative, reduce acidity, leprosy.
11.	*Gmelina arborea* Roxb. (Verbenaceae) Gumardi teku.	Root Leaf Fruit Plant	Stomachic, fever, indigestion, anasarca, increases the secretion of milk. Gonorrhoea, cough, ulcers. Diabetis, leucorrhoea. Snake bite, scorpion sting.
12.	*Holarrhena antidysenterica* (Roxb.ex Fleming) Wall. (Apocynaceae) Kolamuk, pala, Kodaga.	Bark Seed Leaf Root & Leaf	Anti dysentric, astringent, anthelmintic, stomachic, febrifugal, diarrhoea, dropsy, tooth-ache. Erysipelas. Chronic bronchitis, boils, ulcers. Haemorrhagea, bleeding from the nose.
13.	*Madhuca indica* Roxb. (Sapotaceae) Ippa.	Root & bark Seed	Eczema, diabetes, ulcers. Laxative, skin affections, rheumatism.
14.	*Melhania incana* Heyne ex Wt. & Arn. (Sterculiaceae)	Whole Plant	Burns, boils.
15.	*Memecylon umbellatum* burm. f. (Melastomaceae) Kukkalli.	Leaf Root & bark	Leucorrhoea, gonorrhoea, eye troubles. Polyuria, menorrhagia, dysentery.
16.	*Oncoba spinosa* Forsk. (Flacourtiaceae) Line gulabi.	Root	Bladder disorders, dysentery.
17.	*Ruellia prostrata* Poir. (Acanthaceae) Nela nalambramu.	Leaves	Gonorrhoea, ear diseases.
18.	*Schefflera stellata* (Gaertn.) Harms. (Araliaceae) Marrimamidi.	Leaf Berry	Sores, diaphoretic, cathartic, ulcers, lice killer. Rheumatism, abscesses.
19.	*Scutia myrtina* (Burm. f.) Kurz. (Rhamnaceae) Gariki.	Fruit Leaf	Astringent and eaten. Parturition.
20.	*Sesamum indicum* L. (Pedaliaceae) Manchi nuvvulu.	Leaf Seed Root	Dysmenorrhoea, abscess, ulcers. Emollient, diuretic, lactogogue, piles, emmenagogue, aphrodisiac. Growth of the hairs and make it black.
21.	*Shorea talura* Roxb. (Dipterocarpaceae) Chinna jalari.	Resin	Rheumatism, stimulant, astringent.

Contd...

Table 6.1–Contd...

Sl.No.	Scientific Name, Family and Telugu Name	Part Used	Used for Curing Disease
22.	*Shorea tumbuggaia* Roxb. (Dipterocarpaceae) Tumba jalari.	Resin	Duodenal ulcers, stimulant.
23.	*Sida acuta* Burm. f. (Malvaceae) Mutuvapulagam, Neelabenda.	Leaf	Ulcers, diuretic, testicular swellings, elephantiasis, abortifacient, piles, chest pain, anthelmintic.
		Root	Astringent, colling, tonic, stomachic, diaphoretic, anti-pyretic, nervous and urinary diseases, wounds, ulcers, vermifuge.
24.	*Sida cordifolia* L. (Malvaceae) Chirubenda, Tella gorra.	Plant	Elephantiasis.
		Leaf	Demulcent, fever, febrifuge, dysentery, cardiotonic, gonorrhoea, haematuria, hemiplegia.
		Root	Leucorrhoea, urinary troubles, diabetes.
		Seed	Demulcent, laxative, piles, colic, tenesmus.
25.	*Talinum cuneifolium* Willd. (Portulacaceae) Thota pavall.	Leaf	Diabetes, mouth ulcers.
26.	*Tephrosia purpurea* (L.) Pers. (Fabaceae) Vempali.	Herb	Tonic, laxative, diuretic, deobstruent, bronchitis, boils, pimples, bleeding piles, cough, kidney disorders.
		Leaf	Jaundice.
		Root	Dyspepsia, diarrhoea, rheumatism, asthma, urinary disorders, leprous wounds, skin eruptions, elephantiasis, cough.
27	*Terminalia chebula* Retz. (Combretaceae) Nalla karaka.	Fruit	Laxative, stomachic, dentifrice to cure bleeding, ulceration of gums, asthma, dullness of the cattle.
28	*Terminalia paniculata* roth. (Combretaceae) Neemari	Bark	Diuretic, cardiotonic, parotitis, diabetes.
29	*Vernonia cinerea* (L.) Less. (Asteraceae) Sahadevi.	Whole Plant	Skin diseases, leucorrhoea, vomiting, coryza.
		Root	Anthelmintic, blood disorders, fever, eye disease.
30	*Ziziphus jujuba* (L.) Gaertn. (Rhamnaceae) Regu, Bhogi pallu.	Bark	Powerfully astringent.
		Leaf	Cooling, diaphoretic, hair tonic.
		Fruit	Abdominal pains, sedative.
		Seed	Diarrhoea.

75-80 per cent of the world population, mainly in the developing countries, depend on traditional medicine for primary health care because of better cultural acceptability, better compatibility with the human body and lesser side effects in the developed world. In Germany and France, many herbs and herbal extracts are used as prescription drugs and their sales in the countries of European Union were around $ 6 billion in 1991 and may be over $ 20 billion now. In USA, herbal drugs are currently sold in health food stores with a turnover of about $ 4 billion in 1996 which is anticipated to double by the turn of the century (Rawls 1996). In India, the herbal drug market is about $ one billion and the export of plant based crude drugs is around $ 80 million (Biotech Consortium India Ltd., New Delhi 1996). Current market is estimated at about $ 80-250 billion in USA and also in Europe (Brower 1998).

The World Health Organization (WHO) has recently defined traditional medicine (including herbal drugs) as comprising therapeutic practices that have been in existence, often for hundreds of years, before the development and spread of modern medicine and are still in use today (WHO 1991).

The World Health Organization also has recognized the importance of traditional medicine and has been active in creating strategies, guidelines and standards for botanical medicine (WHO 2002).

Traditional knowledge will serve as a powerful search engine and most importantly, will greatly facilitate international, focused and safe natural products research to rediscover the drug discovery process. 30 per cent of the world wide sales of drugs is based on natural products (Bhushan *et al.* 2004).

The most fundamental cure was associated with medicinal plants. That's why bioprospecting of medicinal plants for novel medicines and agrochemical values are under way throughout the world. Botanically and Economically interesting medicinal plants and their uses have been mentioned here.

References

Anon, 2007. *IUCN Red List*, www.iucnredlist.org.

Bhushan Patwardhan, Ashok, D.B.Vaidya and Mukund Chorghade, 2004. Ayurveda and natural products drug discovery, Curr. Sci. 86 (6): 789-799.

Biotech Consortium India Ltd., New Delhi, 1990. Secondary study on Indian Medicinal Plants – status, perspective and strategy for growth.

Brower, V.1998. Nat.Biotechnol 16: 728 – 731.

Kamboj, V.P., 2000. Herbal Medicine, Cur. Sci., 78 (1): 35-39.

Karthikeyan, S. and Mudgal, V., 1995. In Bharat Ki Vnaspati Vividhata, (eds Mudgal, V. and Hajra, P.K.), Botanical Survey of India, Culcutta; 301-318.

Myers, N., Mittermeier, R.A., Mittermeier, C.G., da Fenseca, G.A.B. and Kent, J., 2000. Nature 403 : 853-858.

Nagaraju, N. and Rao, K.N. 1990. Locally scarce and endangered medicinal plants from Tirumala Hills, Chittoor district of A.P. Vegetos 3 (1): 108-122.

Nayar, M.P. 1977. Bull. Bot. Surv. India, 19 : 145-155.

Pullaiah, T., 1998. Taxonomy of Angiosperms. Regency Publication, New Delhi.

Rana, T.S. and Ranade, S.A. 2009. Curr. Sci. 96 (2): 219-229.

Rawls, R., C & EN, Washington, 23 September 1996, 53-60.

Savithramma, N., Rao, K.N. and Ramamurthy, N. 2002. Phyto-Diversity, Monitoring and Conservation, Biodiversity: 29-46.

Savithramma, N. and Sulochana Ch. 1998. Endemic medicinal plants from Tirumala Hills. Andhra Pradesh. India. *Fitoterapia* 69 (3); 253-254.

Thammanna and Narayana Rao, K. 1990. Medicinal Plants of Tirumala.

WHO. Geneva, 1991 in Progress Report by the Director General, Document No. A44/ 20, 22 March, 1991.

WHO, Geneva, 2002. Traditional Medicine Strategy – 2002 – 2005.

Chapter 7

Ethnomedicinal Practices Among Tribals and Local Herbal Healers of Eswaramala Forest of Anantapur District of Andhra Pradesh

B. Thirupal Reddy[1], *B. Aswarthappa*[2], *D. Alimoulali*[3] *and T. Pullaiah*[4]

[1]*Department of Botany, Government Degree College, Pakala, Chittor Dist.*
[2]*Department if Botany, Government Degree College, Penukonda, Anantapur Dist.*
[3]*Department of Botany, Government Degree College, Anantapur – 515 001*
[4]*Departmnt of Botany, Sri Krishnadevaraya University, Anantapur – 515 003*

Introduction

From ancient times, plants have been used as sources of medicine by tribals. They have been used both in the prevention and cure of various diseases of humans and their animals. With the advent of human civilization, many systems of therapy have been developed primarily based on plants. Ayurveda, Homeopathy, Sidda, Unani, etc. are our traditional systems of medicines. The plant-based traditional medical systems continue to provide the primary health care to more than three-quarters of the world's populace. The World Health Organisation has estimated that over 80 per cent of the global population rely chiefly on traditional medicine (Akerele, 1992).

Indigenous herbal treatment is a part of the culture and dominant mode of therapy in most of the developing countries. These traditional phytoremedies, with a considerable extent of effectiveness, are socially accepted, economically viable and mostly are the only available means. Still, one-third of the modern pharmaceutical preparations have botanical origin. International trade on medicinal plants is, therefore, increasing rapidly mainly as result of intensified adoption of crude extracts for self-medication by the general public in the developed countries. In India, the use of plants for medicinal treatment dates back to 5000 years. It was officially recognized that 2500 plant species have medicinal value while over 6000 plants are estimated to be explored in traditional, folk and herbal medicine (Huxley, 1984).

The indigenous people of south India relied for centuries on herbal medicine for all aspects of their primary health care. It is estimated that 7.56 million people belonging to about 20 different communities of South India still use traditional remedies derived from plant based origin (Ravikumar and Ved, 2000).

The tribals of Andhra Pradesh use locally available herbs for a variety of ailments. They have 'Chekka Mandu'- a medicine made of various barks to protect the child and 'Talli mandu' (Mother's medicine) to save the mother from all the diseases connected with child birth (Hemadri, 2003). They have herbs to improve appetite of the mother, galactogogues for increasing breast milk, anthelmintics to kill intestinal worms and antidiarrhoeals to save parts for healing injuries, bone-setting, recipes for controlling fevers, leucorrhoea, menorrhagia, dysmenorrhoea, dysentery and tonics and aphrodisiacs for enjoying life and sex. Their crude pills, decoctions and extracts alleviate body ache, joint pains, heal skin diseases, cure anaemic and dropsical conditions.

They have herbal cures for jaundice, diabetes, paralysis, abortifacient and fertility. They make use of several herbal smokes to "ward off evil spirits", presently the subject of psychiatrists. Thus, there is no dearth of tribal medicine to take care of primary health, particularly in rural areas. In the present work quite a good number of plant extracts used by tribal people and Natuvydyas for prevention and cure of various diseases and manifestations which generally trouble the mother, child and the youth, have been given together with mode of preparation and administration with prescribed doses.

Ethnobotany in Andhra Pradesh

The first recorded study on Ethnobotany in Andhra Pradesh was that of Pal and Banerjee (1971) who reported some less known plants used by tribals, which has triggered the attention of many workers in ethnobotanical studies of Andhra Pradesh. Jain *et al.* (1973) recorded 32 medicinal plants used by adivasis in Andhra Pradesh and Orissa. Harasreeramulu (1980) reported the medicinal importance of 40 wild species used by the tribals of Srikakulam district. Nisteswar and Kumar (1980) published a critical note on the utilitarian value of medical-lore of Rampa agency, while Sudhakar (1980) mentioned the medicinal value of 30 plant species used by Konda Reddis and Koyas of East Godavari District. Reddy (1980) recorded 34 medicinally important plants used by the aboriginals of West Godavari district. Venkaiah (1980) enumerated 35 medicinal plants used by the local Girijans and adivasis of Visakhapatnam district, while Hemadri (1981) reported 22 folklore medicines on rheumatism. Singh *et al.* (1981) published 29 medicinal plants used by the local tribes of Mannanur and also studied on the biological activity of these plants. Ellis (1982) reported wild plant resources of Nallamalais in the Eastern Ghats. Hemadri *et al.* (1987) reported 211 species of medicinal plants from the state. Nagaraju and Rao (1990) conducted a survey on plant crude drugs of Rayalaseema. Aruneekumar and Nisteswar (1990) recorded 188 medicinal plant species of Kakinada region, Vedavathy *et al.* (1991) reported 25 plants species from Rayalaseema used for family planning and birth control. Ravishankar and Henry (1992) reported 30 potential plant species from Adilabad district.

Reddy (1995) conducted ethnobotanical and phytochemical studies on medicinal plants of Cuddapah district. Goud and Pullaiah (1996) reported 40 plants used in ethno-veterinary medicine by the Chenchus, Sugalis and Erukala tribes. Vedavathy and Mrudula (1996) examined the traditional medicines practiced by the Yanadi tribes of Andhra Pradesh. Kumar and Pullaiah (1998) conducted ethno-medico-botanical studies in Mahaboobnagar district. Sudhakar (1998) reported medicinal importance of 100 angiospermic weeds of Chittoor district. Venkata Raju and Reddy (1998) examined medicinal properties of certain rare and interesting plants from Kadapa hills. Six endemic medicinal plants used by the tribal inhabitants of Tirumala hills of Chittoor district were reported by Savithramma and Sulochana (1998). Muralidhar Rao and Pullaiah (2001) conducted ethno-medico-botanical survey of Guntur district. Reddy *et al.* (2007) published a paper on Ehnobotanical studies on medicinal plants of Seshachalam hills in Cuddapah district.

Ethnobotany of Andhra Pradesh

A review was published by Reddy during 2008. Yasodamma *et al.* (2009) wrote an article on Threat assessment (IUCN categorization) for ethnomedicinal plants used by Chencu tribe of Gundlabrahmeswaram in Nallamalai hills in Andhra Pradesh. Suneetha *et al.* (2009) reported 45 ethonomedicinal plants used for bites by the tribals in East Godavari District, Andhra Pradesh.

Ethnobotanical Studies in Anantapur District

The flora of Anantapur district has been worked out by Pullaiah and Yesoda (1989), in which 707 vascular taxa belonging to 397 genera and 105 families were reported. Later Ravi Prasad Rao *et al.* (2003), reported 96 plant taxa as additions to the flora of Anantapur district after Pullaiah and Yesoda (1989). Earlier records on the plants encountered in the district include Beddome (1868-74), Gamble (1915-36).

A thorough review of literature of the ethno-botanical studies in Andhra Pradesh, comes to a conclusion that very little attention was received on the ethnobotanical studies of Anantapur district. Reddy and Venkata Raju (1999) described about 86 plants used in Ethnoveterinary practices in Anantapur district. Jeevan Ram *et al.*

(2002) reported 48 medicinal plants used by the sugali tribes of Gooty forests of Anantapur district, A.P. Aswarthappa *et al.* (2004) carried out preliminary survey of certain ethnomedicinal plants from the forests of Anantapur District, Andhra Pradesh. The present region of study is not surveyed extensively which possesses diversified vegetation of dry deciduous and scrub elements. More over as biotic interference is more in this area, there is a constant threat to the biodiversity. So, the present investigation was undertaken in this area, to study the diversified medicinal plants.

So, it has become the need of the hour to collect, preserve and document valuable information on Medicinal plants from the forests of Eswaramala hills of Anantapur district, Andhra Pradesh, for future use, before their depletion.

Reasons for Undertaking the Present Study

The traditional knowledge system in India is fast eroding due to steady decline in human expertise capable of recognizing various medicinal plants. Much of this wealth of knowledge is totally becoming lost as traditional culture is gradually disappearing because it is mostly oral (Hamilton, 1995). Other reasons for the gradual loss of this traditional knowledge are adaptation of global products by the younger people, extinction of species, urbanization and destruction of habitat, breakdown in traditional structure and certain natural causes like famine, flood, wars *etc*. Therefore, required documentation and computerization is very necessary (Mehrotra and Mehrotra, 2005). There is urgent need to inventorise and record all ethnobotanical information among the diverse ethnic communities before the traditional cultures are completely lost.

There is an increasing demand for Ayurvedic Medicine and tribal medicine by a large section of people because there are no side effects when compared with synthetic drugs. The efficiency of these systems of medicine has been realized especially for certain ailments for which allopathy was not so much effective. The successful introduction of various drugs of plant origin in recent years as well as the continuing efforts of pharmaceutical industry in search of natural source of new drugs from all plants of the globe furnish evidence for the increasing significance being attached to medicines from the plant kingdom. In India despite its rich traditional knowledge and large biodiversity, the marketable drugs are very few. The reason for this is that there are constraints in developing them into successful drugs to test their efficacy and safety through pharmacological and toxicological evaluation.

The area under present study *viz.*, the forests of Eswaramala in Anantapur district are fast dwindling due to biotic interference and scanty rainfall. The major factors like indiscriminate cutting of the forest area for agriculture purposes, industrialization, over grazing, forest fires *etc.*, are responsible for the loss of some rare medicinal plants in this area. In future, degradation of the area will hasten up due to steep increase in human population and live stock along with man's avaricious activities. As most of the tribal people are migrating to urban areas, their precious knowledge of the medicinal plants may not be available to the future generations. In view of the present circumstances the urgent need of the hour is to collect, document and preserve some valuable information on certain medicinal plants from this area before they are lost permanently. The conservation of this plant wealth has to be made available for the benefit of future generations and also for the benefit of researchers in the field of botany, chemistry and pharmacology *etc.* A thorough perusal of botanical literature reveals that less attention has been paid to the present area of study.

Objectives of the Present Investigation

The present investigation is mainly focused on documentation of medicinal properties of plants used by tribal people and local informants.

- ☆ Identification of hotspots in which the tribal inhabitation is rich with valuable information regarding the therapeutic properties of crude drugs.
- ☆ Intensive and extensive field explorations to collect the medicinal plant species from the Eswaramala forest range in consultation with tribal communities and local informants.
- ☆ Taxonomic analysis and systematic evaluation of drug yielding plants used for both human health and care purposes along with the methods of preparation of drugs, disease and mode of administration.
- ☆ Preservation of medicinal plant species in the form of herbarium.
- ☆ Creating environmental awareness among tribals and local people about the precious value of plants and making an endeavour for the declaration of some areas as Medicinal Plants Conservation Areas.

Anantapur district situated in south-western corner of Andhra Pradesh, lies between 13°14′–15°14′ N, 76°47′–78°26′ E, covering an area of 19,130 sq.km. Eswaramala forest covers on area of 6329.42 ha of reserve forest with vast biodiversity. This area mainly consists of sedimentary and archaen group of rocks consisting of granites. The soils are predominantly black cotton, red and gravelly and the climate is tropical semiarid with more sunny days. The average maximum and minimum temperatures are 38.5°C and 16.8°C respectively and the average rainfall is 53.8cm. Pennar is the major river in the area. The forests are dry tropical south Indian mixed deciduous and southern cutch thorn types (Champion and Seth, 1968).

Present Area of Study

Eswaramala forest range possesses the following villages *viz.,* 1. Dhaniancheruvu, 2. Kothapalli, 3. Maddinayanipalem, 4. Edurudona, 5.Gutibailu, 6. Tsamalagondi, 7. Kalavapalli, 8. Pedaballi, 9. Govindarajupalli, 10.Kurumamidi, 11.Gollapalli, 12.Yeddulavandlapalli, 13.Gangannavaripalli 14. Reddivaripalli, 15. Bandameedipalli, 16. Maddivandlapalli.

The forests are inhabited by Sugalis, a predominant tribe in Anantapur district. The 'adivasi' tribes inhabiting the forests use a number of wild plants which are unknown to the urban elite. The present study was aimed at finding out the important medicinal plants used by Sugalis and other herbal healers (natuvaidyas) from Eswaramala forest of Anantapur district Andhra Pradesh.

Ethnology of the Tribal People

Tribals, the distinct groups, usually confined to definite geographical areas speak common dialect. They form a considerable percentage of total population in certain regions of the subcontinent. There are over 550 native or tribal communities under 227 ethnic groups covering 5000 villages in India (Anonymous, 1994).

Tribal people of India mostly live in forests with plateaus and naturally in isolated regions. The tribes are most popularly termed as Adivasi (Anusuchit Janjati – Scheduled tribe). There are altogether 360 tribal communities recognised in India, out of which 8 are predominant in Andhra Pradesh (Shashi, 1994). In and around Eswaramala Hills some tribal communities such as Sugalis, Erukalas, and Yanadis live who still practise the age-old traditional medicines and their applications. As the forests in this area are fast dwindling, the tribals are rapidly being lured into the modern society. Due to this, knowledge about tribal medicine is fast disappearing. Hence, the main aim of this study is to properly document and preserve the unwritten folklore medicinal practices before they suffer further dissipation and disappearance.

As per the Handbook of Statistics of Anantapur District published by the Chief Planning Officer in 2004-2005, the main tribal group inhabiting the district is Sugalis. Major chunk of Sugali tribes in the district exist in the mandals of Kadiri, Puttaparthi, Bukkapatnam, Gorantla and Nallamada. Nearly a total population of 1,18,321 Sugalis are spread in 433 thandas in the Anantapur district. Out of this total, nearly 9,000 Sugalis/lambadies live in Gandlapenta (2144), N.P. Kunta (1703) and Tanakal (5029) mandals which come under the present area of study in Eswaramala Hills. The Sugali settlements are called Thandas. These Thandas are seen in places like Thummalabailu thanda, Pedda thanda, Nallagutta thanda, Pathacheruvu thanda, Sugalipalli P.thanda, Kallubavi thanda, Velichalamala thanda, Pedaballi thanda etc., all present in Gandlapenta, N.P.Kunta and Tanakallu Mandals, which are under the present area of study.

Materials and Methods

Intensive explorations were conducted in the tribal pockets during 2003-2006. The information was collected through personal interviews during field trips and discussions with chieftains, tribal healers, aged men, barefoot doctors and housewives. 94 important crude drugs were found to be used for various ailments, some for more than one purpose. The information on the mode of preparation of the drugs and methods of their administration along with the doses was recorded. An attempt has been made to document possibly a correct terminology of diseases based on the symptoms in consultation with allopathic physicians. The information was cross-checked with the help of available literature in order to confirm the usage and authenticity of the crude drugs (Kirtikar and Basu, 1935; Jain, 1991; Rama Rao and Henry, 1996). Specimens were identified with the help of local/regional floras (Gamble, 1935; Pullaiah and Yesoda, 1989, Pullaiah and Ali Moulali, 1997) and further confirmed by comparison with authentic specimens at Madras Herbarium (MH), Coimbatore and Central National Herbarium (CAL), Kolkata. Voucher specimens have been deposited in Ethnobotany research centre Government Degree and PG College, Anantapur and also one set of specimens were submitted to BSI, Deccan circle, Hyderabad.

Enumeration

Species are enumerated alphabetically with their botanical name followed by family in parentheses, local name(s) and voucher specimen number. Medicinal uses are described with details of part(s) used, method of preparation, dosage and combination with other plant, if any.

Abelmoschus moschatus **Medicus, (Malvaceae), 'Kondabenda', BTReddy 680.**
Aphrodisiac: The root extract is given orally once in a day for 7 days.

Abrus precatorius **L. (Fabaceae), 'Guruvinda', BTReddy 624.**
Antidote: The leaf paste mixed with goat milk, one teaspoonful twice a day is given orally for 7 days.

Baldness: The seed paste mixed with mustard oil, applied over the head for 30 days.

Snake bite: Root paste is applied on bitten part.

Headache: Seed powder is taken as snuff.

Abortifacient: One teaspoonful of seed paste once in a day is taken internally for 7 days.

Ringworm & Scabies: Leaf juice is mixed with common salt and water (1:1 ratio) is applied externally.

Leucorrhoea: The leaf paste mixed with curds and one teaspoonful is given orally for 7 days.

Abutilon indicum **(L.) Sweet, (Malvaceae), 'Thuthurubenda', BTReddy 556.**
Asthma: One teaspoonful of the leaf paste is given orally once in a day for 10 days.

Aphrodisiac: One teaspoonful of the seed paste is mixed with jaggery, given orally once a day for 30 days.

Burning micturition: The leaves chewed and sap swallowed.

Fever: The roots, pepper and garlic ground, 1-2 gms of the extract is given orally once a day for 3 days.

Fertility: The leaves along with *Piper betle* chewed and sap is swallowed once daily for 30 days.

Jaundice: The leaves, pepper and garlic ground, boiled in water and two spoonful of decoction is given orally once a day for 3 days.

Rheumatism: One teaspoonful of the leaf extract is given orally for 30 days.

Scorpion sting: The leaf paste is rubbed over the bitten area.

Acacia chundra **(Roxb. ex Rottl.) Willd. (Mimosaceae), 'Sandra', BTReddy 683.**
Toothache: The tender tip is used to clean teeth and bleeding gums.

Dandruff: The shade dried leaves are made into powder and mixed with albumen and applied on the head before half an hour of taking headbath.

Acacia farnesiana **(L.) Willd. (Mimosaceae), 'Dhaggutumma', BTReddy 684.**

Cough: Fruit powder mixed with honey and 2 spoonful once a day given orally for 3 days.

Acalypha indica **L. (Euphorbiaceae), 'Pippiaku', BTReddy 575.**

Constipation: The leaf paste mixed with cow milk and one teaspoonful in the early morning given orally for 10 days.

Scorpion sting, Eczema: The leaf paste as application.

Malaria: The leaf extract as eye drops. The leaves, pepper and garlic ground, and one teaspoonful of extract given orally for 5 days.

Toothache: Stem is used as tooth brush.

Piles: 5 gms of seed powder mixed with water is used for 7 days two times a day.

Menstrual disorder: The leaves, cardamom and cloves ground and a teaspoonful of extract given orally two times a day for 10 days.

Achyranthes aspera **L. (Amaranthaceae), 'Uttareni', BTReddy 640.**

Asthma: Leaves crushed and 2-3 gms of extract given orally once in a day for 30 days.

Tooth ache: Dried stem used to brush the teeth.

Abortifacient: Whole plant is pounded into fine paste, applied on abdomen to induce abortion.

Diuresis: One teaspoonful of the root bark decoction is given orally twice a day for 10 days.

Dermal diseases: The leaf paste as application

Gastrictrouble: Seed paste boiled with goat milk and one teaspoonful of the extract is given orally for 10 days.

Malaria: The leaves, pepper and garlic ground, 2 gms of the paste is given orally twice in a day for 7 days.

Snake bite: The whole plant, pepper and garlic ground and 1-2 gms of extract is given orally for 5 days.

Anti-fertility: Root is made into a paste, mixed with butter milk and one spoonful is given internally on empty stomach for 7 days, from the fourth day of menstruation as its anti-fertility action will be effective for one cycle.

Piles: Seed powder with water is used for 7 days, two times in a day.

Bronchitis: One spoonful of decoction of roots given internally twice in a day for 30 days

Aegle marmelos **(L.) Correa (Rutaceae), 'Maredu/Velaga', BTReddy 688.**

Dysentery: 5 gms of the pulp of the fruit is given orally along with jaggery in the early morning on empty stomach.

Ulcers: Pulp of fruit is given along with jaggery in the early morning on empty stomach.

Fever: 2 spoonful of stem bark decoction is given orally twice a day for 3 days.

Pains: 2 spoonful of root decoction twice a day is given orally for abdominal pains.

Fracture: The leaves along with blackgram, turmeric, calcium and albumen, ground and the paste is applied over fractured area and bandaged.

***Aerva lanata* (L.) Juss. (Amaranthaceae), 'Kondapindi', BTReddy 643.**

Tongue sore: Fresh inflorescence crushed into a paste and applied locally.

Earache: Leaves are wrapped in a castor oil, heated over fire and the juice obtained by crushing and pressing the leaves is poured in to the ear.

Skin diseases: The leaves ground with ginger oil and turmeric as application.

Stones in bladder: Whole plant juice given orally two spoonful everyday for 30 days.

***Albizia lebbeck* (L.) Willd. (Mimosaceae), 'Dirisena', BTReddy 380.**

Conception: The stem bark and onion ground and one spoonful of the extract is given orally once a day for 7 days.

Epilepsy: The leaves, pepper, garlic and Gorojana ground and one spoonful of the extract once a day is given orally for 7 days.

Sore eyes: The leaf juice is used as eye drops for 3 days.

Rheumatism: The leaves along with the leaves of *Azadirachta indica, Annona squamosa, Cassia auriculata* and *Vitex negundo,* ground, mixed with castor oil and ghee and applied locally.

***Allmania nodiflora* (L.) R. Br. ex Wt. (Amaranthaceae), 'Tellayerraboddiaku', BTReddy 636.**

Rheumatism: 2-3 gms of the leaf paste given once in a day orally for 30 days.

Wounds: The leaf paste mixed with turmeric applied locally.

Scorpion sting: The leaf paste mixed with jaggery, applied over sting area.

***Anisochilus carnosus* (L.f.) Wall. ex Benth. (Lamiaceae), 'Karpuravalli', BTReddy 330.**

Cold, Cough: 2 spoonful of leaf decoction twice in a day is given orally for 3 days.

Itches: Leaf paste is used as an application.

***Argemone mexicana* L. (Papaveraceae), 'Mysurpaala', BTReddy 555.**

Gonorrhoea: The milky latex mixed with sugar one teaspoonful is given twice a day orally for 7 days.

Itching: The milky latex, applied locally.

Leucoderma: The latex is applied on the skin daily once for 30 days.

Malaria: The leaves, pepper and garlic ground, and one gm of mixture is given orally twice a day for 3 days.

Skin diseases: The leaf juice mixed with calcium and common salt and is applied locally.

Wounds: The plant paste is application.

Aristolochia bracteolata **Lam. (Aristolochiaceae), 'Gadidagadapaku', BTReddy 660.**
Snake bite: Whole plant, macerated into paste and applied on bitten area and slightly warmed. Decoction prepared with a teaspoonful of plant powder, human urine and 3 spoonful is taken orally to dilute the poison in the body.

Abortifacient: One teaspoonful of root powder along with jaggery given orally twice a day for 3 to 4 days to induce abortion.

Pimples: The leaf paste as application.

Wounds: The whole plant paste as application.

Aristolochia indica **L. (Aristolochiaceae), 'Nallaeeswari', BTReddy 666.**
Leucorrhoea: The leaf paste mixed with jaggery and two pills are given orally for 5 days.

Rhumatic pains: The whole plant and turmeric made into paste and 2-3 gms taken orally and as an application.

Skin diseases: The leaf paste mixed with goat urine, rubbed over the body.

Snake bite: A spoonful of root decoction given orally and paste as an application on the spot of bite.

Abortifacient: Dry roots ground and boiled, 50 to 100ml decoction is given orally to induce abortion.

Bauhinia racemosa **Lam. (Caesalpiniaceae), 'Arechettu', BTReddy 354.**
Wounds: The leaf paste as application.

Temporary birth control: Whole plant extract of one teaspoonful is given orally for 15 days.

Blepharis repens **(Vahl) Roth (Acanthaceae), 'Nelaparikalu' BTReddy 665.**
Sterility: Whole plant made into paste and one gram once in a day given orally for 15 days.

Buchanania lanzan **Spreng. (Anacardiaceae), 'Sara', BTReddy 370.**
Back Pain: Whole plant made into a powder and 2 gms of it is taken internally three times a day after food for 20 days.

Caesalpinia bonduc **(L.) Roxb. (Caesalpiniaceae) BTReddy 647.**
Allergy: The leaf paste boiled in water and the fumes are inhaled.

Asthma: The seed paste mixed with honey and one teaspoonful per day is given orally for 15 days.

Contraceptive: 2 gms of seed powder is given orally for 7 days after menses.

Hydrocele: Leaves roasted in castor oil as poultice.

Intermittent fever: The leaves, pepper and garlic ground and 2 gms of the paste given orally for 3 days.

Calotropis gigantea **(L.) R.Br. (Asclepiadaceae), 'Tellajilledu', BTReddy 659.**

Abortion: Flower buds / root and camphor ground and 2 gms of the paste once in a day is given orally for 3 days.

Oedema: Latex is applied on the inflamed parts of the body to reduce pains.

Scorpion sting: Root is made into paste and paste is applied on the spot.

Leprosy: The root is pounded with water and liquor made out of *Madhuca indica*, applied externally on leprosy and also taken internally.

Asthma, Cold, Cough: The flower extract of 1 gm once in a day is given orally for 10 days.

Calotropis procera **(Ait.) R.Br. (Asclepiadaceae), 'Jilledu', BTReddy 658.**

Scabies: Latex mixed with castor oil is heated and applied on the skin.

Wounds: Roots are burnt and pounded into powder. The powder is mixed with the oil of *Sesamum orientale* and mixture is applied on the shoulder of bullock to cure wounds.

Jaundice: The tender tips and *Piper betel* leaves ground and one pill of extract given orally for 3 days.

Migraine: The ripe leaf extract as eardrops.

Scorpion sting: Stem is pounded and the powder is made into a paste with hot mustard oil, the paste is applied.

Stomach pain: Latex of the stem is mixed with soil from the plant base and smeared on the belly.

Canthium parviflorum **Lam. (Rubiaceae), 'Balusu', BTReddy 663.**

Dysentery: The unripe fruit extract mixed with curds and one spoonful once a day is given orally for 3 days.

Snake bite: The stem bark paste mixed with cow urine and the extract is dropped into nostrils or given orally for 7 days.

Caralluma umbellata **Haw. (Asclepiadaceae), 'Kundetikommulu', BTReddy 664.**

Headache: Flowers dried, powdered, mixed with honey and ½ spoon given orally for 3 days.

Cardiospermum halicacabum **L. (Sapindaceae), 'Buddakakara', BTReddy 540.**

Skin diseases: Whole plant paste is applied on the effected parts.

Capparis zeylanica **L. (Capparaceae), 'Adavinimma', BTReddy 337.**

Fertility: The leaf paste mixed with cow ghee and prepared pills, given orally twice in a day for 15 days.

Cassia auriculata L. (Caesalpiniaceae), 'Tangedu', BTReddy 626.

Anthelmintic: One teaspoonful of fruit extract given orally for 3 days.

Diabetes: The stem tips eaten daily once for 30 days.

Diarrhoea: The stem tips ground with curds and a teaspoonful twice a day is given orally for 3 days.

Skin diseases: The leaves along with stem bark of *Pongamia pinnata* and gingelly oil ground, applied locally.

Cassia fistula L. (Caesalpiniaceae), 'Rela', BTReddy 669.

Mad dog bite: Leaf juice mixed with milk and 2-3 spoonful twice a day is given orally for 5 days.

Indigestion: Mature leaves are ground and mixed with buttermilk given orally.

Burns: Stem bark ground with coconut oil and the paste is applied externally for week days.

Laxative: One spoonful of the stem bark decoction, twice a day given orally for 5 days.

Stomach ache: One teaspoonful of the fruit paste with ginger and honey given orally for 2 days.

Ulcers: The flowers ground with roots of *Decalepis hamiltonii, Bacopa monnieri* and turmeric, applied locally.

Cassia italica (Mill.) Lam. ex Andr. (Caesalpiniaceae), 'Nelatangedu', BTReddy 654.

Elephantiasis: The whole plant is made into powder and one gram is given orally for 30 days.

Gonorrhoea: The whole plant paste mixed with butter milk and a teaspoonful once in a day is given orally for 7 days.

Leucoderma: The leaves along with roots of *Plumbago zeylanica* and turmeric, ground, extract applied locally.

Syphilis: The leaves along with roots of *Plumbago zeylanica*, Sulphur, brinjal fruit and onion, ground, boiled in gingelly oil and applied locally.

Fractures: Leaves ground and paste applied as plaster.

Stomach ache: Leaf powder (1-2 gms) is given orally twice a day for 3 days.

Constipency: Leaf powder (1-2 gms) is given orally twice a day for 3 days.

Piles: Fresh leaf made into paste and a teaspoonful once a day is given orally for 7 days along with jaggery.

Cassia montana Heyne (Caesalpiniaceae), 'Pagidi tangedu', BTReddy 347.

Leucorrhoea: One teaspoonful of the leaf decoction twice in a day is given orally for 7 days.

Syphilis: The tender stem tips, dry coconut and gingelly oil ground, extract mixed with honey and one spoonful once a day is given orally for 10 days.

Cassia occidentalis **L. (Caesalpiniaceae), 'Kasintha', BTReddy 651.**

Ring worm and wounds: The fresh root juice of one teaspoonful once a day is given orally for 5 days.

Convulsions in children: ½ teaspoon of the seed paste is given orally twice a day for 3 days.

Skin diseases: Plant paste is used as external ointment.

Wounds, Swellings: Fresh leaves are ground, applied externally.

Catunaregum spinosa **(Thunb.) Tirveng. (Rubiaceae), 'Manga', BTReddy 665.**

Scorpion sting: The fruit paste mixed with turmeric is applied over the bitten area.

Tonsilitis: The leaves are chewed in early morning and the extract is used for smearing

Malaria: 2 spoonful of the root decoction twice a day is given orally for 5 days.

Dandruff: Fresh and dried fruits are used in head bath.

Celosia argentea **L. (Amaranthaceae), 'Gurugaku', BTReddy 644.**

Aphrodisiac: The seeds ground with jaggery, onion and made into pills, one pill is given orally every day for 15 days.

Diarrhoea: The seed paste mixed with butter milk, and one spoonful twice a day given orally for 3 days.

Eye diseases: The leaf juice as eye drops.

Minimises thirsting: The root chewed and sap swallowed

Cleome gynandra **L. (Cleomaceae), 'Vavitaku', BTReddy 672.**

Fever: One teaspoonful of root decoction is given in a day orally for 3 days.

Head ache: One teaspoonful of a leaf extract is given orally and plant paste applied on forehead.

Ear ache: Leaves and garlic ground and the extract is used as ear drops.

Consciousness: The root paste is applied on the lips to restore consciousness.

Coccinia grandis **(L.) Voigt (Cucurbitaceae), 'Donda', BTReddy 623.**

Allergy: The leaf paste as application.

Fracture: The leaves along with black gram and albumen ground, paste applied over fractured area and banded.

Dandruff: The leaf paste mixed with castor oil, applied over the head once 7 days for 30 days.

Opacity of Cornea: The leaf juice mixed with common salt and the extract is used as eye drops.

Rabies: The leaves, pepper and garlic ground and one teaspoonful twice a day of the extract given orally for 3 days.

Skin diseases: The leaves along with stem bark of *Pongamia pinnata* and cow urine, ground, applied locally.

Cocculus hirsutus **(L.) Diels (Menispermaceae), 'Dusarateega', BTReddy 653.**

Cooling tonic: Leaf juice is mixed with sugar candy and a teaspoonful is given orally once a day for 7 days.

Menstruation: Whole plant decoction of one spoonful once a day is given orally for 3 days.

Leucorrhoea: Leaves ground with garlic and pepper and made into pills and given 2 pills are given once a day for 7 days.

Urinary disorders: One teaspoonful of leaf decoction is given orally for 5 days.

Anaemia: One teaspoonful of the root paste is given orally on every alternate day.

Commelina benghalensis **L. (Commelinaceae), 'Eennayeddulaku', BTReddy 621.**

Insect bite: Root mixed with turmeric is ground and one teaspoonful of infusion given orally for 3 days.

Leprosy: Plant decoction is used as lotion, drink and bath.

Scorpion sting: The leaves, turmeric and onion ground, paste applied locally.

Corallocarpus epigaeus **(Rottl. & Willd.) C.B. Clarke (Cucurbitaceae), 'Pamudonda', BTReddy 629.**

Snake bite: The tuber/root paste mixed with children's Urine and one spoonful twice a day is given orally for 3 days.

Blisters: The leaf paste mixed with castor oil, slightly warmed and applied extenally.

Poisonous bite: The tubers, pepper and garlic ground, a spoonful of the paste is given orally for 3 days.

Croton bonplandianum **Bail. (Euphorbiaceae), 'Galivanachettu', BTReddy 638.**

Promote vomiting: Leaves mixed with salt and one teaspoonful once in a day is given orally for 3 days.

Delivery bleeding: 2 teaspoonful of root decoction is given orally to stop abnormal bleeding after childbirth.

Cucumis callosus **(Rottl.) Cong. (Cucurbitaceae), 'Adavibudama', BTReddy 593.**

Skin diseases: The leaves, turmeric and onion, ground, paste applied locally.

Chronic wounds: The leaf paste as application.

Curcuma neilgherrensis **Wt. (Zingiberaceae), 'Kondapasupu', BTReddy 325.**

Sneezing: Fresh leaves are ground and the juice is dropped into nostrils.

Skin diseases: Rhizome is shade dried and made into a paste and applied on the effected parts.

Cyanotis tuberosa **(Roxb.) Schult. & Schult f. (Commelinaceae), 'Eggogulagaddalu', BTReddy 606.**

Allergy: The plant, turmeric and castor oil ground, paste applied over the body.

Insect bite: The leaves, pepper, garlic and onion ground, one teaspoonful of extract twice in a day is given orally for 3 days.

Ulcers: Tubers/roots dried, ground and one teaspoonful of the infusion mixed with goat milk is given orally for 7 days.

Cymbopogon coloratus **(Nees) Stapf (Poaceae), 'Bodagaddi', BTReddy 509.**
Rheumatism: 2 spoonful of root decoction is given orally for 15 days.

Dog bite: The aerial parts along with *Bacopa monnieri* paste applied over the bitten area.

Cymbopogon flexuosus **(Nees ex Steud.) Wats. (Poaceae), 'Kasigaddi', BTReddy 686.**
Mosquito repellent: The plant is dried inflamed. The smoke repells mosquitoes.

Skin diseases: The inflorescence is made into a paste and is applied externally.

Cynodon dactylon **(L.) Pers. (Poaceae), 'Garika', BTReddy 620.**
Birth control: One spoonful of the fresh juice of the plant is given orally after delivery to women for 5 days.

Aphrodisiac: The basal parts ground with jaggery and 3 gms once a day is given orally for 15 days.

Decalepis hamiltonii **Wight & Arn. (Asclepiadaceae), 'Maredugaddalu', BTReddy 336.**
Cooling: The roots along with the roots of *Curculigo orchioides* and *Hemidesmus indicus* boiled in water and decoction given orally.

Aphrodisiac: The roots along with those of *Hemidesmus indicus* and *Withania somnifera* and true date palm ground, warmed in ghee, the extract mixed in cow milk and two spoonful is given orally every day for 15 days.

Diospyros melanoxylon **Roxb. (Ebenaceae), 'Tumki/ Beediaku, BTReddy 576.**
Astringent: Decoction of bark is used in Diarrhoea, Dyspepsia, as astringent lotion for the eyes.

Laxative: The leaf powder made into pills, one gm each is given orally for 15 days.

Urinary disorders: One teaspoonful of the decoction of dried flowers is given orally for 7 days.

Skin diseases: The flowers dried in shade and made into powder along with coconut oil is applied on affected parts.

Dipteracanthus patulus **(Jacq.) Nees (Acanthaceae), 'Bellapachuaku', BTReddy 655.**
Skin diseases: The leaves, turmeric and a pinch of calcium ground and applied locally as poultice.

Wounds: The leaf paste as application.

Eclipta prostrata **(L.) L. (Asteraceae), 'Guntagalagara', BTReddy 622.**
Jaundice: Whole plant ground, infusion mixed with salt/sugar, 2 spoonful given orally twice a day for 3 days.

Hair tonic: Fresh plant paste is applied with sesame oil.

Diarrhoea: Juice of fresh leaves is taken two teaspoonful for 3 times a day for 3 days.

Itches: Whole plant is made into paste with mustard seeds applied to itches caused due to excessive contact of the skin with water.

Dog bite: The leaf paste mixed with a pinch of calcium, applied locally.

Muscular disease, Skin diseases: The leaves along with those of *Argemone mexicana*, roots of *Plumbago zeylanica* ground and extract applied locally.

Oedema, Malaria: The leaves, pepper and garlic ground and a teaspoonful of the paste given orally for 3 days.

***Euphorbia acaulis* Roxb. (Euphorbiaceae), 'BarresepuGadda', BTReddy 674.**

Galactogogue: One gm of root powder mixed with sugar (1:1) is given orally for 3 days.

***Euphorbia hirta* L. (Euphorbiaceae), 'Palaku', BTReddy 635.**

Asthma: One teaspoonful of the plant decoction is given orally twice a day for 15 days.

Galactogogue: The leaf paste mixed with jaggery and one teaspoonful once a day is given orally for 15 days.

Burns: Juice of the leaves is applied externally.

Rheumatism: Leaves are crushed and applied as a paste on painful parts.

Warts: Latex of the plants is used as application.

Eye sore: Latex is applied on the lower side of eyelid.

***Euphorbia tirucalli* L. (Euphorbiaceae), 'Kallichettu', BTReddy 646.**

Rheumatid arthritis: Along with the leaves of *Enicostemma axillare* roasted with cow urine is applied over joints.

***Gloriosa superba* L. (Liliaceae), 'Kodijuttugaddalu', BTReddy 379.**

Abortifacient: One spoonful of the root tuber extract once a day is given orally for 15 days.

Leprosy, Wounds: The root tuber paste is used as an application.

***Gymnema sylvestre* (Retz.) R. Br. ex Schult. (Asclepiadaceae), 'Podapatri', BTReddy 667.**

Urinary disorders: Leaves mixed with turmeric ground, applied externally.

Diabetes: One leaf consumed daily at early morning for 40 days.

Diarrhoea: The leaf paste mixed with curds given orally.

Poisonous bite: One teaspoonful of the root extract is given orally twice a day for 3 days.

Hemidesmus indicus **(L.) R.Br. (Asclepiadaceae), 'Sugandhapala', BTReddy 386.**

Debility, Body pains, Lazyness: The roots along with those of *Curculigo orchioides* and *Withania somnifera* musk and garlic ground, 2 pills of the paste once in a day is given orally for 7 days.

Indigestion: The roots, pepper and garlic ground and two pills of the paste given orally for 3 days.

Dysentery: The root paste mixed with water is given in small quantity to children.

Anaemia: Root and leaf decoction of one spoonful is given twice a day for 15 days.

Diarrhoea: A decoction of 5 gms of the root of *Curculigo orchioides, Cyperus rotundus,* Fennel and Cumin is given to children 2 – 3 times a day for 3 days.

Hair tonic: 50 gms of dried roots are soaked in coconut oil and applied.

Hibiscus ovalifolius **(Forsk.) Vahl, (Malvaceae), 'Chittibitti', BTReddy 681.**

Leucorrhoea: One teaspoonful of the leaf and fruit paste is given orally once a day for 7 days.

Fracture: The leaf paste is mixed with goat milk, applied over the fractured area and bandaged.

Fever: The leaves along with the leaves of *Cocculus hirsutus* and tubers of *Cyperus rotundus,* ground one teaspoonful of the extract given orally once a day for 3 days.

Hygrophila auriculata **(Schum.) Heine (Acanthaceae), 'Mudigubbaku', BTReddy 628.**

Fever: The roots, pepper and garlic ground and 2–3 gms of the paste twice in a day given orally for 3 days.

Rheumatism: The leaf paste with turmeric applied locally and leaf decoction of two spoonful twice in a day given orally for 15 days.

Indigofera linnaei **Ali (Fabaceae), 'Nelajanumu', BTReddy 515.**

Antifertility: One teaspoonful of root decoction once in a day is given orally for 6 days.

Diarrhoea, Dysentery, Stomach ache: 1-2 spoonful of plant decoction once in a day is given orally for 3 days.

Ipomoea pes-tigridis **L. (Convolvulaceae), 'Chikunuvvu', BTReddy 673.**

Dog bite: One teaspoonful of the leaf juice is given orally for 10 days.

Boils: Leaf paste is applied.

Carbuncles: Leaf paste is applied.

Laxative: The root paste of 2-5 gms twice in a day is given orally for 3 days.

Jatropha gossypiifolia **L. (Euphorbiaceae), 'Adaviamudam', BTReddy 675.**

Toothache: The stem is used to clean the teeth early in the morning.

Antidote: Seeds fried and powdered and 1 gm is given with molasses as an antidote against poisoning.

Wounds : The latex as application.

Kedrostis foetidissima **(Jacq.) Cogn. (Cucurbitaceae), 'Muradandaku', BTReddy 333.**
Indigestion: Stem bark, pepper and garlic ground, one spoonful of the paste, twice a day given orally for 3 days.

Laxative: The leaf paste mixed in castor oil, rubbed over the head.

Snake bite: The tubers along with tubers of the *Curculigo orchioides* and *Withania somnifera*, pepper and garlic ground, one teaspoonful of the paste twice a day given orally for 3 days.

Lepidagathis cristata **Willd. (Acanthaceae), 'Nakkapentigadda', BTReddy 512.**
Fever: 10gms of fresh leaves are to be taken and made into an extract of 100 ml and given orally twice in a day for 3 days.

Itches: Leaf paste applied externally.

Sores: Whole plant shade dried and burnt to ash and ash mixed with coconut oil and made into an emulsion, paste is to be applied.

Wounds: Whole plant ground, powder mixed with coconut oil, applied externally.

Toothache: Decoction of the inflorescence is used as mouthwash.

Leucas aspera **(Willd.) Link, (Lamiaceae), 'Thummi', BTReddy 645.**
Wounds: Leaf paste is mixed with turmeric and applied.

Anaemia: Flowers boiled with cow milk and one spoonful of the cooled content is given orally to patient for 15 days.

Snake bite: The plant juice is applied externally and given internally.

Fever: The leaves along with fruits of *Citrus* (lemon), pepper and garlic ground and one spoonful of the extract twice a day is given orally for 3 days.

Headache: Leaf paste is applied on the forehead.

Mimosa pudica **L. (Mimosaceae), 'Atthipatthi', BTReddy 561.**
Piles: Leaves and roots are made into powder and one gm is administered with milk daily for 30 days.

Hydrocoele: Leaves are rubbed into a paste and applied.

Conjunctivitis: Leaf paste mixed with horse urine is applied.

Contraceptive: 2 gms of root powder is given orally empty stomach from the 1st day of menses to 6th day to check the conception.

Oldenlandia umbellata **L. (Rubiaceae), 'Chorivelu', BTReddy 351.**
Asthma, Bronchitis: 2 spoonful of the decoction of the roots and leaves is given orally once a day for 10 days.

Snake bite: The plant along with that of *Enicostemma axillare,* pepper and garlic, ground and 5 gms of the paste twice a day is given orally for 3 days.

Passiflora foetida **L. (Passifloraceae), 'Jukamalli', BTReddy 685.**
Skin diseases: Leaf paste is applied externally.

Pedalium murex **L. (Pedaliaceae), 'Yenugupalleru', BTReddy 641.**
Aphrodisiac: The fruit decoction is given orally one teaspoonful once a day for 15 days.

Malaria: The plant along with the leaves of *Andrographis paniculata*, pepper and garlic, ground and the paste of one gm is given orally for 3 days.

Tooth ache: The leaf paste is applied. The leaf decoction is used as mouth gargle.

Pergularia daemia **(Forssk.) Chiov. (Asclepiadaceae), 'Dustaputeega', BTReddy 652.**
Asthma: The leaves along with the leaves of *Wrightia tinctoria*, pepper and garlic ground and 2 pills of the extract given orally for 15 days.

Dysentery: The leaf paste mixed with curd and a spoonful twice a day given orally for 3 days.

Fever, Heart stroke: The leaves, pepper and garlic ground, 1-2 spoonful of extract given orally for 3 days.

Whooping Cough: The tender tips along with the leaves of *Melia azadirachta* and *Ocimum tenuiflorum*, pepper and garlic ground, one spoonful twice a day given orally for 2 days.

Abortifacient: 100ml of root decoction is given orally in empty stamach for 5 days.

Phyllanthus amarus **Schum. & Thonn. (Euphorbiaceae), 'Nelausiri', BTReddy 676.**
Aphrodisiac: One teaspoonful of the leaf decoction once a day is given orally for 15 days.

Jaundice: The whole plant shade dried made into powder, 1 gm once a day is given orally for 3 days.

Dog bite: The leaves, pepper and garlic, ground, 2-3 gms of the paste is given orally, for 10 days.

Diabetes: The shade dried plant powder of one teaspoonful once a day is given orally for 7 days.

Gonorrhoea: Fresh plant parts are ground with buttermilk and made into pills, 3 pills per day are taken orally for 3 days.

Sores: Young shoot milky juice is applied on the body.

Phyllanthus reticulatus **Poir. (Euphorbiaceae), 'Nallapuli', BTReddy 633.**
Bleeding gums: Stem powder is applied.

Insect bite: The leaf extract mixed with goat milk and one teaspoonful once in a day is given orally for 3 days.

Wounds: The leaf paste as an application.

Plumbago zeylanica **L. (Plumbaginaceae), 'Chitramulamu', BTReddy 580.**
Abortion: Root paste is applied on the lower part of belly to induce abortion.

Leucoderma: One teaspoonful of the leaf extract is given orally for 30 days.

Rheumatism: The roots along with leaves of *Cassia italica* made into a paste and 2-3 gms of this extract is given orally for 30 days.

Skin diseases: The root paste mixed with turmeric powder is applied locally.

Polycarpaea corymbosa **(L.) Lam. (Caryophyllaceae), 'Gaddipuvu', BTReddy 642.**
Boils and ulcers: Whole plant is made into paste and two gms in a day is given orally for 5 days.

Polygala elongata **Klein ex Willd. (Polygalaceae), 'Needajanumu', BTReddy 332.**
Chronic bronchites: One spoonful of the decoction of the roots is given orally once a day for 15 days.

Spermatorrhoea: One spoonful of the leaf juice once a day is given orally for 30 days.

Pongamia pinnata **(L). Pierre (Fabaceae), 'Kanuga', BTReddy 678.**
Dysentery: The stem bark decoction mixed with buttermilk, a teaspoonful once in a day is given orally for 3 days.

Fertility: The root bark, garlic and curds ground, a spoonful of paste given orally for 30 days.

Wounds: The seed paste mixed with calcium, applied locally.

Diabetes: The flowers shade dried and the paste of one gm is given orally once a day in the morning for 40 days.

Prosopis cineraria **(L.) Druce, (Mimosaceae), 'Jammi', BTReddy 627.**
Gonorrhoea: The leaves, lemon juice and coconut milk ground, 1-2 spoonful of the extract once a day is given orally for 5 days.

Leucorrhoea: The leaves are ground with sugar and butter milk and the mixture is given orally for 3 days.

Scorpion sting: The leaves are ground with buttermilk and the paste is applied over the bitten area.

Pseudarthria viscida **(L.) Wight & Arn (Fabaceae), 'Nayakuponna', BTReddy 326.**
Rheumatism, Fever, Diarrhoea, Asthma, Heart stroke and Piles: Root powder is used in the form of decoction or powder.

Sarcostemma acidum **(Roxb.) Voigt (Asclepiadaceae), 'Kolakali', BTReddy 662.**
Cough: The leaves with common salt ground, 1-2 spoon ful of the extract given orally for 3 days.

Rabies: Stem mixed with turmeric, ground and the paste is applied externally.

Securinega leucopyrus **(Willd.) Muell.-Arg. (Euphorbiaceae), 'Tellapuli', BTReddy 656.**
Burning Micturition: One teaspoonful of the leaf extract is given orally for 5 days.

Cough: The leaves along with the roots of *Citrullus colocynthis, Tylophora indica* and pepper, ground and one teaspoonful of extract once in day is given orally for 5 days.

Fever: The stem bark, pepper and garlic ground and one teaspoonful of extract given orally for 3 days.

Food: Fresh fruits edible, Yanadis take as food in conditions of drought.

Gonorrhoea: The tender tips along with those of *Tragia involucrata* and sugar ground and 2-3 gms of the paste is given orally for 7 days.

Sores: Leaf paste as application.

Solanum surattense **Burm. f. (Solanaceae), 'Ramamulaga', BTReddy 528.**

Asthma: 2-5 gms of the plant exact is given orally once a day for 15 days.

Chronic cough: 2-5 gms of the fruit extract is given orally twice a day for 10 days.

Epilepsy: The roots, gorojanam and saffron ground and mixed with goat milk and one spoonful once a day is given orally for 5 days.

Jaundice: Juice of raw fruit mixed with turmeric is used as eye drops and administered three times daily for 3 days.

Tarenna asiatica **(L.) O. Kuntze (Rubiaceae), 'Kommi', BTReddy 687.**

Leprosy: The leaves along with the seeds of *Pongamia pinnata,* neem oil and turmeric, ground and paste applied locally.

Skin diseases: Leaves made into a paste is applied on the spot.

Scorpion sting, wounds: The leaf paste is applied on the wounds.

Tinospora cordifolia **(Willd.) Miers ex Hook. f. & Thoms, (Menispermaceae), 'Thippateega', BTReddy 682.**

Snake bite: 1-2 spoonful of stem juice twice a day is given orally for 3 days.

Diabetes: Leaf powder mixed with one teaspoonful of milk is given twice a day for 60 days.

Fever: 2 spoonful of the water extract of the plant is given internally for 3 days.

Piles: Leaf juice mixed with sugar and salt, a teaspoonful of decoction is given twice a day orally for 30 days.

Jaundice: Leaves ground as a paste, made into pills and 2 pills are given orally once a day along with buttermilk for 3 days.

Trianthema portulacastrum **L. (Aizoaceae), 'Ganjeraaku', BTReddy 657.**

Fracture: The leaves mixed with black gram, egg albumen and calcium ground and the paste used as poultice.

Tribulus terrestris **L. (Zygophyllaceae), 'Palleru', BTReddy 639.**

Epilepsy: The leaves, gorojana, ground and 2 gms of the mixture is given orally twice a day for 5 days.

Boils, wounds: The leaf paste is applied locally.

Skin diseases: The plant paste mixed with turmeric is applied locally.

Tridax procumbens **L. (Asteraceae), 'Gaddichamanthi', BTReddy 578.**

Injuries: Leaf juice as application.

Scorpion sting: The leaf paste mixed with jaggery, applied over bitten area.

Warts: Leaf paste applied over the wart for ten minutes.

Wounds: The leaf paste as application.

Triumfetta rhomboidea **Jacq. (Tiliaceae), 'Marlabenda', BTReddy 329.**

Dysentery: The bark and fresh leaves are made into a paste and 2 gms of it once a day is given orally for 3 days.

Diarrhoea: Pounded roots are given for the treatment of intestinal ulcers.

Vitex negundo **L. (Verbenaceae), 'Vavili', BTReddy 693.**

Orchitis: Leaf along with pepper is made into a paste and is applied.

Arthritis: 2 spoonful of leaf decoction is given orally twice a day for 15 days.

Withania somnifera **(L.) Dunal (Solanaceae), 'Asvagandhi', BTReddy 691.**

Aphrodisiac: 100 gms shade dried roots along with 50 gms of the root of *Curculigo orchioides* and *Ipomoea mauritiana* are powdered and one spoon powder is mixed with a cup of goat milk is given orally for 30 days.

Asthma: 10 gms of fresh roots pounded in water, filtered and the extract of 10ml is given in the early morning every day for 15 days.

Wattakaka volubilis **(L.f.) Stapf (Asclepiadaceae), 'Pedakadisaku', BTReddy 592.**

Rheumatism: The leaves, turmeric and calcium ground, paste applied locally.

Wrightia tinctoria **R.Br. (Apocynaceae), 'Palavara', BTReddy 649.**

Asthma : The leaf juice as nasal drops for 7 days.

Toothache : The milky latex as application.

Wounds: Stem bark is made into paste and applied.

Skin diseases: The milky latex as application.

Summary and Conclusions

Eswaramala Hills in Anantapur District spread in an area of 6329.42 Ha in Kadiri range, Anantapur District. The forests are of dry deciduous type. The soils are red and the topography is hilly. In this region at Gutibailu, the great Banyan tree, Thimmamma Marrinanu (a Sacred grove) is present which was recorded in the Guinness Book of World Records as the largest banyan tree in the world with a current age of 570 years and a canopy of 5.3 acres.

Periodical ethnobotanical field trips were conducted in this area during the year 2003-2005, and collected more than Hundred plant species with immense medicinal value in consultation with the Yanadis, Sugalis and other local informants of this area. Voucher specimes were made into herbarium specimens which were deposited in the Ethnobotany Research Centre, Government College, Anantapur. The specimens are compared with the authenticated specimens housed at Sri Krishnadevaraya

University, Anantapur, Madras Herbarium Coimbatore(MH). Among the medicinally useful plants, the data was collected on the plant parts used, its method of collection, processing, preparation of the drug, its dosage, mode of administration, duration of treatment etc. The medicinal properties of plant species are enumerated systematically.

The scientific identification of the native drug source secured from determinable voucher specimen forms the indispensable hinge between field observation and experimental evaluation. The ultimate aim of ethnobotany is to evaluate the traditional preparations through pharmacological findings and to identify the lead compounds through phytochemical evaluation. The information gathered about indigenous drugs is used as feed back to traditional medicine. Knowledge of active constituents of indigenous drugs may lead to substantial improvements in traditional therapy.

Further, the investigation made the following contributions.

1. Good number of endemic plants, such as *Curcuma neilgherrensis,* and endangered plants like *Aristolochia indica, Gymnema sylvestre, Plumbago zeylanica, Hemidesmus indicus, Euphorbia acaulis* are recorded in the present work.
2. Rare taxa *viz., Buchanania lanzan, Decalepis hamiltonii* having medicinal properties are recorded.
3. Some of the medicinal plants hither to not reported with new medicinal properties were identified.

The results are useful to convert crude drugs into refined medicine convenient for immediate consumption and also as having commercial value. Due to rapid acculturation and depletion of forest resources, many tribal herbalists have given up the tradition and taken up other occupations. In view of these facts, it is felt necessary to record the indigenous knowledge before it disappears completely.

The thorough ethnomedicobotanical studies and the knowledge on the indigenous plants provide valuable data for assessment of plant resources. Extensive and intensive ethnobotanical studies of Eswaramala forest range, Anantapur district will provide planning strategies on conservation and the sustainable utilization of available plant resources for the benefit of local tribes in future.

The present studies also help the graduate, postgraduate and research scholars of the departments of Botany, Zoology, Chemistry, Pharmacy and Environmental Sciences to undertake chemical and pharmaceutical work on medicinal plants of Eswaramala forest.

References

Akerele, O.1992 WHO guidelines for assessment of herbal medicines Fitoterapia 63:99-118.

Anonymous, 1994. Ethnology in India a Status report. All India Coordinated project in Ethnobotany. Ministry of Environmental Forests. Govt. of India, New Delhi.

Aruneekumar and Nisteswar, K. 1990. Medicinal plants of Kakinada (East Godavari District, Andhra Pradesh). Ind. Med. 2(2-4): 229-232.

Aswartha, B. Ali Moulali, D., & Pullaiah, T. 2004. Preliminary survey of certain Ethnomedicinal plants from the forests of Anantapur district, Andhra Pradesh. National Symposium on Recent Trends in Plant Sciences, October 29-31, 2004, S.K.University. 118.

Beddome, R.H. 1868-74. Icones Plantarum Indiae Orientalis, Madras.

Champion, H.G. and Seth, S.K. 1968. A revised survey of forest types of India. Government of India. New Delhi.

Ellis, J.L. 1982. Wild plant resources of Nallamalais on the Eastern Ghats of India–1- A preliminary list. Bull. Bot. Surv. India. 10: 140-160.

Gamble, J.S. and Fischer, C.E.C. 1915-35. Flora of Presidency of Madras Vol. I -III. BSI, Calcutta.

Goud, P.S. and T. Pullaiah. 1996. Ethnobotany of Kurnool district Andhra Pradesh. In S.K. Jain (ed.) Ethnobiology of Human Welfare. 410-412. Deep Publications, New Delhi.

Hamilton, A. 1995. The people and plants Initiative. In: Martin, G.J. (Ed.).Ethnobotany a methods manual. WWF International Chapman & Hall, London, pp. X-XI.

Harasreeramulu, S. 1980. Some useful medicinal plants from Srikakulam district, Andhra Pradesh. J. Indian Bot. Soc. 59 (suppl.): 168 (abstract).

Hemadri, K. 1981. Rheumatism : tribal medicine. Anc. Sci. Life. 1: 117-120.

Hemadri, K. 2003. Tribal medicine in primary health care. Int. conference on medicinal and aromatic plants, Hyderabad (March 15-17): 106.

Hemadri, K., Sarma, C.R.R. and Rao, S.S. 1987. Medicinal plant wealth of Andhra Pradesh, Part- II. Anc. Sci. Life 7(1): 55-64.

Huxley, A. 1984. Green inhertise the world wide life found book of India. Collirs/ harvels London.

Jain, S.K. 1991. Dictionary of Indian Folk medicine and Ethnobotany. Deep Publications. New Delhi.

Jain, S.K., Benerjee, D.K. and Pal, D.C. 1973. Medicinal plants among certain adivasis in India. Bull. Bot. Surv. Ind. 15: 85-91.

Jeevan Ram, A. and Venkata Raju, R.R. 2002. Certain potential crude drugs used by tribals of Nallamalais, Andhra Pradesh for Skin diseases. Ethnobotany 13(1&2): 110-115.

Kirtikar, K.R. and Basu, B.D. 1935. Indian Medicinal Plants Vol. I-IV.

Kumar, D.C.T. and Pullaiah, T. 1998. Ethnomedical uses of some plants of Mahaboobnagar district, Andhra Pradesh, India. J. Econ. Taxon. Bot. 23: 341-345.

Mehrotra, S. and Mehrotra, B.N. 2005. Role of traditional and folk lore herbals in the development of new drugs. Ethnobotany 17: 104–111.

Muralidhar Rao, D. and Pullaiah, T. 2001. Ethno-medico-botanical studies in Guntur district of Andhra Pradesh, India. Ethnobotany 13(1&2): 40-44.

Nagaraju, N. and Rao, K.N. 1990. A survey of plant crude drugs of Rayalaseema, Andhra Pradesh, India. J. Ethnopharmacol. 29: 137-158.

Nisteswar, K. and Kumar, K.A. 1980. Utilitarian values of medicinal-lore of Ramapa agency (Andhra Pradesh) in primary health care. Sachitra Ayurveda. 33(3): 210-212.

Pal, D.C. and D.K. Banerjee 1971. Some less known plant foods among the Tribals of Andhra Pradesh and Orissa state. Bull. Bot. Surv. India. 13: 221-223.

Pullaiah, T and Ali Moulali, D. 1997. Flora of Andhra Pradesh Vol.-II Scientific publishers, Jodhpur, India.

Pullaiah, T. and Yesoda, N. 1989. Flora of Anantapur district, Andhra Pradesh. Bishen Singh Mahendra Pal Singh, Dehradun.

Rama Rao, N. and Henry, N. 1996. The ethnobotany of Eastern Ghats in Andhra Pradesh, India. Botanical Survey of India. Calcutta.

Ravi Kumar, K. and Ved, D.K. 2000. 100 Red listed medicinal plants of South India. FRLHT. Banglore.

Ravi Prasad Rao, B., Jayaprada, C., Sandhya Rani, S., Sri Rama Murthy, K., Pullaiah, T., Sreenivasulu, V. and Madhusudana Reddy, A.2003. Additions to the flora of Anantapur District, Andhra Pradesh. J. Econ. Taxon. Bot. 27 (Suppl.) pp.1151–1159.

Ravisankar, T. and Henery, A.N. 1992. Ethnobotany of Adilabad district, Andhra Pradesh. Ethnobotany 4: 45-52.

Reddy, K.N. and Raju, R.R.V. 1999. Plants in Ethnoveterinary practices in Anantapur District, Andhra Pradesh. J. Econ. Taxon. Bot. 23(2): 347-357.

Reddy, R.V. 1995. Ethnobotanical and phytochemical studies on medicinal plant resources of Cuddapah district, A.P., India. Ph.D. thesis, S.K. University. Anantapur, India.

Reddy, T.A. 1980. Notes on some medicinal plants of Polavaram agency tracts, West Godavari District. Andhra Pradesh. J. Indian Bot. Soc. 59 (suppl.): 169 (abstract).

Reddy, K.N. 2008. Ethnobotany of Andhra Pradesh: A Review. Ethnobotanical leaflets, 12:305-310.

Savitramma, N. and Sulochana, C. 1998. Endemic medicinal plants from Tirumala hills, Andhra Pradesh, India. Fitoterapia. 69(3): 253-254.

Shashi, S.S. 1994. Encyclopedia of Indian tribes Vol-2. Andhra Pradesh. Anmol Pub. Pvt. Ltd., New Delhi.

Singh, K.K. Palvi, S.K. and Singh, H.B. 1981. Survey and biological activity of some medicinal plants of Mannanur forests, Andhra Pradesh, Indian J. For 4:115-117.

Sudhakar, A. 1998. Medicinal importance of some angiospermic weeds used by the rural people of Chittor district of Andhra Pradesh. Fitoterapia. 69(5): 390-400.

Sudhakar, S. 1980. Studies on medicinal plants of Upper East Godavari district, Andhra Pradesh. J. Indian Bot. Soc. 59(suppl.): 168(abstract).

Suneetha, J. Seetharami Reddi, T.V.V. and Prasanthi, S. 2009. Traditional phytotherapy for bites in East Godavari district (Andhra Pradesh). Ehnobotany, 21: 75-79.

Vedavathy, S., and Mrudula, V. 1996. Herbal folk medicine of Yanadis of Andhra Pradesh. Ethnobotany 8:109-111.

Vedavathy, S., Rao, K.N., and Nagaraju, N. 1991. Folklore information from Rayalaseema region, Andhra Pradesh for Family Planning and Birth Control. Int. J. Pharmacagnosy 29(2): 113 – 116.

Venkaiah, M. 1980. Some medicinal plants of Vizianagaram district, Andhra Pradesh. J. Indian Bot. Soc. 59(suupl.): 168 (abstract).

Venkataraju, R.R. and Reddy, R.V. 1998. Ethnomedicinal properties of certain rare and interesting plants from Cuddaph hills, Andhra Pradesh, Golden Jubilee. National symposium on Spices, Medicinal and Aromatic Plants Biodiversity, Conservation and Utilization, Calicut, India. 10-12 August. pp.29.

Yasodamma, N. Santosh Kumar Mehar and Paramageetham, Ch. 2009. Threat assessment (IUCN categorization) for ethnomedicinal plants used by Chenchu tribe of Gundlabramheswaram in Nallamalai hills in Andhra Pradesh, Ethnobotany, 21: 51-60.

Chapter 8

Ethnobotanical Plants Used for General Purposes by the Tribes of Komarada Mandal of Vizianagaram District

N. Saradamani and K. Jhansi*

Department of Botany, Andhra University, Visakhapatnam, A.P, India

ABSTRACT

The purpose of the present study is to document the traditional knowledge of the plants used for the general purposes possessed by the tribal people of Komarada mandal of Vizianagaram district. A total of 39 plant species of 37 genera belonging to 27 families were documented that are being used either for house hold or commercial purpose. Among them Amaranthaceae family occupies first with more number of species. Many of these plants have more than one purpose. Thus plants as bioresource are responsible for the socio-economic uplift of the people. The knowledge has been gained by trial and error over the centuries and is priceless and replaceable, but it is being lost by the destruction of Natural ecosystem and acculturation of primitive tribal people.

Keywords: *Tribal, Species, Genera, Families, Amaranthaceae, Bioresource, Ecosystem, Acculturation.*

* Corresponding Author: E-mail: drnsaradamani@gmail.com

Introduction

Ethnobotany is the study of interaction, association and interrelationships of primitive communities especially tribal with the surrounding flora *i.e.*, how they make use of indigenous plants. The tribal have been dependent on the plants for survival, material and emotional needs. The knowledge of ethnic groups on the cultural, spiritual, social and economic values of plants is of immense value to humankind.

India has the largest concentration of tribal in the world. They constitute 8.14 per cent of total population of the country numbering 84.8 million belonging to various communities (Census of India, 2001). The study area Komarada mandal has five types of tribes. They are Mannadora, Mukadora, Kondadora, Jatapu and Gadaba. Their habitats are scattered homesteads. Each habitat is surrounded by fruit bearing trees, agricultural fields and forests. The economy is agro-forest based and in general forests and tribal are inseperable and existence and development of one depend on the other. They prepare various items like ropes, bags, brooms, hand fans, umbrellas etc., using plant parts and sell them in markets as occupation besides using them for domestic purposes also. Thus they earn their livelihood by undertaking occupations like traditional commerce of handicrafts beside settled and shifting agriculture, labour, and animal husbandry.

Materials and Methods

Data were collected by means of semi structural interviews. Information like local names of the plants, part used, habitats are collected along with the general use. Only people who were born and had always lived in the area were taken into consideration. Older people, women are mostly selected. Nearly 15 informants were selected and involved in the study. Voucher specimens of the cited plant species were prepared and referred with Gamble and Fischer (1915-1936) Flora of Presidency of Madras and Flora of Andhra Pradesh (Pullaiah *et al.*, 1997).

Results and Discussion

Ethnobotanical uses are tabulated in Tables 8.1 and 8.2

There comes 39 species belonging to 37 genera and 27 families. Among them Amaranthaceae occupies first with 4 species then Euphorbiaceae, Arecaceae with 3 species each, then Araceae, Fabaceae, Poaceae, Mimosaceae and Moraceae with 2 species each. Most of them are trees which numbers 19, herbs 15, shrubs 4 and climbers only one. The leaf is mostly useful part nearly 12 plants have shown the leaves as useful part, wood of 8 plants fruits of 6, plants without root 5, whole plants 3, stem bark of 3, flowers of 2, root of 1, root bark of 1, rhizome of 1, corms of 2, phloem sap of 1 are useful.

The tribes have a treasure of knowledge on utilization of plant products on the basis of commercial and domestic value. Besides using for household purposes, they also sell the products in the markets. They have a good indigenous creation of art like handicrafts. They make combs, drums, plates, mats, bags, ropes, greeting cards, baskets and various other household articles.

Table 8.1

Sl.No.	Plant Name	Local Name	Family	Habitat	Part Used	Use
1.	*Acacia chundra*	Chundra	Mimosaceae	Tree	Wood Stem bark	Timber Brushing teeth
2.	*Ageratum conyzoides*	Kampurodda	Asteraceae	Herb	Leaf	Insect repellent
3.	*Alangium salvifolium*	Oodika	Alangiaceae	Small tree	Wood	Fencing & Ornamental work
4.	*Alternanthera sessilis*	Pindikura	Amaranthaceae	Herb	Plant	Vegetable
5.	*Amaranthus spinosus*	Mulla thotakura	Amaranthaceae	Herb	Leaf	Vegetable
6.	*Amaranthus viridis*	Chilaka thotakura	Amaranthaceae	Herb	Leaf	Vegetable
7.	*Amorphophallus paeonifolius*	Adavidumpa	Araceae	Herb	Corm	Vegetable
8.	*Bambusa arundinacea*	Veduru	Poaceae	Tree	Culm, leaf Tender tillers, grains	Basket making Edible
9.	*Bixa orellana*	Jabarukaya	Bixaceae	Tree	Seed	Dye
10.	*Blumea mollis*	Kukka pogaku	Asteraceae	Herb	Leaves	Prevent lice in hens & bedbugs
11.	*Bombax ceiba*	Erra buruga	Bombacaceae	Tree	Seed lint	Matresses and pillows
12.	*Borassus flabellifer*	Thati	Arecaceae	Tree	Leaf	Mats, Hand fans, Umbrellas
13.	*Butea frondosa*	Moduga	Fabaceae	Tree	Leaf Stem	Umbrellas Gum
14.	*Canavalia gladiata*	Karu samba	Fabaceae	Climber	Whole plant	Green manure
15.	*Caryota urens*	Jeeluga	Arecaceae	Tree	Inflorescence Plant	Toddy Fodder
16.	*Cassia fistula*	Rela	Caesalpiniaceae	Tree	Wood	Agricultural implements & Roofing
17.	*Cipadessa baccifera*	Rondabilli	Meliaceae	Shrub	Wood	Fuel

Contd...

Table 8.1–Contd...

Sl.No.	Plant Name	Local Name	Family	Habitat	Part Used	Use
18.	*Colocasia esculenta*	Chama	Araceae	Herb	Leaf & corm Corm juice	Vegetable Hair wash
19.	*Cordia dichotoma*	Nakkiri	Boraginaceae	Tree	Fruit	Edible
20.	*Dioscorea bulbifera*	Karukanda	Dioscoreaceae	Herb	Rhizome	Vegetable
21.	*Digera muricata*	Chilaka thotakura	Amaranthaceae	Herb	Leaf	Vegetable
22.	*Eleusine coracana*	Ragulu	Poaceae	Herb	Grains Plant	Food Fodder
23.	*Euphobia hirta*	Pachanumi	Euphorbiaceae	Herb	Plant	Vegetable
24.	*Ficus benghalensis*	Marri	Moraceae	Tree	Leaves & branches Leaf	Religious practices Meal plates
25.	*Ficus religiosa*	Raavi	Moraceae	Tree	Leaf	Greeting cards
26.	*Gmelina arborea*	Peggummudu	Verbenaceae	Tree	Wood	Combs, Drums,Timber
27.	*Helicteres isora*	Chemal nara	Sterculiaceae	Shrub	Plant fibre	Roofing
28	*Hemidesmus indicus*	Suchandhipala	Periplocaceae	Shrub	Root bark	Cool drinks
29.	*Hyptis sauveolens*	S ma tulasi	Lamiaceae	Herb	Stem	Brushing
30.	*Madhuca indica*	Ippa	Sapotaceae	Tree	Flower	Intoxicating beverage
31.	*Mallotus philippensis*	Kampilikam	Euphorbiaceae	Tree	Fruit	Dye
32.	*Phoenix sylvestris*	Eeta	Arecaceae	Tree	Leaf Fruit Phloem sap	Mat, Baskets,Brooms Edible Intoxicating beverage
33.	*Phyllanthus reticulatus*	Pula guvva	Euphorbiaceae	Shrub	Stem bark	Brushing
34.	*Plumbago zeylanica*	Tella chitra mulam	Plumbaginaceae	Herb	Root	Intoxicating beverage (Arrack)
35.	*Prosopis spicigera*	Juvvi	Mimosaceae	Tree	Wood Fruit	Fuel Cattle food

Contd...

Table 8.1–Contd...

Sl.No.	Plant Name	Local Name	Family	Habitat	Part Used	Use
36.	*Shorea robusta*	Guggilam	Dipterocarpaceae	Tree	Resin Timber	Stem Wood
36.	*Solanum nigrum*	Kamanchi	Solanaceae	Herb	Leaf Fruit	Vegetable Edible
37.	*Strychnos potatorum*	Indupu	Loganiaceae	Tree	Seed	Water purification
38.	*Terminalia arjuna*	Tella maddi	Combretaceae	Tree	Fruit Wood	Tannin Timber
39.	*Urena lobata*	Anturasam	Malvaceae	Herb	Plant fibre	Bags & Ropes

Table 8.2: Plants and their Utilities

Sl.No.	Utility/Purpose	Name of the Species
1.	Baskets	1) *Bambusa arundinacea* 2) *Phoenix sylvestris*
2.	Brooms	1) *Phoenix sylvestris*
3.	Combs & Drums	1) *Gmelina arborea*
4.	Dye for clothes Dye for hair	1) *Mallotus philippensis* 2) *Bixa orellana*
5.	Fodder	1) *Caryota urens* 2) *Eleusine coracana* 3) *Prosopis spicigera*
6.	Fencing	1) *Alangium salvifolium* 2) *Borassus flabellifer*
7.	Fuel	1) *Alangium salvifolium* 2) *Cipadessa baccifera* 3) *Prosopis spicigera* 4) *Gmelina arborea*
8.	Greeting cards	1) *Ficus religiosa*
9.	Hand fans	1) *Borassus flabellifer*
10.	Insect repellent	1) *Ageratum conyzoides*
11.	Intoxicating drinks	1) *Borassus flabellifer* 2) *Caryota urens* 3) *Madhuca indica* 4) *Phoenix sylvestris*
12.	Mats	1) *Bambusa arundinacea* 2) *Borassus flabellifer* 3) *Phoenix sylvestris*
13.	Matresses and Pillows	1) *Bombax ceiba*
14.	Meal plates	1) *Ficus benghalensis* 2) *Shorea robusta*
15.	Roofing	1) *Helicteres isora* 2) *Cassia fistula*
16.	Ropes and bags	1) *Urena lobata*
17.	Tannin	1) *Acacia chundra* 2) *Terminalia arjuna*
18.	Timber–For house construction	1) *Borassus flabellifer* 2) *Prosopis spicigera*
	For furniture	1) *Gmelina arborea* 2) *Shorea robusta*
	For ornamental work	1) *Alangium salvifolium* 2) *Gmelina arborea*
	For Agricultural implements	1) *Acacia chundra* 2) *Cassia fistula* 3) *Prosopis spicigera* 4) *Shorea robusta*

Contd...

Table 8.2–Contd...

Sl.No.	*Utility/Purpose*	*Name of the Species*
19.	Water purification	1) *Strychnos potatorum*
20.	Edible vegetables	1) *Amaranthus spinosus* 2) *Amaranthus viridis* 3) *Alternanthera sessilis* 4) *Amorpophallus paeonifolius* 5) *Bambusa arundinacea* 6) *Colacasia esculenta* 7) *Digera muricata* 8) *Euphorbia hirta*
	Edible fruits	1) *Borassus flabellifer* 2) *Cordia dichotoma* 3) *Phoenix sylvestris* 4) *Gmelina arborea* 5) *Solanum nigrum*
	Edible grains	1) *Bambusa arundinacea* 2) *Eleusine coracana*

They use *Gmelina arborea* wood for combs and drums, fibre of *Urena lobata* for bags and ropes, leaves of *Phoenix sylvestris* for brooms, leaf of *Ficus benghalensis* for meal plates, wood of *Alangium salvifolium* for ornamental work etc., these tribes use cotton fibres extracted from dehisced fruits of *Bombax ceiba* for preparation of pillows and mattresses but the tribal of Tarai region, Uttar Pradesh use the calyx as vegetable which is treated as highly nutritive besides using the fibre for pillows (Bajpayee and Dixit, 1996). Wood of *Cassia auriculata* is used by the tribal of Komarada mandal for agricultural implements, however it was reported that the flowers of the same are used as vegetable (Reddy *et al.*, 2007). In this mandal flowers of *Madhuca indica* are being used for making Ippa saara, an intoxicating drink but the same are being used by the people of Orissa as food. When food scarcity of food occurs it forms an important

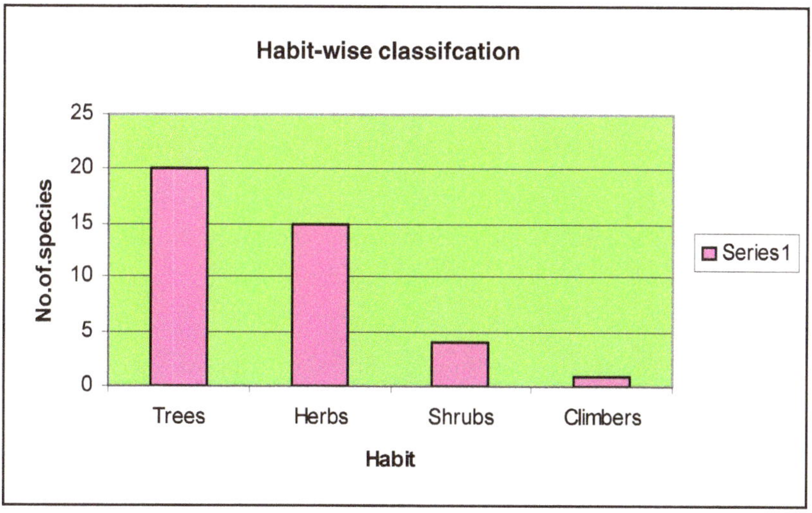

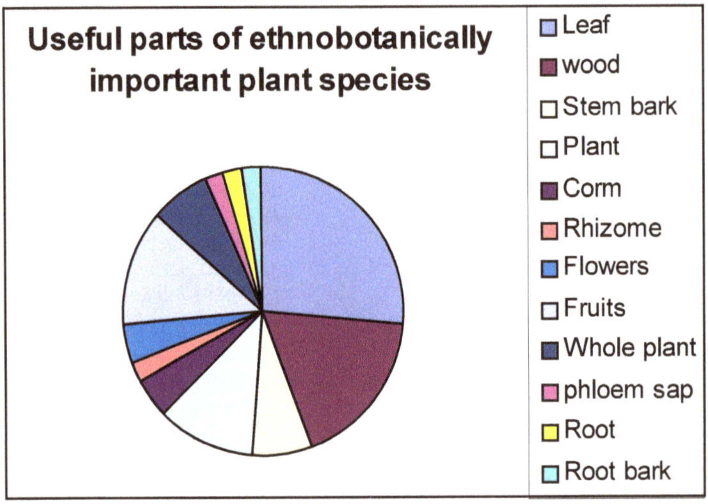

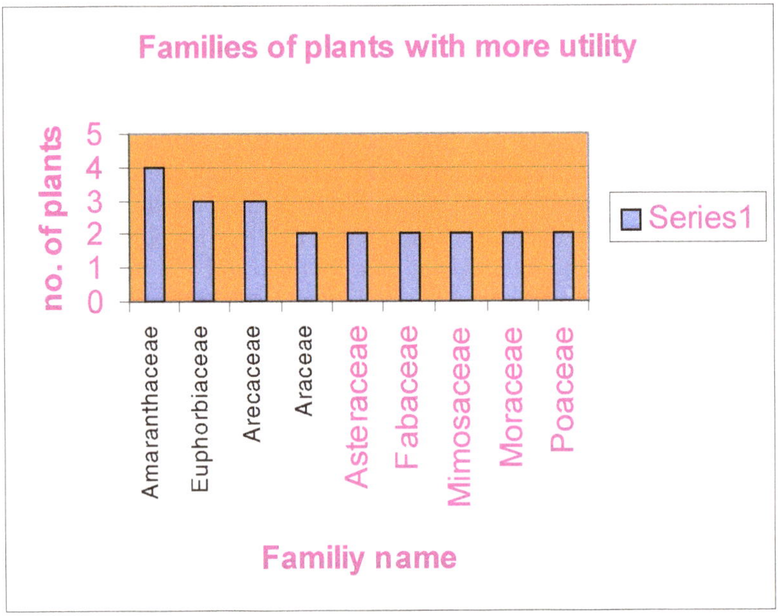

diet (Sinha and Valeria, 2005). Here *Eluesine coracana* grains are used for ambali, but the same are used for pillow making (Reddy *et al.*, 2008).

Conclusions

The plants have so many uses, their products are of low cost and free of pollution and durable. So, more importance should be given in their utilization. But due to urbanization and industrialization, eco globalization forests are degraded and knowledge of tribal is being devasted. So, it is crucial to record the fast disappearing

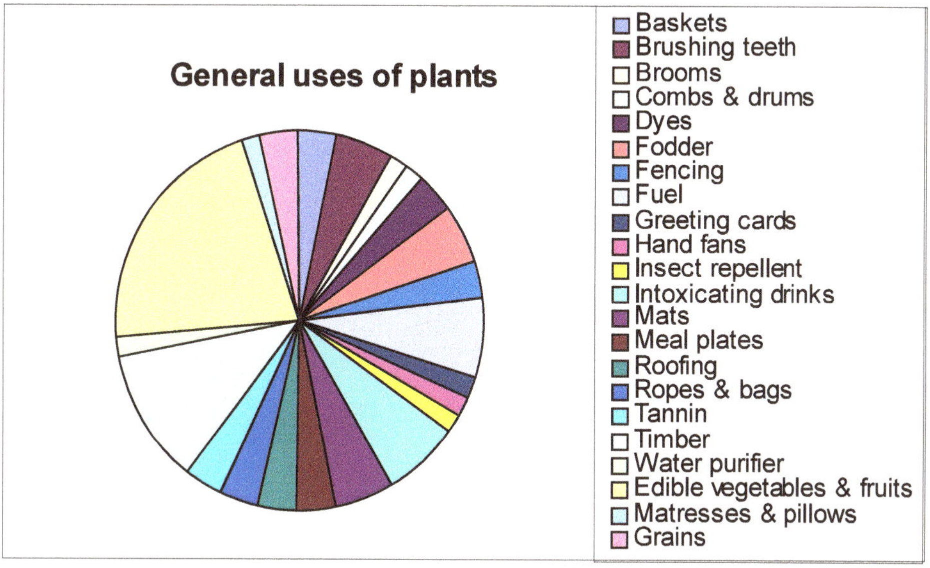

knowledge before it vanishes definitely along with the present generation of elderly persons. Afforestation and horticultural programmes should be intensified and suitable plant species should be germinated and planted to regulate the degraded forests in agency tracts. Social forestry should be developed. Village level committees should form and help the village panchayats in development of forests and cultivation of various plants for their products. Government should advertise their products through media. Due to burgeoning developments in cities the handicrafts have now in demand, but they get very less income by selling the products in the weekly markets. Special funds should be released from State Government to establish training and development centers at block level. TRIFED (Tribal cooperative marketing development federation) should play more role for promotion of tribal products by opening display centers or sales outlets at important places or tourist destinations.

References

Anonymous. 2001. Census of India. Directorate of census operations, Registrar, General of India, New Delhi.

Bajpayee, K. and Dixit, G,1997. Ethnobotanical studies on Food –stuffs of Tribal of Tarai region, Uttar Pradesh. J.Econ. Taxon Bot. Additional series, 12:128-132

Gamble, J.S. and Fischer, C.E.C., 1915-1936. Flora of Presidency of Madras Adland & Son Ltd, London.

Pullaiah, T. *et al.*, 1997. Flora of Andhra Pradesh. 5 Vols. Scientific publishers, Jodhpur.

Rekha, S. and Valeria, L. 2005. Wild tribal food plants of Orissa. Indian J. Traditional Knowledge, 4(3): 246-252.

Reddy, K.N., Pattanaik,C., Reddy, C.S., Murty, E.N. and Raju, V.S. 2008. Plants used in traditional handicrafts in Northern Andhra Pradesh. Indian traditional knowledge 7(1): 162-165.

Reddy, K.N., Reddy, C.S., Pattanaik, C. and Raju, V.S. 2007. Traditional knowledge on wild food plants in Andhra Pradesh. Indian traditional knowledge 6 (1): 223-228.

Chapter 9

A Conspectus of the Tribe: Eragrostideae (Poaceae) in Andhra Pradesh, India

B. Ravi Prasad Rao[1]*, *B. Sadasivaiah*[1], *P. Priyadarsini*[1], *A. Madhusudhana Reddy*[2], *S. Sunitha*[1], *A. Lakshmaiah*[1], *K. Venkata Subbaiah*[1], *J. Narahari Prasad*[1], *S. Khadar Basha*[1], *M.V. Suresh Babu*[1], *and P. Venkateswara Prasanna*[3]

[1]*Biodiversity Conservation Division, Department of Botany, Sri Krishnadevaraya University, Anantapur – 515 003, Andhra Pradesh*

[2]*Department of Botany, Yogi Vemana University, Kadapa, Andhra Pradesh*

[3]*Scientist, Botanical Survey of India, Deccan Regional Centre, Hyderabad, Andhra Pradesh.*

ABSTRACT

The tribe Eragrostideae is represented by 58 taxa (14 genera) in Andhra Pradesh. Nine of them are endemic to India, of which 4 are exclusive to Peninsular India, viz. *Eragrostis ciliaris* var. *clarckei, Eragrostis deccanensis, Tripogon jacquemontii* and *Tripogon wightii*. The largest genus is *Eragrostis* with 24 species. Of the 58 taxa, only one species, *Eleusine coracana* is domesticated. In the present chapter, all these taxa are systematically enumerated and provided dichotomous bracketed keys for easy identification.

* Corresponding Author: E-mail: rao_rp@rediffmail.com

Introduction

Poaceae (*nom. cons.* Gramineae), the family of grasses, are one of the largest families of flowering plants represented by over 10,000 species and 651 genera (Clayton & Renvoize, 1986). It has a major economic and ecological importance and the grasslands occupy third of the land's surface (Schantz, 1954). Grasses are vital for the sustenance of mankind through the production of cereal crops upon which much of the world's population depends for its food. According to an estimate about 70% of the farmlands of the world are cultivated with crop grasses. Grasses are good soil binders and moisture conservers and make good turfs and lawns. Some grasses are of medicinal value, certain other yield essential oils. Grasses are also used for paper making and for thatching, matting, making ropes, furniture, stuffing for pillows, brooms, cigarette and cheroot wrappers, Hindu ritual and musical instruments etc. The grasses are one of the most successful terrestrial life forms on the earth due to their adaptability to changeable environments, ability to coexist with grazing animals and with man and endless variations with distinct life forms (Clayton & Renvoize, 1986). Grasses inhabit the earth in greater abundance than any comparable group of plants. They are cosmopolitan in distribution and are recorded from the warm, humid, tropical climates, polar-regions and deserts. Despite of its importance to mankind, the study on grasses especially of wild ones with reference to taxonomy is neglected subject especially in India, owing to its complexity of floral morphology and erosion of expertise in grass studies over time.

Several classifications have been proposed for grasses. Brown (1810) has clearly recognized the two subdivisions of Gramineae, the Panicoideae and Pooideae, which he referred to as the Paniceae and Poaceae. The natural classification system of Bentham (1881) classified the family into two subfamilies, Panicoideae and Festucoideae. His classification was used with modifications by Hackel (1887), Stapf (1917-1934), Hitchcock (1920-1935) and Bews (1929). During 1950-1960 there was a great activity in the field of grass systematics especially in respect to evidence for phylogenetic arrangement of the major grass groups. New sub-family groupings were proposed by Beetle (1955), Stebbins (1956) and Tateoka (1957). Although in recent past it has been established that many of the characters defining major groups of grasses are associated with difference in photosynthetic metabolism (Clayton and Renvoize, 1986), the classification of grasses based on vegetative and floral morphology is considered easier for all practical purposes. For the present study, the concept of Clayton and Renvoize (1986) has been followed in delimitation of taxa. Accordingly Poaceae comprises 6 subfamilies and 40 tribes. The subfamily Bambusoideae comprise 13 tribes; Pooideae, 10; Centothecoideae, 1; Arundinoideae, 4; Chloridoideae, 5 and Panicoideae, 7. The tribe Eragrostideae is treated under sub-family Chloridoideae.

In the present study, the update records of members of tribe Eragrostideae occurring in the state of Andhra Pradesh has been enumerated. This work is the culmination of the past and current works, but primarily includes the results of our team explorations in the state of Andhra Pradesh for the past 15 years.

Study Area

Andhra Pradesh is the fifth largest state in India. The state lies between NL 12°37' 19°54' and EL 76°46'–84°46' in the southern part of India. The total area of the state is 2, 75,068 sq.km. Topographically 35 per cent of the state area lies below 150m altitude forming coastal plains, 15 per cent area lies between 150-300m; 37 per cent area between 300-600m; 10 per cent area between 600-900m and the remaining area lies +900m MSL. The state has about 968km of coast line. The state comprises twenty three districts which are grouped in to three distinct regions, Rayalaseema, Telangana and Coastal Andhra Pradesh. A total 64,000 sq.km of the total geographical area is under forests and 50 per cent of it covered being in Eastern Ghats (Rao *et al.*, 2008). The state has monsoon type of tropical climate. The mean annual temperature in summer ranges from 20-41°C and in winter 13-32°C. The average annual rainfall is around 940mm. The study area have a wide variety of soils *i.e.*, red, black, deltaic alluvial and coastal alluvial. The major rivers flowing through the state are Godavari, Krishna, Penna and Thungabhadra. According to Champion & Seth (1968), six vegetation types are observed in the state, viz. tropical semi-evergreen, moist deciduous, dry deciduous, dry savannahs, dry ever green and scrub forests. The study area comprises one centre of plant diversity, namely Nallamalais and one endemic centre, Tirupati-Kadapa hills (Davis *et al.*, 1994 & 95; Gadgil & Meher Homjii, 1986). A total of 20 wild life sanctuaries and four national parks are known to present in the state. Among 27 tiger reserves in India, Nagarjuna Sagar-Srisailam Tiger Reserve is situated in Nallamalai hills of Eastern Ghats. Over 600 sacred groves are also known to encounter in the state (Rao, 1996).

A comprehensive account on flora of Andhra Pradesh was documented by Pullaiah *et al.* (1997a, b, and c) and additional data was published by Pullaiah & Karuppasamy (2008) and Reddy *et al.* (2008). Accordingly, the state harbors 2803 species belonging to 1051 genera and 185 families of flowering plants; grasses comprises 294 species. Except Suxena (1947) much of the work on grasses has been carried out by the research teams working in Sri Krishandevarya University, Anantapur. This includes Prasanna (1988), Gayathri (1997), Sunitha (2002), Reddy (2002), Reddy *et al.* (2002, 2004, 2005, and 2009), Narahari Prasad (2006), Subbaiah (2008), Sadasivaiah *et al.* (2008, 2010), Rao *et al.* (2009a, b, 2010) and Reddy and Rao (2009). Other significant studies include Subba Rao & Kumari (2008).

Eragrostideae in Andhra Pradesh

The tribe Eragrostideae has 5 sub tribes namely Monanthochloineae, Eleusinineae, Triodiinae, Uniolinae and Sporobolinae. Representatives of Monanthochloineae, Eleusinineae and Sporobolinae are found in Andhra Pradesh. The tribe comprise about 1000 species belongs to 77 genera (Clayton & Renvoize, 1986). Presently, India harbors 106 species belonging to 24 genera.

The current study recorded 58 taxa of the tribe Eragrostideae belonging to 14 genera. Genus-wise representation of taxa: *Acrachne* (1), *Aeluropus* (1), *Coelachyrum* (1), *Dactyloctenium* (2), *Desmostachya* (1), *Dinebra* (1), *Eleusine* (2), *Eragrostiella* (4), *Eragrostis* (24), *Leptochloa* (4), *Myriostachya* (1), *Oropetium* (2), *Sporobolus* (10) and *Tripogon* (4). In the present study, the genus *Diplachne* P.Beauv. is merged with

Leptochloa P. Beauv. following Clayton & Renvoize (1986). Nine of the 58 taxa are endemic to India, of which 4 are strict endemics to Peninsular India, viz. *Eragrostis ciliaris* var. *clarckei*, *E. deccanensis*, *Tripogon jacquemontii* and *Tripogon wightii*. *Eragrostiella walkeri* is recollected in the state after lapse of a century. Of the 58 taxa, only one species, *Eleusine coracana* is cultivated.

All the taxa of the tribe Eragrostideae are systematically enumerated. The keys are dichotomous bracketed type. The taxa were provided with full citation and description. General world-wide distribution, distribution in India and Andhra Pradesh are provided. Representative specimens cited are from the collections of the authors, unless otherwise mentioned and in many cases they represent all the three regions of the state. All the specimens unless otherwise mentioned are deposites of SKU (herbarium).

Abbreviations Used in the Text

Districts in Andhra Pradesh: ADB- Adilabad; ATP- Anantapur; CTR-Chittoor; EG-East Godavari; GNT-Guntur; HYD-Hyderabad; KDP-Kadapa; KMM-Khammam; KNL-Kurnool; KSN-Krishna; MBNR-Mahaboobnagar; MDK-Medak; NLG- Nalgonda; NLR-Nellore; NZB-Nizamabad; PKM-Prakasam; SKLM-Srikakulam; VSKP-Visakhapatnam; VZN-Vizianagaram. *Plant collectors*: AL-A. Lakshmaiah; AMR- A. Madhusudana Reddy; BR-B. Ravi Prasad Rao; BSS-B. Sadasivaiah; CAB-C.A. Barber; CECF-C.E.C. Fischer; CP-C. Prabhakar Raju; GO-G. Obulesu; GVS-G.V.Subba Rao; JNP- J. Narahari Prasad; JSG-J.S. Gamble; KH-K. Hanumanthappa; KJ-K Jacobs; KVS – K.V. Subbaiah; MRS- M.R. Suxena; MSB-M.V. Suresh Babu; MSG-M.S. Gayathri; MSR-M. Sridhar Reddy; NPBK- N.P. Balakrishnan; PMR-PM. Rajeswaramma; PP- P. Priyadarsini; PVP- P.V. Prasanna; SKB-S. Khadar Basha; SS-S. Sunitha; TP-T. Pullaiah; VNS-V.Narayana Swamy.

Herbaria: CAL-Central national Herbarium; DD-Dehradun Herbarium; MH-Madras Herbarium; SKU-Sri Krishandevaraya University, Herbarium.

Systematic Enumeration of Taxa

Key to Genera

1. Lemmas 5-13-nerved .... *Aeluropus*
1. Lemmas 1-3-nerved .... (2)
2. Spikelet with one fertile floret, sunk in the rachis .... *Oropetium*
2. Combination of characters not as above .... (3)
3. Spikelets 2 -or more-flowered (1-flowered in *Leptochloa neesi*) .... (4)
3. Spikelets strictly 1-flowered .... *Sporobolus*
4. Lemmas usually entire at apex, obtuse, acute or acuminate or 2-toothed, then glabrous near the margins and along the side nerves .... (5)
4. Lemmas usually emarginate or 2-4 lobed or toothed at apex, rarely entire and then hairy along the nerves, freequently mucronate or awned .... (13)

5. Glumes aristate or aristate-acuminate .... (6)
5. Glumes otherwise .... (7)
6. Spikelets 1-3-flowered .... *Dinebra*
6. Spikelets many-flowered .... *Myriostachya*
7. Spikelets in open, contracted or panicled spikes or rarely in solitary spikes .... (8)
7. Spikelets in digitately or racemosely arranged spikes .... (10)
8. Inflorescence panicled .... (9)
8. Inflorescence a solitary secund spike .... *Eragrostiella*
9. Lemmas rounded on the back; spikes crowded in globose heads .... *Coelachyrum*
9. Lemmas acutely keeled on the back; spikes seated on open panicles .... *Eragrostis*
10. Axis and branches of the inflorescence ending with a spikelet .... (11)
10. Axis of spikes terminating in a sharp point; spikes digitate .... *Dactyloctenium*
11. Spikelets falling entire at maturity .... *Desmostachya*
11. Spikelets breaking at maturity .... (12)
12. Spikes digitate or sub digitate; spikelets 3-6-flowered .... *Eleusine*
12. Spikes usually in pseudo-whorls or scattered; spikelets 8-20-flowered .... *Acrachne*
13. Spikelets in terminal, solitary secund spikes .... *Tripogon*
13. Spikelets otherwise .... *Leptochloa*

ACRACHNE Wight & Arn.

Acrachne racemosa (Heyne) Ohwi in Bull. Tokyo Sci. Mus. N. 18:1. 1947; Bor, Grass. India 487. 1960; Moulik, Grass. Bamb. India 2:584. 1997. *Eleusine racemosa* Heyne ex Roem. & Schult., Syst. Veg. 2:583. 1817. *Eleusine verticillata* Roxb., Fl. Ind. 1:346. 1820; Hook. f., Fl. Brit. India 7:295. 1897; Fischer, Fl. Madras. 3:1840. 1934.

Annuals, culms erect, tufted, to 70cm; nodes glabrous. Leaf sheaths glabrous. Ligules hairy. Leaf blades linear-lanceolate, ovate-acuminate, base rounded, acuminate, margins scaberulous. Racemes spiciform. Spikelets ovate, 6-18-flowered. Lower glume ovate, keeled, 3-nerved, acute and mucronate. Upper glume ovate-lanceolate, keeled, acuminate, awned. Lower lemma broadly ovate, 3-nerved, awned. Upper lemma ovate, 2-nerved, awned. Palea minute. Stamens 3. Grains oblong, transversely rugose.

Distribution: *World*: Australia, South-East Asia, Sri Lanka, and Tropical Africa. *India*: Throughout. *Andhra Pradesh*: Throughout.

Common near streams in forests. *Fl. & Fr.*: July - Dec. *Rep. Spec.*: Upper Ahobilam (KNL), BR & AMR 20074; Rollapenta (PKM), BSS & SKB 29393; Bhajaranga kona (KDP), KVS 29716.

AELUROPUS Trinius

Aeluropus lagopoides (L.) Trin. ex Thw. Bnum. Pl. Zeyl. 374. 1864, *'lagopodioides'*; Fischer, Fl. Madras. 3:1843. 1934; Bor, Grass. India 380. 1960. *Dactylis lagopoides* L. Mant. 33.1767. *Aeluropus villosus* Trin. ex C. Meyer, Verz. Pfl. Cauc. Merr. 18. 1831; Fl. Brit. India 7:334. 1896.

Perennials, culms creeping, tufted, to 20cm; nodes glabrous. Leaf sheaths glabrous. Ligule shortly hairy ridge. Leaf blades linear, base rounded, margin convolute, apex acuminate, glabrous. Heads subglobose pedunculate, terminal. Spikelets 4–8-flowered, crowded. Glumes ovate-oblong, subacute, ciliate. Lemmas ovate, obtuse, many-nerved, ciliate. Palea broad, 3-lobed. Stamens 3. Grains obovoid-oblong, dorsally compressed.

Distribution: *World*: Afaghanistan, Arabia, Bangladesh, Caspian region, Iran, Mediterranean Regions, Pakistan, Russia, and Sri Lanka; *India*: Maharashtra, Tamil Nadu, and West-Benghal; *Andhra Pradesh*: East Godavari, Guntur, Krishna and Nellore districts.

Common along sea coast and on alkaline soils. *Fl & Fr.*: Mar.-Aug. *Repr. Spec.*: Krishnapatnam (NLR), MSG 10885 (SKU).

COELACHYRUM Hochst. & Nees

Coelachyrum lagopoides (Burm.f.) Senaratna, Grass. Sri Lanka 79.1956; Bor, Grass. India 488. 1960. *Coelochyropsis lagopoides* (Burm.f.) Bor in Ann. Natur. Hist. Mus. Wien.75:25. 1972; Moulik, Grass. Bamb. India 2:585.1997. *Eleusine lagopoides* (Burm. f.) Merr. In Philipp. J. Sci. 19:339.1921; Fischer, Fl. Madras 3:1840.1934. *Eleusine brevifolia* (Willd.) R. Br. ex Hook. f., Fl. Brit. India 7:294.1896.

Perennials, culms creeping or erect, to 50cm, rooting at nodes; nodes glabrous. Leaf sheaths glabrous. Ligules membranous. Leaf blades oblong-lanceolate, margins ciliate. Spikes numerous, crowded into a globose head. Spikelets elliptic-ovate, many–flowered. Lower glume lanceolate, 1-nerved, keeled, chartaceous. Upper glume ovate-lanceolate, chartaceous, acute, 1-nerved. Upper lemma elliptic-lanceolate, 3-nerved, ciliate, hyaline, keels winged. Palea oblong, coriaceous, base bearded. Stamens 3. Grains trigonous or ovate, transversely rugose.

Distribution: *World*: Sri Lanka; *India*: South India; *Andhra Pradesh*: East and West Godavari districts, Krishna, Nalgonda, Nellore, Rayalaseema.

Occasional in plains and forest peripheries. *Fl. & Fr.*: July-Dec. *Repr. Spec.*: Peapully (KNL), PVP 3189; Varikuntapadu (NLR), PMR 16472; Kalasamudram (ATP), AMR 24521; Konapuram (ATP), JNP & AL 26659; Kalasamudram (ATP), KVS & BSS 28992; Ahobilam (KNL), BR & BSS 32265.

Note: The genus *Coelochyropsis* Bor (1960) was treated as synonym to *Coelachyrum* by Clayton & Renovoize (1986), same concept is followed in the present study.

DACTYLOCTENIUM Willd.

1. Stoloniferous; spikes 2-6 .... *D. aegyptium*
1. Non-stoloniferous; spikes solitary .... *D. aristatum*

Dactyloctenium aegyptium (L.) P. Beauv., Ess. Agrost. Expl. Pl. 15.1812; Fischer, Fl. Madras 3:1840.1934; Bor, Grass. India 489. 1960; Moulik, Grass. Bamb. India 2:585.1997. *Eleusine aegyptia* (L.) Desf., Fl. Atlant. 1:85. 1798; Hook. f., Fl. Brit. India 7:295. 1896. *Cynosurus aegyptius* L., Sp. Pl. 72.1753.

Annuals, culms stoloniferous, to 70cm, erect or ascending, rooting at nodes; nodes glabrous. Leaf sheaths glabrous. Ligules membranous. Leaf blades linear-lanceolate, acuminate, glabrous, sparsely ciliate on the margins. Spikes digitate. Spikelets oblong 3-5-flowered. Lower glume ovate, boat-shaped, 1-nerved, keeled, scabrid. Upper glume obovate, 1-nerved, mucronate, scabrid. Lemma ovate-lanceolate, mucronate, keeled. Palea lanceolate, acuminate, 2-keeled. Stamens 3. Grains obovate, transversely rugose.

Distribution: *World*: Afghanistan, Africa, America, Myanmar and Sri Lanka; *India*: Throughout; *Andhra Pradesh*: Throughout.

Common in forest plains, margins and found as a weed in crop fields and waste lands. *Fl.* & *Fr.*: June – Sept. *Repr. Spec.*: Mosra Forest Nursery (NZB), TP & BR 6069; Leguntapadu (NLR), BR & PMR 16432. Kambagiri konda (KNL), BR & SS 20010; Kalasamudram (ATP), BR & AMR 21898; Tatipudi (VZN), AMR 24526; Lambasangi (VSKP), AMR 25630; Konakondla (ATP), JNP & AL 26602; Chendrayuni kona (ATP), JNP 26730; Kannekal (ATP), KVS 28055; Krishnapuram (CTR), AL & BSS 28567; Borapuram (MBNR), BSS & SKB 29567 & 32291.

Dactyloctenium aristatum Link, Horty. Berol. 1:59. 1827; Bor, Grass. India 489. 1960; Moulik, Grass. Bamb. India 2:587. 1997. *Eleusine aristata* Hook. f., Fl. Brit. India 7:296.1896.

Annuals, culms erect or geneculately ascending, tufted, to 10cm, rooting at nodes; nodes glabrous. Leaf sheaths hairy. Ligules fimbriate, membranous. Leaf blades linear-lanceolate, base cordate or rounded, acuminate, glabrous. Spike solitary. Spikelets ovate, 2-3-flowered. Lower glume ovate, 1-nerved, keeled, chartaceous. Upper glume ovate-elliptic, aristate, scabrid, chartaceous. Lower lemma ovate-lanceolate, membranous, keeled, acute. Upper lemma akin to lower lemma. Palea ovate-lanceolate, chartaceous, keeled. Stamens 3. Grains globose, rugose.

Distribution: *World*: North-East Africa; *India*: North West and South India; *Andhra Pradesh*: Adilabad, Anantapur, Guntur and forests of Kurnool districts.

Rare in open forests and peripheries. *Fl.* & *Fr.*: June – Nov. *Repr. Spec.*: Rudrakod (KNL), SS & AMR 20197; Hanimireddi palli (ATP), JNP & AL 26892; Nagalooty (KNL), BR & BSS 30149; Velgode (KNL), BSS & SKB 30230.

DESMOSTACHYA Stapf

Desmostachya bipinnata (L.) Stapf in Dyer. Fl. Cap. 7:632.1900; Fischer, Fl. Madras 3:1819.1934; Bor, Grass. India 491.1960; Moulik, Grass. Bamb. India 2:588.1997.

Briza bipinnata L., Syst. Nat. ed. 10:875.1759. *Eragrostis cynosuroides* (Retz.) P.Beauv. Ess. Agrost. 162.1812; Hook.f. Fl. Brit. India 7:324.1896.

Perennials, stoloniferous, culms erect, to 90cm, branches from base. Leaf sheaths glabrous. Ligule a ring of hairs. Leaf blades linear, margin hispid, acuminate. Panicle narrow. Spikelets ovate-lanceolate, sessile, secund, 2 seriate, many-flowered. Lower glume ovate, small, 1-nerved, membranous, scabrid. Upper glume ovate-lanceolate, 1-nerved, membranous. Lemmas ovate, acute, glabrous, faintly 3-nerved, chartaceous. Palea linear keeled, shorter than the lemma, scabrid. Grains ovoid.

Distribution: *World*: Middle East to Indo-China, Tropical Africa; *India*: Throughout; *Andhra Pradesh*: Anantapur, Krishna, Nellore, Visakhapatnam and West Godavari districts.

Common in marshy habitats, in plains. *Fl.* & *Fr.*: Very infrequent. *Repr. Spec.*: Kanekal (ATP), BR & KVS 28052; Hottebetta (ATP), JNP, AL & KVS 28690.

DINEBRA Jacq.

Dinebra retroflexa (Vahl) Panz. in Denksehr., Acad. Wiss. Mynchen. 270. t. 12. 1814; Fischer, Fl. Madras 3:1841. 1934; Bor, Grass. India 491.1960; Moulik, Grass. Bamb. India 2:589.1997. *Cynosurus retroflexus* Vahl, Symb. Bot. 2:20.1791. *Dinebra arabica* Jacq., Fragm. 77.t.121. f. 1.1809; Hook. f., Fl. Brit. India 7:297. 1896.

Annuals, culms erect or ascending, tufted, leafy, to 1.2m; nodes slender, glabrous. Leaf sheaths glabrous. Ligules membranous. Leaf blades linear-lanceolate, base cordate, apex acuminate, glabrous. Inflorescence elliptic-oblong to pyramidal. Spikelets 1-3-flowered. Lower glume elliptic-lanceolate, 1-nerved, keeled, coriaceous, scaberulous, awned. Upper glume elliptic-lanceolate, 1-nerved, keeled, awned or cuspidate, persistent. Lemmas narrow, ovate, hyaline, sub acute or obtuse, 1-nerved. Paleas ovate, hyaline, 2-keeled, sparsely ciliate on the keels. Stamens 3. Grains ovoid, trigonous.

Distribution: *World*: Africa, Pakistan, and Sri Lanka; *India*: Peninsular & Central India; *Andhra Pradesh*: Throughout.

Occasional in forests and common in crop fields. *Fl.* & *Fr.*: Aug.-Sept. *Repr. Spec.*: Kammavaripalli (KNL), PVP 2786; Narsapur Forest (MDK), BR 1163; Sirikonda RF (NZB), BR 9699; Jagadevipet (NLR), BR & PMR 19521; Gundumallappa kona (KNL), SS & AMR 20285; Kalasamudram (ATP), BR & AMR 21890; Kadapa (KDP), AL 26579; Konapuram (ATP), JNP & AL 26669; Reddi palli form (ATP), JNP & AL 26755; Mannila (ATP), JNP & AL 26792; Pamidi (ATP), KVS 28015; Jyothi (KDP), BR & BSS 30859; Gundlabrah-meswaram (KNL), BSS & SKB 32972.

ELEUSINE Gaertner

1. Spikes slender, narrow, straight; wild species .... *E. indica*
1. Spikes stout, broad, curved at apex; cultivated species .... *E. coracana*

Eleusine coracana (L.) Gaertner, Fruct. Pl. 1:8. 1789; Hook. f., Fl. Brit. India 7:294. 1896; Fl. Madras 3:1839; Bor, Grass. Ind. 492. 1960. *Cynosurus corocanus* L. Syst. Nat. (ed. 10) 875. 1759.

Annuals, erect, tufted, to 1m; nodes glabrous. Leaf sheaths glabrous. Ligule hairy. Leaf blades linear, convolute, overlapping the culm, acuminate, base truncate, glabrous. Spikes 4-7. Spikelets ovoid, glabrous. Glumes oblong, margins scaberulous, acute. Lemma broadly ovate. Palea ovate. Stamens 3. Caryopsis globose, dark red brown, often black.

Distribution: *World*: Tropical countries; *India, Andhra Pradesh:* Throughout.

Commonly cultivated; oftenly run wild. *Fl.* & *Fr.*: July–Sep. *Repr. Spec.*: Chinnakunta (ATP), PVP 2639.

Eleusine indica (L.) Gaertner, Fruct. 1:8.1789; Hook. f., Fl. Brit. India 7:293.1896; Fischer, Fl. Madras 3:1839.1934; Bor, Grass. India 493. 1960; Moulik, Grass. Bamb. India 2:590.1997. *Cynosurus indicus* L., Sp. ed. 1.72.1753.

Annuals or perennials, culms erect, tufted, to 90cm; nodes glabrous. Leaf sheaths glabrous. Ligules membranous. Leaf blades linear or linear-lanceolate, base cordate, tip blunt. Spikes digitate or sub-digitate. Spikelets 2-5-flowered. Lower glume oblong-lanceolate, chartaceous, keeled, 1-3-nerved, acute, scabrid on the keels. Upper glume ovate-oblong, chartaceous, keeled, 5-7-nerved. Lemmas ovate oblong, chartaceous, 3-nerved. Paleas elliptic-oblong, hyaline, 2-keeled, 2-nerved. Stamens 3. Grains ovate.

Distribution: *World*: Tropics and sub tropics; *India, Andhra Pradesh:* Throughout.

Common in open places usually near water courses. *Fl.* & *Fr.*: July–Dec. *Repr. Spec.*: Mosra Forest Nursery (NZB), TP & BR 6081; Bhogeswaram (KNL), BR & SS 22666; Tirumala hills (CTR), BR & AMR 21801; Talakona (CTR), AMR 21823; Mydukur (KDP), AL 26574; Mannila (ATP), JNP 26787; Gundlabrahmeswaram (KNL), BSS & SKB 32232.

ERAGROSTIELLA Bor

1. Leaves from the basal sheaths reaching to the base of the inflorescence .... (2)
1. Leaves from the basal sheaths very much shorter, never reaching the base of the inflorescence .... (3)
2. Leaves flat; lemmas 2.5mm .... *E. walkeri*
2. Leaves convolute, filiform, with long white hairs; lemmas below 2mm .... *E. nardoides*
3. Keels of the palea broadly-winged .... *E. bifaria*
3. Keels of the palea narrowly-winged .... *E. brachyphylla*

Eragrostiella bifaria (Vahl) Bor, in Indian For. 66:270.1940; Bor, Grass. India 494.1960; Moulik, grass. Bamb. India 2:594.1997. *Poa bifaria* Vahl, sym. Bot. 2:19.1791. *Erogrostis bifria* (Vahl) Wight ex Steudel, Nom. Bot. Ed. 2.1:562.1840; Hook. f., Fl. Brit. India 7:325.1896; Fischer, Fl. Madras 3:1828.1934.

Annuals or perennials, culms erect, tufted, to 90 cm; nodes glabrous. Leaf sheaths glabrous. Ligules membranous or fimbriate. Leaf blades linear or acicular, often rolled, base narrow, acute, glabrous. Inflorescence spikes. Spikelets slightly compressed, ovate-lanceolate or oblong-lanceolate, 5-30-flowered. Lower glume oblong–lanceolate chartaceous, 1-nerved, keel scabrid. Upper glume ovate–lanceolate, chartaceous, 1-nerved. Lemmas broadly ovate, chartaceous, 3-nerved. Paleas elliptic–ovate, hyaline, 2-keeled, winged, ciliate. Stamens 3. Grains globose.

Distribution: *World*: Myanmar, Srilanka; *India*: Madhya Pradesh, Maharastra and South India; *Andhra Pradesh*: Throughout.

Common in grass lands, plains and forest peripheries. *Fl.* & *Fr.*: Oct. – Dec. *Repr. Spec.*: Arsapalli (NZB), BR & KH 6487; Tirumala hills (CTR), BR & AMR 21862; Konapuram (ATP), JNP & AL 26661; Chandrayunikona (ATP), JNP 26743; Penukonda (ATP), JNP 26765; Uravakonda (ATP), AL 26766; Omkaram (KNL), MSR & BSS 29349.

Eragrostiella brachyphylla (Stapf) Bor, Indian For. 66:270. 1940; Bor, Grass. India 494. 1960; Moulik, Grass. Bamb. India 2:594. 1997. *Eragrostis brachyphylla* Stapf in Hook. f., Fl. Brit. India 7:327. 1896; Fischer, Fl. Madras 3:1828. 1934.

Perennials, culms erect, tufted, to 50cm, slender; nodes glabrous. Leaf sheaths glabrous. Ligule a ciliate rim. Leaf blades linear, base rounded, apex acute, glabrous. Inflorescence solitary raceme. Spikelets ovoid-oblong, secund, 5-18-flowered. Lower glume linear-lanceolate, acute, 1-nerved. Upper glume lanceolate, acute, 1-nerved. Lemma broadly ovate, 2-nerved, keeled, margins hyaline. Palea obovate, keels ciliate, margins inflexed, hyaline. Stamens 3. Grains ellipsoid.

Distribution: *World*: Endemic to India; *India*: Bihar, West Bengal, and Peninsular India; *Andhra Pradesh*: Throughout.

Common in waste lands, forest plains and on hill slopes. *Fl.* & *Fr.*: July-Nov. *Repr. Spec.*: Kalasamudram (ATP), PVP 2613; Kamareddy (NZB), BR 9279; Upper Ahobilam (KNL), SS & AMR 22692; Rollapenta (PKM), BR & BSS 29396.

Eragrostiella nardoides (Trin.) Bor, Indian For. 66: 270. 1940; *Eragrostis nardoides* Trin. Mem. Acad. Imp. Sci. St. – Petersbourg 1:415. 1830; Fl. Brit. India 7:326. 1896.

Perennial, culms erect, tufted, to 60cm. Leaf blades filiform, convolute, with very delicate, long hairs. Spikes long, slender. Glumes subequal, acutely keeled. Lemmas turgid. Caryopsis ellipsoid-oblong with oblique base.

Distribution: Endemic to India; *India*: *Andhra Pradesh*, Uttar Pradesh and Bihar.

Rare in open moist grasslands. *Fl.* & *Fr.*: Aug.-Dec.

Note: The taxon is included based on Reddy *et al.* (2008), species is cited without any locality and voucher specimen.

Eragrostiella walkeri (Stapf) Bor, Indian For. 66:270. 1940; Bor, Grass. India 495. 1960; Moulik, Grass. Bamb. India 2:594. 1997. *Eragrostis walkeri* Stapf in Trin., Fl. Ceyl. 5:298. 1900; Fischer, Fl. Madras 3:1828. 1934.

Perennials, culms erect, tufted, to 80cm, tuft of leaf sheaths at the base; nodes glabrous. Leaf sheaths glabrous. Ligules ciliate, membranous. Leaf blades filiform, convolute, base rounded, glabrous, margin scabrous, acuminate. Inflorescence terminal, solitary racemes; rhachis flattened. Spikelets secund, ovate, subsessile, margin scaberulous, 2-8 flowered. Lower glume linear-lanceolate, acute, 1-nerved. Upper glume boat shaped, keeled, acute, 1-nerved. Lemmas broadly ovate, margin hyaline, 3-nerved. Palea spatheolate, keels ciliate, margin inflexed, hyaline. Stamens 3. Grains ellipsoid.

Distribution: *World*: Srilanka; *India*: South India; *Andhra Pradesh*: Anantapur, Kurnool, Mahaboobnagar, Prakasam and Visakapatnam districts.

Common in forest plains and on open slopes. *Fl.* & *Fr.*: July-Dec. *Repr. Spec.*: Gundumallappa sacred grove (KNL), BR & SS 20287; G. Madugula (VSKP), BR & AMR 25607; Srisailam (KNL), AMR 25688; Penukonda (ATP), JNP 26751; Vajrakarur (ATP), AL 28571; Pamidi (ATP), BR & KVS 28023; Amrabad (MBNR), BSS & SKB 32210.

ERAGROSTIS Wolf

1. Spikelets breaking up from above downwards; rachis fragile .... (2)
1. Spikelets breaking up from below upwards; rachis tough .... (13)
2. Lemma ciliate on the margins .... (3)
2. Lemma not ciliate on the margins .... (4)
3. Panicles clavate, dense, shining; keels of the palea narrowly winged .... *E. deccanensis*
3. Panicle cylindric, compact, dull; keels of the palea not winged .... *E. ciliata*
4. Keels of the palea ciliate .... (5)
4. Keels of the palea not ciliate .... (11)
5. Panicle spiciform or compact and dense .... (6)
5. Panicle effuse .... (8)
6. Lemma ciliate on the keels; annuals .... *E. ciliaris* var. *ciliaris*
6. Lemma not ciliate on the keels; perennials .... (7)
7. Panicle very dense, cylindric; lemma rounded at apex .... *E. riparia*
7. Panicle more or less dense, interrupted; lemma acute at apex .... *E. coarctata*
8. Lemma ciliate on the keels .... *E. ciliaris* var. *clarckei*
8. Lemma not ciliate on the keels .... (9)
9. Plants viscous; Panicles often leaden grey .... *E. viscosa*
9. Plants not viscous .... (10)
10. Panicle loose; axis hairy at the nodes .... *E. tenella* var. *tenella*
10. Panicle contracted; axis not hairy at the nodes .... *E. tenella* var. *insularis*
11. Panicle lax, open thyrsiform .... *E. aspera*
11. Panicles not thyrsiform .... (12)
12. Spikelets few-flowered, loose, scattered; lemmas less than 1mm long .... *E. japonica*
12. Spikelets 6-14-flowered, densely clustered; lemmas 1mm or more long .... *E. diarrhena*
13. Annuals .... (14)

13. Perennials .... (20)
14. Lemmas upto 1.5mm long .... (15)
14. Lemmas over 1.5mm long .... (18)
15. Plants glandular .... (16)
15. Plants eglandular .... *E. gangetica*
16. Lowest branches of the panicle fascicled or sub whorled .... *E. pilosa*
16. Lowest branches not whorled .... (17)
17. Spikelets less than 2mm wide; leaf margins eglandular, grains truncate at both ends .... *E. maderaspatana*
17. Spikelets more than 2mm wide; leaf margins glandular, grains globose .... *E. minor*
18. Spikelets 2mm or more wide, ovate or ovate-oblong .... (19)
18. Spikelets 0.75-2mm wide, lanceolate .... *E. tremula*
19. Plants with or without glands; upper glume 3-nerved, paleas persistent .... *E. cilianensis*
19. Plants eglandular; upper glume 1-nerved, paleas decidous .... *E. unioloides*
20. Lemmas upto 1.8mm long .... (21)
20. Lemmas 2mm or longer .... (23)
21. Spikelets 1.7-2.5mm wide, paleas deciduous .... *E. atrovirens*
21. Spikelts 1-1.2mm wide; paleas persistent .... (22)
22. Panicle contracted; spikelets crowded; straight, tip of lemma overlapping base of next lemma .... *E. nutans*
22. Panicles wide spreading; spikelets loose, slightly curved; tip of lemma just touching the base of next lemma .... *E. papposa*
23. Pedicels of the spikelets with glandular band .... *E. tenuifolia*
23. Pedicels of the spikelets devoid of glandular band .... *E. nigra*

Eragrostis aspera (Jacq.) Nees, Fl. Afr. Austr. 911.408.1841; Hook. f., Fl. Brit. India 7:314.1896; Fischer, Fl. Madras 3:1825.1934; Bor, Grass. India 501.1960; Moulik, Grass. Bamb. India 2:597.1997. *Poa aspera* Jacq., Hort. Bot. Vindob 332.t.56.1776.

Annuals, culms erect, tufted, to 70cm. Leaf sheaths glabrous. Ligules ring of hairs. Leaf blades linear-lanceolate, acuminate. Panicles lax, branches neumerous. Spikelets linear-oblong, 4-15-flowered. Lower glume ovate-lanceolate, charactaceous, acute, 1-nerved. Upper glume ovate-lanceolate, acute, chartaceous, bifid, purple at apex, 1-nerved. Lemmas broadly ovate, markedly truncate, not ciliate on margins. Paleas obovate, keeled, scabrid, hyaline, 2-nerved. Grains sub-globose.

Distribution: *World*: Africa, Mascarene Islands; *India*: Peninsular to North India; *Andhra Pradesh*: Anantapur, Kadapa and Kurnool districts.

Occasional in forest plains, waste lands and ground nut fields. *Fl. & Fr.*: July – Nov. *Repr. Spec.*: Pathakothacheruvu (ATP), BR & JNP 26631; Bhajaranga kona (KDP), KVS & AL 29115; Kuntimaddi (ATP), AL & BSS 26899.

Eragrostis atrovirens (Desf.) Trin. ex Steudel, Nom. Bot. Ed. 2.1:562.1840; Bor, Grass. India 503.1960; Moulik, Grass. Bamb. India 2:597.1997. *Poa atrovirens* Desf., Fl. Atlant. 1:73.t.14.1798.

Perennials, culms erect, tufted, to 1.3m; nodes glabrous. Leaf sheaths glabrous. Ligules membranous. Leaf blades linear–lanceolate, base rounded or cordate, apex acuminate. Panicles contracted, often lax. Spikelets short pedicelled, ovate, 3-30-flowered. Lower glume lanceolate, chartaceous, 1-nerved. Upper glume ovate–lanceolate, chartaceous, 1-nerved. Lemmas ovate, chartaceous, acute, 3-nerved. Paleas lanceoloate keeled, hyaline, keels ciliate, 2-nerved. Stamens 3. Grains ellipsoid.

Distribution: *World*: Africa, Malaysia, Myanmar, Sri Lanka; *India*: Bihar, Himachal Pradesh, Madhya Pradesh, North-East, Rajasthan and South India; *Andhra Pradesh*: Throughout.

Common along stream banks and in marshy areas. *Fl. & Fr.*: July - Feb. *Repr. Spec.*: Sirnapalli river (NZB), BR & GO 9077; Pochammralu (MDK), BR & CP 11497; Kalasamudram (ATP), BR & AMR 24227; Tirumala hills (CTR), AMR 21849, 21863; Kambakkam (CTR), AMR 24288; Appapur (MBNR), AMR 25700; Kalasamudram (ATP), BR & AL 26895; Batrepalli (ATP), KVS & MSB 28114. Sunnipenta (KNL), BR & BSS 29309; Rollapenta (PKM), BSS & SKB 29392; Velgode (KNL), BR & BSS 30231; Isukagundam (PKM), BSS & SKB 32983.

Eragrostis cilianensis (Alt.) Vignolo-Lutatiin Malapighia 18:386.1900; Fischer, Fl. Madras. 3:1827.1934; Bor, Grass. India 503.1960; Moulik, Grass. Bamb. India 2:598.1997. *Poa cilianensis* Alt., Fl. Pedem. 2:246. t. 91. f. 2. 1785. *Eragrostis major* (L). Host., Gram. Austr. 4:14 & 24.1809; Hook f., Fl. Brit. India 7:320.1896.

Annuals, culms erect or ascending, tufted, to 70cm, leafy; nodes glabours. Leaf sheath hairy at mouth. Ligule ciliate. Leaf blades linear-lanceolate, base cordate, apex acuminate, margins glandular. Panicles contracted, often lax, green. Spikelets oblong, 10-25-flowered. Lower glume oblong–lanceolate, chartaceous, 1-nerved. Upper glume ovate-lanceolate, acute, chartaceous, 3-nerved. Lemmas ovate, chartaceous, 3-nerved, glands on the keels. Paleas oblong, 2-keeled, 2-nerved, keels ciliate. Stamens 3. Grains sub-globose.

Distribution: *World*: Tropics; *India*: Bihar, Gujarat, Himachal Pradesh, Maharashtra, Punjab, Rajasthan, Uttar Pradesh and South India; *Andhra Pradesh*: Adilabad, Anantapur, East Godavari, Kurnool, Mahaboobnagar, Ranga Reddy, and Srikakulam districts.

Occasional along the forest margins in shady localities. *Fl. & Fr.*: Dec. – June. *Repr. Spec.*: Gundumallappa kona (KNL), SS & AMR 20288; Kalasamudram (ATP), BR & AMR 24237; Konakondla (ATP), JNP 26613; Penukonda (ATP), JNP 26763; Guntakal (ATP), JNP 26634; Timmapuram (ATP), AL 26634; Pathikonda (KNL), KVS & AL 29141; Ahobilam (KNL), BR & BSS 29307, 29319.

Eragrostis ciliaris (L.) R. Br. in Tuckey Narr. Exp. Congo, App. 478.1818; var. *ciliaris;* Hook.f., Fl. Brit. India 7:314.1896; Fischer, Fl. Madras 3:1825.1934; Bor, Grass. India 506.1960; Moulik, Grass. Bamb. India 2:598.1997. *Poa ciliaris* L., Syst. Nat. ed. 102:875.1759.

Annuals, clums erect, tufted, to 60cm, slender; nodes glabrous. Leaf sheaths ciliate at mouth. Ligules short silky hairy. Leaf blades linear, base rounded, acuminate, glabrous. Panicles spiciform or compact. Spikelets ovate. Lower glume ovate-acute, chartaceous, 1-nerved. Upper glume ovate-lanceolate, chartaceous, scabrid, 1-nerved. Lemmas broadly ovate, chartaceous, 3-nerved, ciliate. Palea obovate, truncate, soft ciliate, 2-keeled, 2-nerved. Stamens 3. Grains ovoid.

Distribution: *World*: Tropics and Subtropics. *India*: Daman, Madhya Pradesh, Maharashtra, Rajasthan, Uttar Pradesh and South India. *Andhra Pradesh*: Anantapur, Chittoor, Nellore, Srikakulam, Visakhapatnam and West Godavari districts.

Common in plains and forest margins. *Fl. & Fr.*: Aug. - Nov. *Repr. Spec.*: Tirumala hills (CTR), BR & AMR 21808; Sitampeta (SKLM), AMR 25653; Chennekothapalli (ATP), BR & JNP 27339; Guntakal (ATP), JNP & AL 26605; Hottebetta (ATP), BR & KVS 28085;

Eragrostis ciliaris (L.) R.Br. in Tuckey, Narr. Exp.Congo, APP. 478 (1818) var. *clarckei* Stapf ex Hook.f., Fl. Brit. Ind. 7:315 1896.

Annuals, culms erect, tufted, to 30cm; nodes glabrous. Leafsheaths ciliate at mouth. Ligule ciliate. Leaf blades lanceolate, flat, base rounded, glabrous, acuminate. Spikelets ovate, 4-6-flowered. Lower glume ovate-acute, chartaceous, 1-nerved, scabrid on upper side. Upper glume ovate-lanceolate, chartaceous, 1-nerved, scabrid on upper side. Lemmas broadly ovate, 3-nerved, 2-keeled, ciliate on the keels. Paleas oblong, 2-keeled, 2-nerved, long ciliate on the keels. Grains ovoid.

Distribution: Endemic to India; *India*: Delhi; *Andhra Pradesh*: Kurnool district.

Rare in plains. *Fl. & Fr.*: Aug. - Nov. *Repr. Spec.*: Gundlabrahmeswaram (KNL), BR & BSS 32964.

Eragrostis ciliata (Roxb.) Nees, Agrost. Bras. 512.1829; Fl. Brit. India 7:373.1896; Fischer, Fl. Madras. 3:1825.1934; Bor, Grass. India 506. 1960. *Poa ciliata* Roxb. Fl. Ind. 2:256. 1820.

Annuals or perennials, tufted, erect, to 80cm, geniculate; nodes glabrous. Leaf sheaths glabrous, bearded at mouth. Ligule a few hairs. Leaf blades linear-lanceolate, flat or convolute, apex acuminate, glabrous. Panicles compact, cylindrical. Spikelets compressed, ovate-lanceolate, 6-12-flowered. Lower glume ovate, chartaceous, 1-nerved, acute. Upper glume ovate, chartaceous, 1-nerved, acute. Lemma obovate, chartaceous, 3-nerved, ciliate. Palea ovate-lanceolate, hyaline, 2-keeled, 2-nerved, long ciliate. Grain ovate-elliptic.

Distribution: *World*: Burma and Indo-China; *India*: Bihar and Tamil Nadu; *Andhra Pradesh*: North circars, Karimnagar, Kurnool, Mahaboobnagar, Nalgonda and Prakasam districts.

Occasional in open places. *Fl. & Fr.*: May-Sep. *Repr. Spec.*: Betludatha konda (PKM) BSS & SKB 32281; Gundlabrahmeswaram (KNL), BR & BSS 32963.

Eragrostis coarctata Stapf in Hook. f., Fl. Brit. India. 7:313.1896; Fischer, Fl. Madras 3:1825.1934; Bor, Grass. India 507.1960; Moulik, Grass. Bamb. India 2:601.1997.

Perennials, culms erect, tufted, to 1.1m, geniculate; nodes glabrous. Leaf sheaths ciliate at mouth. Ligules hairy. Leaf blades linear–lanceolate, base rounded, apex acuminate. Panicles compact, dense, cylindrical. Spikelets short pediceled, oblong, grey colour. Lower glume lanceolate, chartaceous, 1-nerved. Upper glume ovate-lanceolate, chartaceous, 1-nerved. Lemmas broadly ovate, chartaceous, 3-nerved, ciliate on the margins. Paleas oblong, 2-keeled, 2-nerved, ciliate on the keels. Stamens 3. Grains ovoid.

Distribution: *World*: Bangladesh, Myanmar; *India*: Throughout; *Andhra Pradesh:* Guntur, Hyderabad, Karimnagar, Kurnool, Prakasam and North Eastern Ghat districts.

Occasional in stream beds and marshy areas in forests. *Fl.* & *Fr.*: Dec. - July. *Repr. Spec.*: Omkaram (KNL), SS 21348; Kasumuru (NLR), BR & PMR 22334; Punyagiri (VZN), BR & AMR 24538; Bukkapatnam (ATP), BR & JNP 27349; Hottebetta (ATP), BR & KVS 28100; Pathikonda (KNL), KVS & AL 29125; Isukagundam (PKM), BR & BSS 29291; Perayya kunta (PKM), BSS & SKB 32953.

Eragrostis deccanensis Bor, Grass. India 507.1960; Moulik, Grass, Bamb. India 2:602, 1997. *E. spicata* Jedwaln. in Bot. Arch. 5:185.1924, non Varsey 1891; Fischer, Fl. Madras 3:1825.1934. *E. phleoides* Stapf in Hook. f., Fl. Brit. India 7:313.1896.

Annuals, culms erect or geniculate, tufted, to 75cm. Leaf sheaths glabrous, ciliate at mouth. Ligule ciliate. Leaf blades linear-lanceolate, flat, base rounded, apex acuminate, glabrous. Panicle clavellate, dense, interrupted at base, shining; branches very short, villous at base. Spikelets compressed. Glumes sub-equal, obtuse, chartaceous, 1-nerved. Lemma ovate, membranous, minutely scaberulous, 3-nerved, nerves prominent. Palea keels narrowly winged, long ciliate, hyaline, 2-nerved. Stamens 3. Grains ellipsoid-globose.

Distribution: *World*: Endemic to Peninsular India. *India*: Karnataka and Tamilnadu. *Andhra Pradesh*: In forests of Chittoor, East & West Godavari, Srikakulam and Vizayanagaram districts.

Occasional in forest plains and margins. *Fl.* & *Fr.*: March-June. *Repr. Spec.*: Rampa hills (EG), TP & MSG 12279; Tirumala hills (CTR), KCJ 471 (MH).

Eragrostis diarrhena (Schultes) Steudel, Syn. Pl. Glum. 1:266.1854; Bor, Grass. India 507.1960; Moulik, Grass. Bamb. India 2:602.1997. *Poa diarrhena* Schult., Syst. Veg. 2:Mant. 615.1827. *Eragrostis interrupta* P. Beauv. var. *diarrhena* Stapf in Hook.f., Fl. Brit. India 7:316.1896.

Perennials, culms erect, tufted, to 1.3m; nodes glabrous. Leaf sheaths glabrous. Ligules membranous. Leaf blades linear-lanceolate, base cordate, apex acuminate, glabrous. Panicles contracted, interrupted, branches dense. Spikelets ovate, 6-15-flowered. Lower glume lanceolate, chartaceous, 1-nerved. Upper glume oblong-lanceolate, chartaceous, 1-nerved. Lemmas broadly ovate, or ovate-elliptic, obtuse, 3-nerved. Paleas oblong, hyaline, 2-keeled, 2-nerved, keels scaberulous. Stamens 3. Grains ovate.

Distribution: *World*: South-East Asia; *India, Andhra Pradesh*: Througuout.

Common in stream beds and near water cources. *Fl. & Fr.*: Jan.-Dec. *Repr. Spec.*: Gandhari (NZB), TP & BR 6347; Kottala cheruvu (KNL), SS 22608; Bitragunta (NLR), PMR 23026; Kalasamudram (ATP), BR & AMR 24518; Tadipatri railway cross fields (ATP), JNP 27391; Pamidi (ATP), AL & JNP 28001; Kannekal (ATP), KVS 28034; Hottebetta (ATP), KVS & AL 28094; Rollapenta (PKM), BR & BSS 29386; Gundlamotu Reservoir (PKM), BSS & SKB 32933.

Eragrostis gangetica (Roxb.) Steudel, Syn. Pl. Glum. 1:266.1854; Fischer, Fl. Madras 3:1826.1934; Bor, Grass. India 508. 1960; Moulik, Grass. Bamb. India 2:602.1997. *Poa gangetica* Roxb., Fl. Ind. 1:341.1820. *Eragrostis stenophylla* Hochst. ex Miq. in Ver. Konink. Nederl. Inst. 3,4,39.1851; Hook. f., Fl. Brit. India 7:318.1896.

Annuals, culms erect, tufted, to 80cm, slender; nodes glabrous. Leaf sheaths hairy at mouth. Ligules membranous. Leaf blades linear-lanceolate or linear, base cordate, acute or acuminate. Panicles lax, branches slender. Spikelets ovate-oblong, 6-15-flowered, grey. Lower glume ovate, chartaceous, 1-nerved. Upper glume oblong-lanceolate, chartaceous, 1-nerved. Lemmas ovate-lanceolate, chartaceous, 3-nerved. Paleas elliptic-lanceolate, 2-keeled, 2-nerved, keels ciliate. Stamens 3. Grains orbicular.

Distribution: *World*: Tropical Asia and Africa; *India, Andhra Pradesh:* Throughout.

Common in forest plains. *Fl. & Fr.*: July – Jan. *Repr. Spec.*: Saemella (ADB), GO & PVP 4222; Mudheli RF (NZB), BR & CP 7206; Gundlabrahmeswaram (KNL), SS & AMR 22436; Gundumallappa kona (KNL), BR & SS 20093; Gulapalyam hills (ATP), JNP & AL 26648; Hanimireddipalli (ATP), JNP & KVS 27368; Bhajaranga kona (KDP), KVS & BSS 29119; Gangutla RF (NLR), BR & BSS 29304; Saleswaram (MBNR), BSS & SKB 30584; Appapur (MBNR), BSS & SKB 32297; Perayya kunta (PKM), BSS & SKB 32952.

Eragrostis japonica (Thunb.) Trin., in Mem. Acad. Sci. Peterst., Ser. 6.1:405.1831; Fischer, Fl. Madras 3:1826.1934; Bor, Grass. India 509.1960; Moulik, Grass. Bamb. India 2:603. 1997. *Poa japonica* Thunb., Fl. Jap. 51. 1784. *Eragrostis interrupta* P.Beauv. var. *tenuissima* Stapf in Hook. f., Fl. Brit. India. 7:316.1896.

Perennials, culms erect, tufted, to 1.3m, ascending; nodes glabrous. Leaf sheaths glabrous. Ligules membranous. Leaf blades linear-lanceolate, involute, base rounded, acuminate, glaucous. Panicles contracted, dense, spiciform. Spikelets oblong-lanceolate, 5-15-flowered. Lower glume ovate-lanceolate, chartaceous, 1-nerved. Upper glume obovate, chartaceous, 1-nerved. Lemmas obovate, chartaceous, 3-nerved. Paleas oblong-lanceolate, 2-keeled, 2-nerved, keels ciliate. Stamens 3. Grains ellipsoid.

Distribution: *World*: Tropical Asia. *India*: Throughout; *Andhra Pradesh*: Medak, Nellore, Nizamabad, Prakasam, Srikakulam and West Godavari and Rayalaseema districts.

Occasional in stream beds and moist places of forests. *Fl. & Fr.*: Oct.-June. *Repr. Spec.*: Upper Ahobilam (KNL), SS & AMR 21274; Nandalur (KDP), AL & BSS 26568; Inagalur (KDP), KVS & BSS 29122; B.T. dam (ATP), BR & KVS 29172. Omkaram (KNL), MSR & BSS 29257; Isukagundam (PKM), BR & BSS 29290, 30274; Udayagiri fort (NLR), BR & BSS 29301; Chaparlapalli (KDP), BSS & SKB 29306; Baireni (KNL), BSS & SKB 32947.

Eragrostis maderaspatana Bor, Grass. India 509. 1960. *E. willdenowiana* Nees ex Stapf in Hook.f. Fl. Brit. India 7:322. 1896, non Nees ex Hook. & Arn. 1838; Fischer, Fl. Madras. 3:1827. 1934.

Annuals, culms erect or geniculately ascending, tufted, to 40cm; nodes glabrous. Leaf sheaths glabrous. Ligule a ring of hairs. Leaf blades, linear, margins eglandular, acuminate, glabrous. Panicle oblong, lax, open. Spikelets ovate-laneolate, 4-18-flowered. Lower glume oblong-lanceolate, 1-nerved, chartaceous. Upper glume ovate-lanceolate, 1-nerved, chartaceous. Lemmas broadly elliptic, 3-nerved. Paleas lanceolate, 2-keeled, keels scabrid, persistant. Grains oblong.

Distribution: *World*: Sri Lanka; *India*: Karnataka and Tamilnadu. *Andhra Pradesh*: Guntur and Kurnool districts.

Rare in open areas and near streams. *Fl. & Fr.*: Mar.-Sep. *Repr. Spec.*: Gundlabrahmeswaram (KNL), BR & BSS 32966; Narasaraopet (GNT), CAB 4549 (MH).

Note: Our collection is subsequent after C.A. Barbar's collection.

Eragrostis minor Host., Icon. Descr. Austr. 4. 15. 1809; Fl. Brit. India 7:321. 1896. *E. poaeoides* P. Beauv., Ess. Agrost. 162. 1812; Fischer, Fl. Madras. 3:1827. 1934; Bor, Grass. India 512. 1960.

Annuals, culms erect, tufted, to 45cm, leafy; nodes glabrous. Leafsheaths ciliate at mouth, glabrous. Ligule a pubescent ridge. Laef blades linear, base narrow, usually glandular along the margin, apex acuminate, glabrous. Panicle lax, open, branches filiform, false whorled at base. Spikelets ovate-oblong, 12-flowered. Glumes subequal, ovate, acute, 1-nerved. Lemma broadly ovate, acute, 3-nerved. Palea obovate-oblong, keels scabrid or slightly ciliate. Stamens 3. Grains broadly oblong, dorsally rounded or slightly flattened.

Distribution: *World*: Mediterranean regions and Pakistan; *India*: Kerala and Tamilnadu; *Andhra Pradesh*: Adilabad, East Godavari, Hyderabad, Medak and Nalgonda districts.

Rare in crop fields and waste lands. *Fl. & Fr.*: July – Dec. *Repr. Spec.*: Bellampalli (ADB), PVP 9445; Pocharam RF (MDK), TP & MSG 12061; Golkonda (HYD), MRS 303 (DD); Mandasa mendu (EG), VNS 4624 (MH); Nagarjunakonda valley (NLG), KJ 9687 (CAL).

Eragrostis nigra Nees ex Steudel, Syn. Pl. Glum. 1:267.1854; Hook. f., Fl. Brit. India 7:324.1896; Fischer, Fl. Madras 3:1827. 1934; Bor, Grass. 511.1960; Moulik, Grass. Bamb. India 2:605.1997.

Perennials, culm erect, tufted, to 75cm; nodes glabrous. Leafsheaths ciliate at mouth. Ligule ciliate. Leaf blades linear, convolute, sparsely pilose above. Panicle effuse. Spikelets ovate-lanceolate, 8-flowered; pedicels glandular. Glumes ovate-lanceolate, subequal. Lemmas broadly ovate, eglandular, straight on the back. Palea obtuse, denticulate, persistent. Stamens 3. Grain smaller, subquadrate, dorsaly slightly grooved.

Distribution: *World*: China and Sri Lanka; *India*: Throughout; *Andhra Pradesh*: Anantapur, Chittoor and Visakhapatnam districts.

Common in forest plains and peripheries. *Fl.* & *Fr.*: July – Dec. *Repr. Spec.*: Tirumala hills (CTR), MSG 12136 & 12179; Anantagiri (VSKP), GVS 19470 (CAL).

Eragrostis nutans (Retz.) Nees ex Steudel, Nom. Bot. Ed. 2.563. 1840; Fischer, Fl. Madras 3:1826.1934; Bor, Grass. India 511.1960; Moulik, Grass. Bamb. India 2:605.1997. *Poa nutans* Retz., Obs. Bot. 4:19.1786. *Eragrostis elegantula* Nees ex Steudel, Syn. Pl. Glum. 1:266.1854; Hook. f., Fl. Brit. India 7: 318. 1896.

Perennials, culms erect, tufted, to 90cm, slender; nodes glabrous. Leaf sheaths glabrous. Ligules hairy. Leaf blades linear-lanceolate, base rounded, apex acuminate. Panicles contracted, black or gray colour. Spikelets oblong-lanceolate, 6-15-flowered. Lower glume ovate, chartaceous, 1-nerved. Upper glume oblong, chartaceous, 1-nerved. Lemmas ovate, chartaceous, 3-nerved. Palea, obovate, 2-keeled, 2-nerved, ciliate on the keels. Stamens 3. Grains ellipsoid.

Distribution: *World*: Malaysia and South China; *India*: Bihar, Karnataka, Madhya Pradesh, Maharashtra, Orissa, Rajasthan, Tamilnadu, Uttar Pradesh and West Bengal; *Andhra Pradesh*: Throughout.

Common in moist habitats. *Fl.* & *Fr.*: June – April. *Repr. Spec.*: Gundumallappa kona (KNL), BR & SS 20100, Tirumala hills (CTR), BR & AMR 21847, 21871, 21860, 21866, 21878; G. Madugula (VSKP), AMR 24554; Mahendragiri (SKLM), AMR 25661; Akuthota palli (ATP), JNP & AL 27201; Reddi palli form (ATP), JNP & AL 27220; Aspari (KNL), AL & BSS 27825; Bhajaranga kona (KDP), KVS & AL 28988; Perayya kunta (PKM), BSS 32951; Garnipenta (PKM), BSS & SKB 32959.

Eragrostis papposa (Roemer & Schultes) Steudel, Syn. Pl. Glum. 1:263.1854; Bor, Grass. India 511.1960; *Megastachya papposa* Roem. & Schult., Syst. Veg. 2:585.1817.

Perennials, culms erect, tufted, to 50cm; nodes glabrous. Leaf sheaths glabrous, ciliate at mouth. Ligule ciliate. Leaf blades short, stiff, linear, convolute. Panicle lax, open, filiform; Spikelets linear, 7-23-flowered. Glumes subequal, hyaline, lower nerveless, upper 1-nerved. Lemma broadly ovate, hyaline. Palea persistent, obtuse, keels scabrid. Stamens 3. Grains obovoid, dorsally grooved.

Distribution: *World*: Mediterranean region; *India*: North-West India; *Andhra Pradesh*: East and West Godavari districts.

Occasional in streams and forest shade. *Fl.* & *Fr.*: Nov.-May. *Repr. Spec.*: Rampa hills (EG), TP & MSG 12278.

Eragrostis pilosa (L.) P. Beauv., Ess. Agrost. 71:162.175.1812; Hook. f., Fl. Brit. India 7:323.1896; Fischer, Fl. Madras 3:1827.1934; Bor, Grass. India 512.1960; Moulik, Grass. Bamb. India 2:606. 1997. *Poa pilosa* L., Sp. Pl. 68.1753.

Annuals, culms erect, tufted, to 1.2m, slender; nodes glabrous. Leaf sheaths glabrous. Ligules fimbriate, membranous. Leaf blades linear-lanceolate, base rounded, acuminate. Panicles lax, filiform. Spikelets ovate-lanceolate or linear-oblong, 4-10-flowered. Lower glume ovate-lanceolate, chartaceous, 1-nerved. Upper glume lanceolate, chartaceous, 1-nerved. Lemmas oblong-lanceolate, chartaceous, 3-nerved. Paleas ovate, keeled, 2-nerved, keels scabrid. Stamens 3. Grains ovate.

Distribution: *World*: Tropics; *India*: Throughout; *Andhra Pradesh*: Adilabad, Hyderabad, Mahaboobnagar, Nellore, Srikakulam, Warangal, West Godavari and Rayalaseema districts.

Occasional in forest margins and plains. *Fl. & Fr.*: Aug. – Dec. *Repr. Spec.*: Balapalii RF (KDP), PVP 3461; Kambagiri konda (KNL), BR & SS 20039; Kalasamudram (ATP), BR & AMR 24219; Guvvalacheruvu (KDP), AMR 25606; Appapur (MBNR), AMR 25699; Kalasamudram (ATP), JNP & AL 26896; Hanimireddi palli (ATP), JNP & AL 27379, 27801. Chelama (KNL), MSR & BSS 32243; Gundlamotu Reservoir (PKM), BSS & SKB 32934.

Eragrostis riparia (Willd.) Nees, Agrost. Bras. 512.1829; Fischer, Fl. Madras 3: 1825.1934. Bor, Grass. India 513.1960; Moulik, Grass. Bamb. India 2:606.1997. *Poa riparia* Willd. in Ges., Natur. Freunde Berlin, Neue, Schrift. 4:185.1803. *Eragrostis tenella* P. Beauv. var. *riparia* Stapf in Hook. f., Fl. Brit. India 7:315.1896.

Perennials, culms erect, tufted, to 90cm, ascending; nodes glabrous. Leaf sheaths glabrous, hairy at mouth. Ligules a row of hairs. Leaf blades linear -lanceolate, base rounded, acute, hairy. Panicles spiciform, dense. Spikelets ovate, oblong, 5-10-flowered. Lower glume ovate-linear, chartaceous, 1-nerved. Upper glume oblong, chartaceous, 1-nerved. Lemmas oblong, chartaceous, 3-nerved. Paleas oblong, 2-keeled, 2-nerved, keels ciliate. Stamens 3. Grains elliptic.

Distribution: *World*: Endemic to India; *India*: Maharashtra, Sikkim, Uttar Pradesh, West Bengal and South India. *Andhra Pradesh*: Krishna, Nalgonda, Nellore, Prakasam, Srikakulam, Vizayanagaram, West Godavari and Rayalaseema districts.

Occasional in marshy areas especially near sandy moist soils. *Fl. & Fr.*: May-Nov.

Repr. Spec.: Marripadu (NLR), PMR 16513; Tirumala hills (CTR), AMR 21842; Kalasamudram (ATP), BR & AMR 24256; Gulapalyam hills (ATP), JNP & AL 26651; Penukonda (ATP), JNP 26757; Hottebetta (ATP), KVS & AL 28086; Pamidi (ATP), BR & KVS 29184; Rollapenta (PKM), BR & BSS 29384; Ahobilum-Mattamcheruvu (KNL), MSR & BSS 29315; Gundlamotu reservoir (PKM), BSS & SKB 32935.

Eragrostis tenella (L.) P. Beauv. ex Roemer et Schultes, Syst. Veg. 2:576.1871. var. *tenella*; Bor, Grass. India 513.1960; Moulik, Grass. Bamb. India 2:608.1997. *Poa tenella* L., Sp. Pl. 69. 1753. *Eragrostis plumosa* (Retz.) Link, Hort. Berol. 1:192.1827; Fischer, Fl. Madras 3:1826. 1934. *Eragrostis tenella* (L.)P. Beauv. ex Roem. & Schult. var. *plumosa* (Retz.) Stapf in Hook. f. Fl. Brit. India. 7:315.1896.

Annuals, culms erect, tufted, to 80cm, slender; nodes glabrous. Leaf sheaths hairs at mouth. Ligules membranous. Leaf blades linear-acuminate, base rounded, acuminate. Panciles effuse. Spikelets oblong-ovate, 5-9-flowered. Lower glume lanceolate, chartaceous, 1-nerved. Upper glume ovate-lanceolate, chartaceous, 1-nerved. Lemmas oblong-ovate, chartaceous, 3-nerved. Paleas lanceolate or elliptic-lanceolate, 2-keeled, 2-nerved, keels ciliate. Stamens 3. Grains ovate-elliptic.

Distribution: *World*: Tropics; *India*: Throughout; *Andhra Pradesh*: Throughout.

Common in hill slopes, waste places and plains. *Fl.* & *Fr.*: April – Nov. *Repr. Spec.*: Kalasamudram (ATP), BR & AMR 20040, 24238; Chendrayuni kona (ATP), BR & JNP 26736; Penukonda (ATP), JNP 26762; Gooty hills (ATP), JNP & AL 26775; Pamidi (ATP), KVS & JNP 28031. Sunnipenta (KNL), BR & BSS 32978; Saleswaram (MBNR), BSS & SKB 34801.

Eragrostis tenella (L.) P. Beauv. ex Roemer et Schultes Syst. Veg. 2:576.1871. var. *insularis* C.E.Hubbard in Kew Bull. 654. 1939. Bor, Grass. India 513.1960; Moulik, Grass. Bamb. India 2:608.1997.

Annuals, culms, erect, tufted, to 30cm, slender; nodes glabrous. Leaf sheaths hairy at mouth. Ligules membranous. Leaf blades linear-acuminate, base rounded, acuminate. Panciles contracted. Spikelets oblong-ovate, 5-9- flowered. Lower glume lanceolate, chartaceous, 1-nerved. Upper glume ovate-lanceolate, chartaceous, 1-nerved. Lemmas broadly ovate, chartaceous, 3-nerved. Paleas lanceolate or elliptic-lanceolate, hyaline, 2-keeled, 2-nerved, keels ciliate. Grains ovate-elliptic.

Distribution: *World*: Madagascar, Mascarene Island and Sri Lanka; *India*: Tamil Nadu and West Bengal; *Andhra Pradesh*: Mahaboobnagar district.

Rare in plains in shady localities. *Fl.* & *Fr.*: April – Nov. *Repr. Spec.*: Kollapur (MBNR), BR & BSS 32918.

Eragrostis tenuifolia Hochst. ex Steudel, Syn. Pl. Glum. 1:268. 1854; Hook. f., Fl. Brit. India 7:322.1896; Fischer, Fl. Madras 3:1827. 1934; Bor, Grass. India. 514.1960; Moulik, Grass. Bamb. India 2:609.1997.

Perennials, culms erect, slender, tufted, to 60cm; nodes glabrous. Leaf sheaths glandular, sparsely hairy. Ligules ciliate. Leaf blades linear base rounded, apex acuminate, glabrous. Panicles lax. Spikelets linear-oblong, pedicels glandular. Lower glume minute, ovate, linear, 1-nerved, lobed. Upper glume ovate-oblong, 1-nerved, 2-3-lobed. Lemmas ovate-elliptic, 3-nerved, mucronate. Palea oblanceolate, 2-nerved, 2-keeled, keels scabrid, hyaline, persistant. Grains oblong.

Distribution: *World*: Tropical Africa; *India*: South and North-East India; *Andhra Pradesh*: Adilabad, Guntur, Visakhapatnam and Rayalaseema districts.

Rare in open dry places and along road sides. *Fl.* & *Fr.*: July – Oct. *Repr. Spec.*: Muniswaruni Kunta (ATP), PVP 2617; Bugga (ATP), BR & JNP 27389; S.K.U campus (ATP), KVS & BSS 28985.

Eragrostis tremula Hochst. ex Steudel, Syn. Pl. Glum. 1:269.1854; Hook. f., Fl. Brit. India 7:320.1896; Fischer, Fl. Madras 3:1827.1934; Bor, Grass. India 514.1960; Moulik, Grass. Bamb. India 2:609. 1997.

Annuals or perennials, culms erect, tufted, to 90cm, ascending; nodes glabrous. Leaf sheaths bearded at mouth. Ligules fimbriate. Leaf blades effuse, filiform. Spikelets oblong, 15-50-flowered. Lower glume oblong, chartaceous, 1-nerved. Upper glume oblong, chartaceous 1-nerved. Lemmas ovate-oblong, chartaceous, 3-nerved. Paleas elliptic-oblong, 2-keeled, 2-nerved, keels scabrid. Stamens 3. Grains globose.

Distribution: *World*: Tropical Africa, Myanmar; *India, Andhra Pradesh*: Throughout.

Occasional in forest plains. *Fl. & Fr.*: June - Nov. *Repr. Spec.*: Ramayampeta (MDK), BR & CP 11520; Tirumala hills (CTR), BR & AMR 21863; Bhogeswaram (KNL), BR & SS 22679; Kalasamudram (ATP), BR & AMR 24239; Chintapalli (VSKP), AMR 24577; Pathakothacheruvu (ATP), BR & AL 27212; Baireni (KNL), BR & BSS 32246; Chelama (KNL), BSS & SKB 32948.

Eragrostis unioloides (Retz.) Nees ex Steudel, Syn. Pl. Glum. 1:264. 1854; Fischer, Fl. Madras. 3:1826;1934; Bor, Grass. India 515.1960; Moulik, Grass. Bamb. India 2:609.1997. *Poa unioloides* Retz., Obs. Bot. 5:19.1789. *Eragrostis amabilis* auct. non Wight & Arn. ex Nees 1838: Stapf in Hook. f., Fl. Brit. India. 7:317.1896.

Annuals or perennials, culms erect or ascending, tufted, to 90cm; nodes glabrous. Leaf sheaths glabrous. Ligules membranous. Leaf blades linear-oblong, base cordate, acuminate. Panicles lax, purplish. Spikelets ovate, 10-30-flowered, purplish. Lower glume ovate-oblong, chartaceous, 1-nerved. Upper glume ovate-oblong, chartaceous, 1-nerved. Lemmas broadly ovate, chartaceous, 3-nerved. Paleas oblong-elliptic, 2-keeled, 2-nerved, keels scabrid. Stamens 3. Grains obovoid.

Distribution: *World*: South-East Asia; *India, Andhra Pradesh:* Throughout.

Common in wet lands, streams and in moist habitats. *Fl. & Fr.*: July – April. *Repr. Spec.*: Alampalli (ADB), TP & PVP 4170; Annaram (NZB), BR & CPR 7190; Kambagirikonda (KNL), BR & SS 20042; Kambakkam (CTR), AMR 24283; Punyagiri (VZN), BR & AMR 24537; Chinta palli (VSKP), AMR 24579; Lambasangi (VSKP), AMR 25632; Kadiri (ATP), JNP & AL 27342; Bhajaranga kona (KDP), KVS 29108; Ahobilam (KNL), MSR & BSS 32263.

Eragrostis viscosa (Retz.) Trin., in Men. Acad. Sci. Petersb. Ser. 6.1:397. 1830; Fischer, Fl. Madras 3:1826.1934; Bor, Grass. India 515. 1960; Moulik, Grass. Bamb. India 2: 611.1997. *Poa viscosa* Retz., Obs. Bot. 4:20.1786. *Eragrostis tenella* var. *viscosa* Stapf in Hook.f.,Fl. Brit. India 7:315.1896.

Annuals, culms erect or ascending, tufted viscous below, to 70cm; nodes glabrous. Leaf sheaths glabrous. Ligules membranous. Leaf blades linear, convolute, base rounded, apex acuminate. Panicles narrowly cylindric, glandular. Spikelets oblong-lanceolate, 4-20 flowered. Lower glume linear, chartaceous, 1-nerved. Upper glume oblong, chartaceous, 1-nerved. Lemmas ovate, chartaceous, 3-nerved. Paleas linear, ciliate on the keels, 2-keeled, 2-nerved. Stamens 3. Grains ovoid.

Distribution: *World*: Africa; *India*: Throughout; *Andhra Pradesh*: Throughout.

Common in wastelands and forest plains. *Fl. & Fr.*: April – Nov. *Repr. Spec.*: Sole (ADB), TP & PVP 4020; Leguntapadu (NLR), BR & PMR 16437; Variguntapadu (NLR), PMR 16475; Kottala cheruvu (KNL), BR & SS 21283; Kalasamudram (ATP), BR & AMR 24236; Chendrayuni kona (ATP), JNP 26737; Hanimireddi palli (ATP), JNP & KVS 27384; Kannekal (ATP), BR & KVS 28066; Papavinasanam (CTR), BSS & SKB 34822.

LEPTOCHLOA P. Beauv.

1. Lemmas 3-lobed; racemes loosely spiculate .... *L. fusca*
1. Lemmas not lobed; racemes densely spiculate .... (2)
2. Spikelets 2-7-flowered .... (3)
2. Spikelets 1-flowered .... *L. neesii*
3. Lemmas acuminate; anthers 0.2-0.5mm .... *L. panicea*
3. Lemmas acute-obtuse, anthers 0.5-0.7mm .... *L. chinensis*

Leptochloa chinensis (L.) Nees in Syll. Ratisb. 1:4.1824; Hook. f., Fl. Brit. India 7:299. 1896; Fischer, Fl. Madras 3:1820. 1934; Bor, Grass. India 516. 1960; Moulik, Grass. Bamb. India 2:613.1997. *Poa chinensis* L., Sp. Pl. ed. 1:69.1753.

Annuals, culms erect or ascending, tufted, to 90cm; nodes glabrous. Leaf sheaths glabrous. Ligules membranous. Leaf blades linear or linear-lanceolate, base rounded, acuminate, glabrous. Panicles open. Spikelets ovate-oblong, 3-7-flowered. Lower glume lanceolate, chartaceous, 1-nerved. Upper glume ovate-lanceolate, chartaceous, 1-nerved. Lemmas ovate, chartaceous, 3-nerved. Paleas ovate or oblong, 2-keeled, 2-nerved, keels ciliate. Stamens 3. Grains oblong.

Distribution: *World*: South-East Asia; *India*: West Bengal and South India; *Andhra Pradesh*: Throughout.

Common near water sources. *Fl. & Fr.*: July-Dec. *Repr. Spec.*: Bijjur RF (ADB), GO & PVP 4877; Leguntapadu (NLR), PMR 16430; Maddiletiswamy temple (KNL), BR & SS 20281; Andhra Reservoir (VZN), AMR 24544; Rajampet (KDP), AL & BSS 26570; Pamidi (ATP), JNP & AL 26620; Kothacheruvu (ATP), BR & JNP 26704, 27398; Musalreddypalli (KDP), BSS & SKB, 30823; Gundlabrahmeswaram (KNL), BR & BSS 32967.

Leptochloa fusca (L.) Kunth, Revis. Gramin. 1:91. 1829; *Diplachne fusca* (L.) P.Beauv., Ess. Agrost. 80 & 163. 1812; Hook. f., Fl. Brit. India 7:329. 1896; Fischer, Fl. Madras 3:1829.1934; Bor, Grass. India 492. 1960; *Festuca fusca* L., Sp. Pl. (ed.2) 109.1762. *Poa procera* Roxb., Fl. Ind. 1.344.1820.

Perennials, culms erect, tufted, to 1.2m; nodes glabrous. Leaf sheaths glabrous. Ligules membranous. Leaf blades linear-lanceolate, base rounded, acuminate, glabrous. Panicles densely spiculate. Spikelets oblong-lanceolate, 4-8-flowered. Lower glume oblong, chartaceous, 1-nerved. Upper glume oblong-lanceolate, chartaceous, 1-nerved. Lemmas oblong, 3-lobed, chartaceous, aristate, scabrid. Paleas oblong, hyaline keeled, 2-nerved, keels scabrid. Stamens 3. Grains oblong.

Distribution: *World*: Tropics and sub tropics of the old world and Australia; *India*: Upper gangetic plains to Bengal and Southwards; *Andhra Pradesh*: Throughout.

Common in moist and shady localities of the forests, and in streams and ponds. *Fl. & Fr.*: Oct.-April. *Repr. Spec.*: Kottala cheruvu (KNL), SS & AMR 22637; Pamidi (ATP), JNP & AL 26619; Nagaluty (KNL), BSS & SKB 30150.

Leptochloa neesii (Thw.) Benth. in J. Linn. Soc. (Bot.) 19:108.1881; Bor, Grass. India 517.1960; Moulik, Grass. Bamb. India. 2:613. 1997. *Cynodon neesii* Thw., Enum. Ceyl. Pl. 3:71.1864; *Leptochloa polystachya* Benth., Fl. Austral 7:617.1878; Hook. f., Fl. Brit. India 7:298.1896; Fischer, Fl. Madras 3:1820. 1934.

Annuals, culms erect or geniculately ascending, tufted, to 1.1m; nodes glabrous. Leaf sheaths glabrous. Ligules membranous. Leaf blades linear or linear-lanceolate, base cordate, apex acuminate. Spikelets ovate-oblong, 1-flowered. Lower glume ovate-lanceolate, chartaceous, 1-nerved. Upper glume ovate-lanceolate, chartaceous, 1-nerved. Lemmas lanceolate, membranous, keeled, 3-nerved. Paleas ovate-oblong, hyaline, keeled, 2-nerved. Stamens 3. Grains ovoid.

Distribution: *World*: Sri Lanka, Malaya, Australia and other warm countries; *India*: Peninsular India; *Andhra Pradesh*: Chittoor, East Godavari, Kadapa, Krishna, Kurnool and Srikakulam districts.

Occasional in marshy areas. *Fl.* & *Fr.*: July – Dec. *Repr. Spec.*: Kambakkam (CTR), AMR 24286; Baireni (KNL), BSS & SKB 32239; Salur (SKLM), NPBK 1134(CAL); Mandapenta (EG), CAB 4709(MH).

Leptochloa panicea (Retz.) Ohwi in Bot. Mag. Tokyo 55:311. 1941; Bor, Grass. India 517. 1960. *Poa panacea* Retz. Obs. Bot. 3:11. 1783. *Leptochloa filformia* sensu Hook. f., Fl. Brit. India 7: 298.1896 non Beauv. 1812; Fischer, Fl. Madras. 3:1820. 1934.

Annuals, erect, tufted, to 25cm; nodes glabrous. Leafsheaths tuberculate-hispid. Ligule membranous. Leaf blades linear-lanceolate, glabrous. Panicle interrupted at base. Spikelets linear, 3-flowered. Glumes subequal, persistent, membranous. Lemmas membranous, keeled. Grains broadly ellipsoid.

Distribution: *World*: Warmer parts of Asia & Africa; *India*: Andhra Pradesh, Maharashtra, Madhya Pradesh and Punjab; *Andhra Pradesh*: East Godavari and Visakhapatnam districts.

Rare in wet lands. *Fl.* & *Fr.*: Aug.-Nov. *Repr. Spec.*: Samalkota (EG), Acc.No. 84310 (MH); Gajapatinagaram (VSKP), VNS 8241 (MH).

MYRIOSTACHYA (Bentham) Hook.f.

Myriostachya wightiana (Nees ex Steudel) Hook. f., Fl. Brit. India 7:327. 1896; Fischer, Fl. Madras. 3:1818. 1934; Bor, Grass. India 518. 1960. *Leptochloa wightiana* Nees ex Steud. Syn. Pl. Glum. 1:209. 1854.

Perennials, tall, tufted, marshy grass, culms erect, to 1m, root stock sheathed, sometime floating. Leafsheaths very long, compressed. Ligule absent. Leaf blades several feet long, linear, apex acuminate, glabrous. Spikelets compressed, 4-8-flowered. Glumes unequal, coriaceous, lanceolate, 1-nerved, tapering into an awn, keels scabrid. Lemmas ovate-lanceolate, coriaceous, 3-nerved, keels glabrous, awned. Palea chartaceous, bi-dentate, 2-keeled, keels ciliate. Stamens 3. Grains obliquely ovoid.

Distribution: *World*: Burma, Sri Lanka, Malya & China; *India*: Tamil Nadu and West Benghal; *Andhra Pradesh*: East Godavari, West Godavari & Krishna districts.

Found in estuarine forests. *Fl.* & *Fr.*: Oct. – Jan. *Repr. Spec.*: Coringa canal bank (EG), GVS 68700 A, & KCJ 12505 (MH); Y.D.Palem RF (KSN), PV 5609 (MH).

OROPETIUM Trin.

1. Spikelets 1-flowered; lemma obovate; caryopsis oblong .... *O. thomaeum*
1. Spikelets 1-2-flowered; lemma rounded; caryopsis terete .... *O. roxburghianum*

Oropetium roxburghianum (Steudel) S.M.Phillips in Kew Bull. 36:469. 1975. *Lepturus roxburghianus* Steud. Syn. Pl. Glum. 1:357. 1854; Fl. Brit. India 7:365. 1896. *Tripogon roxburghianus* (Steud.) Bhide in J.Proc. Asiat. Soc. Beng. 1911 (n.s.) 7:515. 1912; Fischer, Fl. Madras. 3: 1834. 1934; Bor, Grass. India 524. 1960.

Annual, culms erect, tufted, slender to 15cm; nodes glabrous. Leafsheaths glabrous. Ligules membranous. Leaf blades filiform, margin ciliate. Spike solitary. Spikelets 1-2-flowered. Lower glume oblique, hyaline, 2-lobed, 1-nerved. Upper glume coriaceous, 3-nerved. Lower lemma membranous, bidentate, awned in the cleft, ciliate in lower part, callus bearded. Upper lemma bisexual or imperfect or 0. Palea 2-keeled, scabrid. Stamens 3. Grains terete, oblong-ellipsoid, pointed at one side.

Distribution: World: Africa. *India*: Maharashtra, Madhya Pradesh, Rajasthan, and Tamil Nadu; *Andhra Pradesh*: Anantapur, Chittoor, Kadapa and Nellore districts.

Rare in moist rock crevices. *Fl. & Fr.*: July – Dec. *Repr. Spec.*: Talakona (CTR), BR & BSS 32929; Puttalur (NLR), CECF 4136 (CAL).

Oropetium thomaeum (L.f.) Trin., Fund. Agrost. 98.t.3.1820; Hook.f., Fl. Brit. India 7:336.1896; Fischer, Fl. Madras 3:1830.1934; Bor, Grass. India 474.1960; Moulik, Grass. Bamb. India 2:617.1997. *Nardus thomaea* L.f., Suppl. 105.1781.

Annuals, culms erect, tufted, to 6cm; nodes glabrous. Leaf sheaths small, compressed. Ligules membranous. Leaf blades linear, filiform, base rounded, acute, ciliate on margins. Spikes solitary, terminal, slender. Spikelets sessile, lanceolate, solitary, sunk in the alternatingly distichous hollows in the rhachis, 1-4-flowered. Lower glume minute. Upper glume linear-lanceolate. Lemmas ovate-oblong, hyaline, 1-3-nerved. Paleas lanceolate, 2-keeled. Stamens 3. Grains oblong.

Distribution: *World*: South-East Asia, Sri Lanka, Myanmar and Tropical Africa; *India*: Throughout; *Andhra Pradesh*: Throughout.

Common in forest plains in dry places. *Fl. & Fr.*: July-Nov. *Repr. Spec.*: Devaragattu (KNL), SS & AMR 22671; Rapur (NLR), AMR 24563; Pachipenta (VZN), AMR 24544; Kasapuram (ATP), JNP & AL 26638; S.K.U Campus (ATP), KVS 28036; Thummalabilu (PKM), BSS & SKB 29383; S.K.U Campus (ATP), BR & PP 36814.

SPOROBOLUS R.Br.

1. Lower glume shorter than the spikelet .... (2)
1. Lower glume as long as the spikelet .... *S. virgnicus*
2. Upper glume conspicuously shorter than the lemma .... (3)
2. Upper glume as long as or longer than the lemma .... (7)
3. Panicle effuse .... (4)
3. Panicle otherwise .... (6)

4. Lemmas more than 1mm long .... (5)
4. Lemmas not morethan 1mm long .... *S. tenuissimus*
5. Stamens 2; spikelets crowded, slatey-grey in colour, lower glume truncate .... *S. diander*
5. Stamens 3; spikelets scattered, lower glume very short, obtuse .... *S. wallichii*
6. Plants grow near shady or saline soils; panicle spicate .... *S. spicatus*
6. Plants otherwise; panicle subspiciform or interrupted .... *S. fertilis*
7. Panicle contracted; branches not whorled .... (8)
7. Panicle effuse; branches truly whorled .... *S. coromandelianus*
8. Stamens 3; perennials .... (9)
8. Stamens 2; annuals .... *S. piliferus*
9. Leaf-margins smooth; spikelets 1.5-1.7mm long; lower glume lanceolate .... *S. tremulus*
9. Leaf-margins spinulose; characters not as above .... *S. maderaspatanus*

Sporobolus coromandelianus (Retz.) Kunth, Rev. Gram. 1:68.1829; Hook. f., Fl. Brit. India 7:252. 1896; Fischer, Fl. Madras, 3:1817.1934; Bor, Grass. India 627. 1960; Moulik, Grass. Bamb. India 2:477.1997. *Agrostis coromandeliana* Retz., Obs. Bot. 4:19.1786. *Sporobolus commutatus* (Trin.) Kunth, Enum. Pl. 1:214.1833. *Vilta commutata* Trin., Diss. Bot. 156.1824.

Perennials, culms erect or decumbent, tufted, to 20cm; nodes glabrous. Leaf sheaths glabrous. Ligules membranous. Leaf blades linear- lanceolate with sparsely tubercle based hairy, base rounded or cordate, margins spinulosely toothed, acute. Panicles pyramidal, decompound, whorled. Spikelets ovate. Lower glume oblong, small. Upper glume ovate, chartaceous, 1-nerved. Lemma ovate, chartaceous, 1-nerved. Palea elliptic, 2-nerved. Stamens 3. Grains ellipsoid.

Distribution: *World*: Afghanistan, Myanmar, Sri Lanka and South Africa; *India*: Throughout. *Andhra Pradesh*: Guntur, Hyderabad, Khammam, Mahaboobnagar, Rayalaseema and West Godavari districts.

Occasional in forest plains and in dry habitats. *Fl.* & *Fr.*: Aug.-Dec. *Repr. Spec.*: Devaragattu (KNL), BR & SS 22678; Chennur (KDP), AL 26577; Kinnerasani (KMM), BR & AMR 26816; Konapuram (ATP), JNP & AL 26662; S.K.U Campus (ATP), BR & PP 36823.

Sporobolus diander (Retz.) P. Beauv., Ess. Agrost. 26.147.178. 1812; Hook. f., Fl. Brit. India 7:247.1896; Fischer, Fl. Madras 3:1817.1934; Bor, Grass. India 629.1960; Moulik, Grass. Bamb. India 2:478.1997. *Agrostis diandra* Retz., Obs. Bot. 5:19.1789.

Annuals, culms erect, tufted, to 90cm; nodes glabrous. Leaf sheaths glabrous. Ligules membranous. Leaf blades linear-lanceolate, base rounded, acuminate, glabrous. Panicles effuse, filiform. Spikelets obovoid, lanceolate. Lower glume lanceolate, truncate, chartaceous. Upper glume ovate, chartaceous, 1-nerved. Lemma lanceolate, chartaceous, 1-nerved. Palea oblong, elliptic, 2-nerved. Stamens 2. Grains obovoid.

Distribution: *World*: Pakistan, Myanmar, Sri Lanka and Australia; *India, Andhra Pradesh*: Throughout.

Occasional in moist and shady localities of the forests. *Fl. & Fr.*: July – Dec. *Repr. Spec.*: Horsely hills (CTR), PVP 4661; Yedpalli (NZB), BR & KH 6473; Rudrakod (KNL), BR & SS 20168; Madiiletiswamy temple (KNL), BR & SS 20266; Tirumala hills (CTR), BR & AMR 21844; Kalasamudram (ATP), BR & AMR 21889; Palakonda (SKLM), AMR 24250; Gudem (VSKP), BR & AMR 24595; Appapur (MBNR), AMR 25686; Kalasamudram (ATP), JNP & AL 27400; Rollapenta (PKM), BSS & SKB 29385.

Note: Jovet & Guedes (1973) treated this taxon as a variety under *S.indicus.* However we followed the concept of Clayton & Renovoize (1986).

Sporobolus fertilis (Steudel) W.D. Clayton in Kew Bull. 19:291. 1965; Moulik, Grass. Bamb. India 2:478.1997. *S. indicus* sensu Hook. f., Fl. Brit. India 7:247. 1896; Fischer, Fl. Madras 3:1817. 1934; Bor, Grass. India 630.1960.

Perennials, culms erect, tufted, to 1.1m; nodes glabrous. Leaf sheaths glabrous. Ligules hairy or membranous. Leaf blades base rounded or cordate, apex acuminate, glabrous. Panicles contracted. Spikelets obovoid. Lower glume ovate, oblong, chartaceous, obtuse. Upper glume ovate, chartaceous, 1-nerved. Lemma oblong-ovate, chartaceous, 1-nerved. Palea oblong-elliptic, hyaline, 2-nerved. Stamens 3. Grains ovate-oblong.

Distribution: *World*: Myanmar, Sri Lanka; *India*: Throughout; *Andhra Pradesh*: Anantapur, Krishna, Mahaboobnagar, Nellore, Visakhapatnam, West Godavari and North circar districts.

Occasional along road sides and forests plains. *Fl. & Fr.*: Sept. – Dec. *Repr. Spec.*: Chinta palli (VSKP), AMR 24580; Penukonda (ATP), JNP 27227; Hottebetta (ATP), KVS & AL 28082, 28097; Gangutla RF (NLR), BR & BSS 29305; Althuru padu (NLR), BSS & SKB 32916; Kollapur (MBNR), BSS & SKB 32996.

Note: Jovet & Guedes (1973) treated this taxon as a variety under *S.indicus.* We followed the concept of Clayton & Renovoize (1986).

Sporobolous maderaspatanus Bor in Kew Bull. 234.1957; Bor, Grass. India 632.1960; Moulik, Grass. Bamb. India 2:480.1997. *Sporobolous orientalis* Kunth, *nom. illegit*; Hook. f., Fl. Brit. India 7:251.1896; Fischer, Fl. Madras 3:1817.1934.

Perennials, culms erect, to 50cm, stoloniferous; nodes glabrous. Leaf sheaths glabrous. Ligules membranous. Leaf blades linear, base rounded, acuminate, convolute. Panicles pyramidal. Spikelets oblong. Lower glume ovate, hyaline. Upper glume ovate-lanceolate, acute. Lemma ovate, lanceolate, chartaceous, acute. Palea ovte-lanceolate, hyaline. Grains ovoid.

Distribution: *World*: Sri Lanka; *India*: Peninsular India; *Andhra Pradesh*: Anantapur, East Godavari and Mahaboobnagar districts.

Rare in forest shade and plains. *Fl. & Fr.*: July-Nov. *Repr. Spec.*: Kollapur (MBNR), BR & BSS 32917; Mallapukota (EG), CAB 8294 (MH); Gooty (ATP), JSG 20845 (MH).

Sporobolus piliferus (Trin.) Kunth, Enum Pl. 1:211.1833; Hook. f., Fl. Brit. India 7:251.1896; Fischer, Fl. Madras 3:1817. 1934; Bor, Grass. India 632.1960; Moulik, Grass. Bamb. India 2:480.1997. *Agrostis spicata* Vahl, Symb. Bot. 1.9.1970.

Annuals, culms erect, tufted, to 30cm; nodes glabrous. Leaf sheaths ciliate at mouth. Ligules ring of hairs. Leaf blades radical, linear, base narrow, acuminate, panicles linear, densely flowered. Lower glume ovate-lanceolate, hyaline. Upper glume oblong-lanceolate. Lemma oblong. Palea oblong. Stamens 2. Grains ellipsoid.

Distribution: *World*: Sri Lanka, East Africa; *India*: Throughout; *Andhra Pradesh*: Guntur, Nellore and Visakhapatnam districts.

Occasional on hill slopes and moist habitats. *Fl.* & *Fr.*: Aug. – Dec. *Repr. Spec.*: Sunkurimetta (VSKP), GVS 21631 (CAL & MH).

Sporobolus spicatus (Vahl) Kunth, Rev. Gram. 1:67.1829. Hook. f., Fl. Brit. India 7:250.1896; Fischer, Fl. Madras 3:1817.1934; Bor, Grass. India 632.1960; Moulik, Grass. Bamb. India 2:481.1997. *Agrostis spicata* Vahl, Symb. Bot. 1:9.1970.

Perennials, stoloniferous, aquatic, culms erect, tufted, to 30cm; nodes glabrous. Leaf sheaths glabrous. Ligules membranous. Leaf blades oblong, base cordate or truncate, acute, glabrous. Panicles spiciform, solitary. Spikelets elliptic. Lower glume ovate-oblong, truncate, hyaline, 1-nerved. Upper glume ovate-oblong, chartaceous, 1-nerved. Lemma linear, lanceolate, chartaceous, hyaline. Palea oblong, truncate, 2-nerved. Stamens 3. Grains obovoid.

Distribution: *World*: Arabia, Africa; *India*: Peninsular India; *Andhra Pradesh:* Anantapur, Kadapa and Kurnool districts.

Occasional along stream banks and moist places. *Fl.* & *Fr.*: April – Dec. *Repr. Spec.*: Peapully (KNL), PVP 2682; Balaraju palli (KDP), AMR 24573; Pamidi (ATP), JNP & AL 27365; B.T.dam (ATP), BR & KVS 29181; Ahobilam (KNL), MSR & BSS 32264.

Sporobolus tenuissimus (Schrank) Kuntze, Rev. Gen. Pl. 3:369.1893; Bor, Grass. India 633.1960; Moulik, Grass. Bamb. India 2:481.1997. *Panicum tenuissimum* Schrank in Denkschr., Bot. Ges. Regensb. 2:26.1822. *Sporobolus minutiflorus* (Trin.) Link., Hort. Berol. 1:88.1827; Hook. f., Fl. Brit. India 7:248.1896; Fischer, Fl. Madras 3:1816.1934.

Annuals, culms erect, tufted, to 1.2m; nodes glabrous. Leaf sheaths glabrous. Ligules hairy, membranous. Leaf blades linear-lanceolate, base rounded, acuminate. Panicles effuse. Spikelets ovate. Lower glume oblong, hyaline. Upper glume ovate, hyaline, 1-nerved. Lemma ovate, chartaceous, 1-nerved. Palea oblong, hyaline, 2-nerved. Stamens 2. Grains obovoid.

Distribution: *World*: Africa, America. *India*: Peninsular India; *Andhra Pradesh*: Anantapur, Chittoor, Kurnool, Khammam, Mahaboobnagar, Medak, Nellore and Rayalaseema districts.

Occasional near water courses. *Fl.* & *Fr.*: Aug. – Dec. *Repr. Spec.*: Balapalli RF (KDP), PVP 3459; Upper Ahobilam (KNL), SS 22698; Kinnerasani (KMM), BR & AMR

26820; Muchukota (ATP), JNP 27397; Sri Kalahasthi (CTR), AL 28700; Gangutla RF (NLR), BR & BSS 29305.

Sporobolus tremulus (Willd.) Kunth, Rev. Gram. 1:67.1829; Hook. f., Fl. Brit. India 7:250.1896; Fischer, Fl. Madras 3:1817. 1934; Bor, Grass. India 633. 1960; Moulik, Grass. Bamb. India 2:481. 1997. *Agrostis tremula* Willd., Sp. Pl. 1:372.1798.

Annuals, culms tufted, to 30cm, stoloniferous; nodes glabrous. Leaf sheaths glabrous. Ligules hairy. Leaf blades linear-lanceolate. Panicles, subspiciform, interrupted. Spikelets ellipsoid. Lower glume lanceolate. Upper glume ovate-lanceolate, membranous, acute. Lemma oblong, chartaceous, 1-nerved. Palea ovate, chartaceous. Stamens 3. Grains ellipsoid.

Distribution: *World*: Sri Lanka; *India*: Throughout; *Andhra Pradesh*: Anantapur, Chittoor, Guntur, Hyderabad, Krishna, Kurnool and Mahaboobnagar districts.

Occasional in forest peripheries. *Fl. & Fr.*: July-Dec. *Repr. Spec.*: Gundlabrahmeswaram (KNL), BR & BSS 32965; Akuthotapalli (ATP), AL 27395.

Sporobolus virginicus (L.) Kunth, Rev. Gram. 1:67. 1829; Hook. f., Fl. Brit. India 7:249. 1896; Fischer, Fl. Madras 3:1818. 1934; Bor, Grass. India 634. 1960; Moulik, Grass. Bamb. India 2:482. 1997. *Agrostis virginica* L., Sp. Pl. ed. 1:63. 1753.

Perennials, culms erect, tufted, to 10cm, stoloniferous. nodes glabrous. Leaf sheaths terete. Ligules row of hairs. Leaf blades linear-lanceolate, base rounded, convolute, apex acute. Panicles sub-spiciform. Spikeltes lanceolate, crowded. Lower glume lanceolate, chartaceous. Upper glume ovate-lanceolate, chartaceous. Lemma ovate-lanceolate, chartaceous, 1-nerved. Palea oblong, elliptic, hyaline. Stamens 3. Grains obovoid.

Distribution: *World*: Myanmar, Sri Lanka; *India*: Peninsular India; *Andhra Pradesh:* Anantapur, Kurnool, Mahaboobnagar, Srikakulam and Vizayanagaram districts.

Occasional in river banks, hill slopes and near streams. *Fl. & Fr.*: Sept. - Nov. *Repr. Spec.*: Velgode (KNL), BSS & SKB 30233; Tatipudi (VZN), AMR 24529; Devaragattu (KNL), SS & AMR 22675; Pamidi (ATP), BR & JNP 26617; Penukonda (ATP), BR & JNP 26749.

Note: This taxon was known earlier recorded only from Srikakulam district (Rao & Harasreeramulu, 1986). The present report extends its distribution to Anantapur, Kurnool and Vizayanagaram.

Sporobolus wallichii Munro ex Trin., in J. Bot. 27:171. 1889; Hook. f., Fl. Brit. India 7:248.1896; Fischer, Fl. Madras 3:1817.1934; Bor, Grass. India 634.1960; Moulik, Grass. Bamb. India 2:482.1997.

Perennials, culms erect, tufted, to 1.2m, slender; nodes glabrous. Leaf sheaths glabrous. Ligules membranous. Leaf blades linear, convolute, base cordate or truncate, apex acuminate, glabrous. Panicles effuse, flexuous. Spikelets oblong-lanceolate. Lower glume ovate, chartaceous, truncate. Upper glume ovate-lanceolate, acute, 1-nerved. Lemma ovate-elliptic, chartaceous, 3-nerved. Palea oblong, chartaceous. Stamens 3. Grains oblong.

Distribution: *World*: Myanmar, Sri Lanka; *India*: South & East India; *Andhra Pradesh*: Anantapur, Chittoor, Kadapa, Krishna, Kurnool, Mahaboobnagar, Prakasam and Rayalaseema districts.

Common in moist localities of the forests. *Fl. & Fr.*: July-Dec. *Repr. Spec.*: Rollapenta (PKM) PVP 2341; Tirumala hills (CTR), BR & AMR 21841, 21845, 21848; Kottala Cheruvu (KNL), SS 22464; Kalasamudram (ATP), BR & AMR 24230; Guvvalacheruvu (KDP), BR & AMR 26074. Nallamada (ATP), BR & JNP 27337; Velgode (KNL), BSS & SKB 30232, 30600; Mannanur (MBNR), BSS & SKB 30601.

TRIPOGON Roemer et Schultes

1. Lemmas 4-lobed .... *T. bromoides*
1. Lemmas 2-lobed .... (2)
2. Median awn as long as or longer than the lemma .... *T. wightii*
2. Median awn shorter than the lemma or absent .... (3)
3. Leaf blades filiform, convolute .... *T. jacquemontii*
3. Leaf blades linear to linear-lanceolate .... *T. purpurascens*

Tripogon bromoides Roemer et Schultes, Syst. Veg. 2:600.1817; Hook. f., Fl. Brit. India 7:287. 1896; Fischer, Fl. Madras 3:1834.1934; Bor, Grass. India 521.1960; Moulik, Grass. Bamb. India 2:622.1997.

Perennials, culms erect, tufted to 30cm, nodes glabrous. Leaf sheaths glabrous. Leaf blades linear, base cordate, acuminate, hairy. Spikelets 6-13- flowered. Lower glume oblong-lanceolate, chartaceous, 1-nerved. Upper glume lanceolate, ovate-oblong, 2-lobed, 1-nerved. Lemmas ovate, chartaceous, 3-nerved, 3-awned. Paleas elliptic-oblong, keeled, 2-nerved, hyaline. Stamens 3. Grains oblong.

Distribution: *World*: Sri Lanka. *India*: Meghalaya, South India and Rajasthan. *Andhra Pradesh*: Mahaboobnagar, Srikakulam, Vizayanagaram and Rayalaseema districts.

Common on hill slopes and forest plains. *Fl. & Fr.*: Aug.-Feb. *Repr. Spec.*: Horseley hills (CTR), PVP 3500; Pachipenta (VZN), AMR 24543; Appapur (MBNR), AMR 25678; Penukonda (ATP), JNP & AL 27348; Irlapenta (MBNR), BR & BSS 32289.

Tripogon jacquemontii Stapf in Kew Bull. 85.1892; Hook. f., Fl. Brit. India 7:286. 1896; Fischer, Fl. Madras 3:1833. 1934; Bor, Grass. India 522.1960; Moulik, Grass. Bamb. India 2:622.1997.

Perennials, culms erect, tufted, to 90cm; nodes glabrous. Leaf sheaths glabrous. Ligules fimbriate. Leaf blades linear, base narrow, acuminate, glabrous. Spikelets oblong, 8-20-flowered. Lower glume ovate-lanceolate, 1-nerved, lobed on one side, sub coriaceous. Upper glume ovate-lanceolate, 1-nerved, chartaceous. Lemma elliptic-obovate, awned. Palea lanceolate, hyaline, 2-keeled, winged. Stamens 3. Grains oblong.

Distribution: *World*: Endemic to Peninsular India; *Andhra Pradesh*: Chittoor and Mahaboobnagar districts.

Rare in barren lands and forest plains. *Fl. & Fr.*: Sept.–Dec. *Repr. Spec.*: Mallelateertham (MBNR), BR & BSS 32217; Horseley hills (CTR), MSG & AA 12221 (SKU).

Note: Although, Mitra (1958) mentioned the taxon occurrence in Assam, but has not been recorded by Shukla (1996). Further, Ahmedullah & Nayar (1986) and Nayar (1996) considered the taxon endemic to Peninsular India.

Tripogon purpurascens Duthie in Ann., Roy. Bot. Gard. Cal. 9:75. 1901; Bor, Grass. India 522. 1960; Moulik, Grass. Bamb. India 2:623. 1997. *T. abyssinicus* Hook. f., Fl. Brit. India. 7:287. 1896 in part non Nees (1854).

Perennials, culms erect, tufted, slender, to 15cm; nodes glabrous. Leaf sheaths glabrous. Ligules ciliate, membranous. Leaf blades filiform, convolute, tip obtuse with purple colour, densely hairy near the base. Inflorescence terminal, contracted. Spikelets 2-3-flowered. Lower glume small, triangular, 1-nerved, slightly awned. Upper glume lanceolate, 1-nerved, membranous, central thickened, coriaceous, margin hyaline, ciliate upper portion. Lemmas 3, 3-nerved, cleft, median nerve slightly awned in the cleft, margin hyaline. Paleas 3, 2-keeled, tip 2-lobed except third palea. Stamens 2-3. Grains linear.

Distribution: *World*: Pakistan; *India*: North-West India, Himalayas, Uttar Pradesh and Rajasthan; *Andhra Pradesh*: Mahaboobnagar, Prakasam, Vizayanagaram and Rayalaseema districts.

Occasional in dry and moist rocky habitats of forests. *Fl. & Fr.*: July-Nov. *Repr. Spec.*: Devaragattu (KNL), BR & SS 21221; Pachipenta (VZN), AMR 24543; Kuntimaddi (ATP), AL & BSS 27357; Kalasamudram (ATP), KVS & BSS 29702; Amrabad (MBNR), BSS & SKB 32201; Appapur (MBNR), BSS & SKB 32220.

Note: New distributional record to Peninsular India (Rao *et al.*, 2009c)

Tripogon wightii Hook. f., Fl. Brit. India 7:286. 1896; Fishcer, Fl. Madras 3:1833. 1934; Bor, Grass. India 524. 1960; Moulik, Grass. Bamb. India 2:623.1997.

Perennials, culms erect, tufted, to 30cm; nodes glabrous. Leaf sheaths glabrous. Leaf blades filiform. Inflorescence spike. Spikelets oblong, many-flowered. Glumes linear-oblong, hyaline, 2-lobed, mucronate. Lemma narrow, base bearded, 2-lobed, awned in the cleft. Palea oblong, winged, scabrid. Stamens 3. Grains narrow.

Distribution: *World*: Endemic to Peninsular India. *India*: Tamilnadu; *Andhra Pradesh*: Eastern Ghats of Chittoor district.

Rare in rock crevices and hill slopes. *Fl. & Fr.*: Aug. – Dec. *Repr. Spec.*: Horseley Konda (CTR), JSG 20920 (CAL & DD).

Note: Till date, the cited specimen is the only record from Andhra Pradesh.

Acknowledgements

The authors are grateful to Department of Science & Technology (SP/SO/A-40/98) and Department of Biotechnology (BT/PR6603/NDB/51/089/2005), New Delhi for financial assistance. Thanks are due to Andhra Pradesh Forest Department authorities and staff for their help during field explorations.

References

Ahmedullah, M and M. P. Nayar. 1986. *Endemic Plants of Indian Region*. Vol.1. (Peninsular India). Botanical Surey of India, Calcutta.

Beetle, A.A. 1955. The four sub-families of Gramineae, *Bull. Torrey Bot. Club*. 82: 196-197.

Bentham, G. 1881. Notes on Gramineae. *J. Linn. Soc. Bot.*, 19: 14. 134.

Bews, J.W. 1929. *The world's Grasses – Their Differentiation, Distribution, Economics and Ecology*. Longmans, Green & Co. London.

Bor, N. L. 1960. *The Grasses of Burma, Ceylon, India and Pakistan*. Pergamon Press. Oxford.

Brown, R. 1810. *Prodomus florae novae Hollandiae*, Vol.1. R Taylor, London, pp. i – viii. 145 – 590.

Champion, H.G. and S.K. Seth. 1968. *A Revised Surey of the forest types of India*. Government of India, New Delhi.

Clayton, W.D and S.A. Renvoize. 1986. *Genera Graminum – Grasses of the world*. Kew Bulletin Additional series XIII, RBG, Kew; HMSO, London.

Davis, S.D., V.H. Heywood and A.C. Hamilton. 1994 & 1995. *Centres of Plant Diversity Volume I: Europe, Africa and South West Asia and Volume II: Asia, Australasia and the Pacific.* WWF and IUCN. IUCN Publications Unit, Cambridge, UK.

Gadgil, M. and V.M. Meher-Homji. 1986. Localities of great significance to conservation of india's biological diversity. *Proc. Indian Acad. Sci. (Animal Sci. Plant Sci. Suppl.* 165-180.

Gayathri, M.S. & T.Pullaiah. 1997. Poaceae. in T.Pullaiah (ed.) *Flora of Andhra Pradesh (India)*. Vol.3. Scientific Publishers, Jodhpur.

Hackel, E. 1889. *Monographia Andropogonearum*. Paris.

Hitchcock, A.S. 1920. *The genera of grasses of the United States, with special reference to the economic species.* U.S.Dep. Arg. Bull. 772.

Hitchcock, A.S. 1935. *Manual of grasses of the United States.* U.S. Dep. Arg. Misc. Publ. 200.

Jovet, P. & M. Guedes. 1973. Validation of names in *Sporobolus. Taxon* 22: 163.

Mitra, J.N. (1958). *Flowering plants of Eastern India*. Vol.1. Monocotyledons. Calcutta.

Nayar, M.P. 1996. *'Hot spots' of endemic plants of India, Nepal and Bhutan*. Tropical Botanical Garden and Research Institute, Palode, Trivandrum.

Narahari Prasad, J. 2006. Grass Diversity in Anantapur District, Andhra Pradesh. M. Phil, dessertation. Sri Krishnadevaraya University, Anantapur.

Prasanna, P.V. 1988. *Poaceae in Rayalaseema, Andhra Pradesh.* Ph.D.thesis submitted to Sri Krishnadevaraya University, Anantapur.

Pullaiah, T and E. Chennaiah. 1997a. *Flora of Andhra Pradesh (India)*. Vol.1. Scientific Publishers, Jodhpur.

Pullaiah, T and D.Ali Moulali. 1997b. *Flora of Andhra Pradesh (India)*. Vol.2. Scientific Publishers, Jodhpur.

Pullaiah, T. 1997c. *Flora of Andhra Pradesh (India)*. Vol.3. Scientific Publishers, Jodhpur.

Pullaiah, T and S. Karuppusamy. 2008. *Flora of Andhra Pradesh (India)*. Vol. 5. Scientific Publishers, Jodhpur.

Rao, Ravi Prasad, B. 1996. in *Sacred and protected groves of Andhra Pradesh*. WWF – India, A. P. State Office, Hyderabad. pp 15-19, 28-31, 46-52, 60-63, 65-70.

Rao, Ravi Prasad, B., M. Sridhar Reddy and T. Pullaiah. 2008. Flora and Vegetation of Andhra Pradesh. *Proc. A.P. Akademi of Sciences* 12 (1&2): 1-13.

Rao, Ravi Prasad, B., B. Sadasivaiah, S. Sunitha, A. Madhusudhana Reddy. 2009a. Occurrence of East Himalayan floral elements in the Eastern Ghats of Andhra Pradesh: I. Grasses. *Pleione* 3(1) 40-44.

Rao, Ravi Prasad, B., A. Madhusudhana Reddy, B. Sadasivaiah, K.V.Subbaiah and S. Sunitha. 2009b. Two new distributional records of Poaceae for Peninsular India. *Journ. Eco. Tax. Bot.* 33(2): 434-436.

Rao, Ravi Prasad. B., A. Madhusudhana Reddy, B. Sadasivaiah, K.V.Subbaiah and S. Sunitha. 2009c. Two New Distributional Records of Poaceae for Peninsular India. *J. Econ. Taxon. Bot.* 33 (2): 434-436.

Rao, Ravi Prasad B., A. MadhusudhanaReddy* and B. Sadasivaiah. 2010 Two endemic grasses of India, additions to flora of Andhra Pradesh, India. *Journ. Eco. Tax. Bot.* 34(2): 394-397.

Rao, R.S. and S. Hara Sreeramulu. 1986. *Flora of Srikakulam district, Andhra Pradesh, India*. Indian Botanical Society, Meerut.

Reddy, Madhusudhana, A. 2002. *Grasses of Eastern Ghats of Andhra Pradesh*. Ph. D., Thesis submitted to Sri Krishnadevaraya University, Anantapur.

Reddy, Madhusudhana, A., S. Sunitha and B. Ravi Prasad Rao. 2002. *Brachiaria brizantha* (A.Rich) Stapf, a new report to Peninsular India. *J. Eco. Tax. Bot.* 26 (1):166-168.

Reddy, Madhusudhana, A., S. Sunitha and B. Ravi Prasad Rao. 2004. *Cyrtococcum deccanense* Bor (Poaceae), a new record to Andhra Pradesh, India. *J. Eco. Tax. Bot.* 28(1):84-86.

Reddy, Madhusudhana, A., S. Sunitha and B. Ravi Prasad Rao. 2005. *Paspalum conjugatum* Berg. (Poaceae)- A new record to Andhra Pradesh. *J. Bomb. Nat. Hist. Soc.* 102(2): 259-260.

Reddy, Madhusudhana, A., B. Sadasivaiah, K.V. Subbaiah, M.V. Suresh Babu and B. Ravi Prasad Rao. 2009. Three new distributional records of Poaceae for Andhra Pradesh, India. *Journ. Eco. Tax. Bot.* 33(2): 411-413.

Reddy, Madhusudhana, A. and B. Ravi Prasad Rao. 2009. *Isachne albens* (Poaceae) - a new distributional record for peninsular India. *Journal of Indian Botanical Society.* 88 (3&4): 158-159.

Reddy, C.S., K.N. Reddy and Vasthavaya S. Raju. 2008. *Supplement to Flora of Andhra Pradesh, India.* Deep Publications, New Delhi.

Sadasivaiah, B., K. Venkata Subbaiah, S. Khadar Basha, M.V. Suresh Babu and B. Ravi Prasad Rao. 2008. *Arundinella tuberculata* Munro ex Lisboa- A new distributional record for Eastern Ghats, India. *Proc. of A.P. Akademi of Sciences* 12 (3):289-291.

Sadasivaiah, B., P. Priyadarshini, S.Khadar Basha and B. Ravi Prasad Rao. 2010. Occurrence of *Urochloa mosambicensis* (Hack.) Dandy (Poaceae) in Andhra Pradesh, India. *Indian Forester* 136 (1): 402-405.

Schantz, H.L. 1954. The place of grasslands in the earths cover of vegetation. *Ecology* 35: 143-145.

Shukla, U. 1996. *The Grasses of North - Eastern India.* Scientific Publishers. Jodhpur.

Stapf, O. 1917-1934. Gramineae. in D.Prain (ed.), *Flora of Tropical Africa,* vol.9. L. Reeve and co., Ltd., Kent.

Stebbins, G.L. 1956. Cytogenetics and the evolution of the grass family. *Amer. J. Bot.* 43: 890-905.

Subbaiah, K.V. 2008. *Grass diversity of Arid Zone of Peninsular India.* Ph.D. thesis submitted to Sri Krishnadevaraya University, Anantapur.

Subba Rao, G.V. and G.R. Kumari. 2008. *Flora of Visakhapatnam (Andhra Pradesh) Vol.2.* Botanical Survey of India, Kolkata.

Sunitha, S. 2002. *Plant Biodiversity of the Sacred Groves of Kurnool district,* Andhra Pradesh. Ph.D. thesis submitted to Sri Krishnadevaraya University, Anantapur.

Suxena, M.R. 1947. A study of the grasses of the H.E.H., the Nizam's dominions, Hyderabad, Deccan. *Osmania Univ. Journ. Res.* 12 :28-65.

Tateoka, T. 1957. Miscellaneous papers on the phylogeny of Poaceae (10). Proposition of a new phylogenetic system of Poaceae. *J. Jap. Bot.* 32: 275-287.

Chapter 10

Cultivated Plants of Andhra Pradesh

P. Priyadarsini and B. Ravi Prasad Rao*

Biodiversity Conservation Division, Department of Botany, Sri Krishnadevaraya University, Anantapur – 515 003, Andhra Pradesh

ABSTRACT

A total of 256 plant species belonging to 71 families are cultivated in Andhra Pradesh. Products from these plants are being utilized for various purposes such as food, medicine, fibre, fodder and aromatic oils. The present paper provides enumeration of all the cultivated species of Andhra Pradesh.

Introduction

Agrobiodiversity is an evolutionary divergent line of biodiversity concerned with agro-ecosystems, and variation in agriculture-related plants. It has multiple, economic, ecological and social benefits, and is a crucial and integral component of sustainable development. Agroecosystems provide various ecological services, such as landscape protection, soil protection, health, water cycle and quality, air quality etc. (Cromwell, 1999). Agriculture is the main sources of food, fodder and is the basis for many industries and therefore forms the important part of economy of the nation. Crop diversity is vital both to the culture and basic food security of India. The Indian region is one of the world's 8 centers of crop plant origin and diversity as recognized

* Corresponding Author: E-mail: rao_rp@rediffmail.com

by Vavilov (1951). According to Rana (1993) at least 166 food crop species and 320 wild relatives of crops have been originated in India. Don Ellison (1999) listed 4,250 internationally grown trees, shrubs, and climbers.

Andhra Pradesh is one of the agriculturally rich and food grain surplus states of the country. Andhra Pradesh is fourth largest state by area, and fifth largest by population in India. The state lies between NL 12° 37′ -19°54' & EL 76°46′-84°46′ in the southern part of India. It is characterized by semi-arid, moist to dry, sub-humid climate. Andhra Pradesh has bountiful natural resources endowed with fertile soils, water and various agroclimatic conditions and hence supports rich agrobiodiversity. Andhra Pradesh state is known as Anna poorna, the rice bowl of the south. About 70 per cent of the total population depends on Agriculture. It is the second largest producer of rice in india. Rice is the large monocrop cultivated in 40 lakh hectares in the state, followed by groundnut, 18 lakh hectares under cultivation (Anon, 2002).

The total geographic area of the state is 2, 75,068 sq. km and has 10.1 million ha cultivated area, constituting 37 per cent of state's total geographical area. The irrigated area is 36 per cent of the total cultivated area (Anon., 2007).The state comprises 23 districts which are grouped into 3 distinct regions namely Rayalaseema, Telangana, and Coastal Andhra. Andhra Pradesh is divided into 7 agro ecological zones (Anon, 2002), Krishna Godavari, North Coastal, Southern, Northern Telangana, Southern Telangana, Low Rainfall Rayalaseema and High Altitude Tribal Zones. Andhra Pradesh has three cropping seasons based on monsoon and availability of water from manmade sources: the Kharif, Rabi and summer.

The soils of the state are ranging from poor coastal sands to highly fertile deltaic alluviums. Red soils occupy 65 per cent of the cultivated area and are mostly situated in Rayalaseema region. Black soils cover 25 per cent of the cultivated area. The alluvial clay soils cover 5 per cent of the cultivated area. The coastal sandy occupies only 3 per cent , while the remaining 1 per cent is covered by lateritic soils. The mean annual temperature in summer varies 20-41°C and winter temperature of 13-32°C. Andhra Pradesh receives rainfall from southeast monsoon as well as the north east monsoon. The average annual rainfall is around 940 mm. Three major Rivers pass through the state are Godavari, Krishna and Pennar.

Data Compilation

Since there is no precise information on the cultivated species of Andhra Pradesh, the present paper aims to fill the gap. The present paper provides a checklist of cultivated species of Andhra Pradesh based on our research team field observations for the past two decades in Andhra Pradesh and extensive referencing of all the published floristic literature pertaining to the region and State level publications include, the flora of Andhra Pradesh (Pullaiah and Chennaiah, 1997; Pullaiah and Ali Moulali, 1997; Pullaiah, 1997; Pullaiah and Sandhya Rani, 1999; Lakshmaiah, 2006; Rao, 2008). The references for district level flora information includes: Srikakulam (Rao and Sreeramulu, 1986), Vizianagaram (Venkaiah, 2004), Visakhapatnam (Venkateswarulu *et al.*, 1972; Subba Rao and Kumari, 2008), East Godavari (Rao *et al.*, 1999), West Godavari (Rao *et al.*,1986), Krishna (Lakshminarayana *et al.*, 1997), Guntur (Pullaiah *et al.*, 2000), Prakasam (Bhairavamurthy and Krishna Mohan, 1991), Nellore

Table 10.1: Cultivated Species of Andhra Pradesh

Sl.No.	Name of the Species	Family	H	Telugu Names	English Names	RS	CA	TN	Use Value
1.	*Adhatoda vasica* Nees	Acanthaceae	S	Addarasam, Vasa	Vasaka, Malabar nut			+	Med
2.	*Andrographis paniculata* (Burm.f.) Wallich ex Nees	Acanthaceae	H	Nelavemu, Punirma	Kalmegh, King of bitters	+	+	+	Med
3.	*Crossandra infundibuliformis* (L.) Nees	Acanthaceae	H	Kanakambaram, Gobbi, Priyadarsi	Tropic flame	+	+	+	Orn-Flowers
4.	*Agave sisalana* Perrine	Agavaceae	S	Kittanara, Kalabanda	Sisal hemp	+	+	+	Fibre
5.	*Amaranthus caudatus* L.	Amaranthaceae	H	Netikura, Kikeera	Amaranth			+	Veg-Leaf
6.	*Amaranthus tricolor* L. (=*A. gangeticus* L.)	Amaranthaceae	H	Perugu thotakura, Tella thitakura	Chinese spinach, Josephs coat	+		+	Veg-Leaf
7.	*Gomphrena globosa* L.	Amaranthaceae	H	Rudrakshi	Globeamaranth	+	+	+	Orn-Flowers
8.	*Crinum defixum* Ker.-Gawl. (=*C.latifolium* L.)	Amaryllidaceae	H	Kesarichettu, Chembugadda	Wild Garlic		VSKP		Med
9.	*Polianthus tuberosa* Willd.	Amaryllidaceae	H	Sampenga, Lilly, Sugandharaja	Lilly, Tuburose		+	+	Orn-Flowers
10.	*Anacardium occidentale* L.	Anacardiaceae	T	Godambi, Jeedipappu, Munthamamidi, Zeedimamidi	Cashewnut, Kaju	+	+	+	Edible-Nuts
11.	*Mangifera indica* L.	Anacardiaceae	T	Mamidi, Amra	Mango, Amra	+	+	+	Edible-Fruits
12.	*Annona cherimola* Miller	Annonaceae	T	Guanabana	Cherimoya	CTR			Edible-Fruits
13.	*Annona muricata* L.	Annonaceae	T	Mundla sithaphalamu	Sour sop				Edible-Fruits
14.	*Annona reticulata* L.	Annonaceae	T	Ramaphalamu, Ramasithaphalamu	Netted custard apple	+	+	+	Edible-Fruits
15.	*Annona squamosa* L.	Annonaceae	T	Sithaphalamu	Custard apple, Sugar apple Gandhagatramu	+	+	+	Edible-Fruits

Contd...

Table 10.1–Contd...

Sl.No.	Name of the Species	Family	H	Telugu Names	English Names	RS	CA	TN	Use Value
16.	*Artabotrys hexapetalus* (L.f.) Bhand.	Annonaceae	C	Tigasampangi, Manoranjitam	Climbing ilang-ilang	+	+	+	Orn-Flowers
17.	*Anethum graveolens* L.	Apiaceae	H	Dill	Dill, European dill		EG		SC-Fruits
18.	*Trachyspermum roxburghianum* (DC.) Craib.(= *Carum roxburghianum* Benth.)	Apiaceae	H	Radhuni	Wild Celery		WG &EG		Edible-Fruits
19.	*Coriandrum sativum* L.	Apiaceae	H	Kottimeera, Daniyalu	Coriandar	+	+	+	SC-Fruits-Leaves
20.	*Cuminum cyminum* L.	Apiaceae	H	Jeelakarra	Cumin	+		+	SC-Fruits
21.	*Daucas carota* L.	Apiaceae	H	Carrot, Erra mullangi	Carrot	+		+	Veg-Tubers
22.	*Ferula asafoetida* L.	Apiaceae	H	Inguva, Hingu	Asafetida		+	+	SC/Resin
23.	*Foeniculum vulgare* Miller	Apiaceae	H	Sompu, Sopu	Fennel, Globet of life	ATP KNL			Med/SC
24.	*Trachyspermum ammi* (L.) Turrill (=*Carum copticum* Hiern.)	Apiaceae	H	Vamu, Omamu	Ammi, Lovage Bishops weed	+	+	+	SC-Fruits
25.	*Catharanthus roseus* (L.) G. Don (=*Vinca rosea* L.)	Apocynaceae	H	Billaganneru, Sadapuspi	Madgaskar periwinkle	ATP			Med
26.	*Rauvolfia serpentina* (L.) Benth. ex Kurz	Apocynaceae	S	Sarpagandhi, Pathalagarudi	Snake root		+	+	Med
27.	*Acorus calamus* L.	Araceae	H	Vasa, Vaja, Ugragandha	Sweet flag	+	+		Med
28.	*Amorphophallus paeoniifolius* (Dennst.) Nicolson	Araceae	H	Manchikanda, Kanda	Telugu potato, Telinga potato	CTR	WG		Veg-Tubers
29.	*Colocasia esculenta* (L.) Schott. (=*C. antiquorum* Schott.)	Araceae	H	Chemagadda	Taro, Dasheen, Cocoyam	+	+	+	Veg-Tubers

Contd...

Table 10.1–Contd...

Sl.No.	Name of the Species	Family	H	Telugu Names	English Names	RS	CA	TN	Use Value
30.	*Amorphophallus campanulatus* Blume	Araceae	H	Kandagadda, Ratnapurigadda	Elephant foot yam		+		Veg-Tubers
31.	*Areca catechu* (L.) P. Beauv.	Arecaceae	T	Vakka, Poka	Areca nut, Betel nut	+	+	+	Narcotics
32.	*Borassus flabellifer* L.	Arecaceae	T	Tati, Tadi	Palmyra palm	+	+	+	Edible-Fruits
33.	*Cocos nucifera* L.	Arecaceae	T	Kobbari, Tenkaya, Narikelam	Coconut	+	+	+	Edible-Fruits
34.	*Elaeis guineensis* Jacq.	Arecaceae	T	Palm oil	Palm oil,		+	+	Oil
35.	*Gymnema sylvestre* (Retz.) R.Br. ex Roemer & Schultes	Asclepiadaceae	C	Podapatri, Meshasrungi	African oil palm		NLR		Med
36.	*Anacyclus pyrethrum* (L.) Link	Asteraceae	H	Akkalakarra, Akharakaraba	Mount Atlas daisy Spilanthus	+		+	Med
37.	*Artemisia pallens* Wallich (=*Artemisia nilagirica* (Cl.) Pampan.)	Asteraceae	H	Davanamu	Davana	+		+	AO
38.	*Carthamus tinctorius* L.	Asteraceae	H	Kusuma	Saf flower	+	+	+	Oil
39.	*Chrysanthemum indicum* L.	Asteraceae	H	Chamanthi	Japanese Chrysanthemum	+	+	+	Orn-Flowers
40.	*Cichorium intybus* L.	Asteraceae	H	Chikori	Chicory		GNT		Med
41.	*Guizotia abyssinica* Cass.	Asteraceae	H	Valiselu, Gaddi nuvvalu	Niger seed oil	+	+	+	Oil
42.	*Helianthus annuus* L.	Asteraceae	H	Poddutirugudu, Adityalu, Kusalu, Gurnulu	Sunflower	+	+	+	Oil
43.	*Lactuca sativa* L.	Asteraceae	H	Kavukura	Lettuce	+	+	+	Veg- Leaf
44.	*Sausserea lappa* C.B.Clarke	Asteraceae	H	Chengalvacost	Costus, Kuth			WGL	Med
45.	*Tagetes erecta* L.	Asteraceae	H	Mudda banti	African/ American marigold	+	+	+	Orn-Flowers

Contd...

Table 10.1–Contd...

Sl.No.	Name of the Species	Family	H	Telugu Names	English Names	RS	CA	TN	Use Value
46.	*Tagetes patula* L.	Asteraceae	H	Banti	French Marigold	+	+	+	Orn-Flowers
47.	*Basella alba* L. (=*B.rubra* L.)	Basellaceae	C	Bachali	Red vine spinach, Ceylon spinach	+	+	+	Veg-Leaf
48.	*Bixa orellana* L.	Bixaceae	T	Jafra, Bixa, American jafra	Annatto, Jafra	+	+	+	Med
49.	*Ceiba pentandra* (L.) Gaertner	Bombacaceae	T	Buruga	Silkcotton	+	+	+	Cotton
50.	*Brassica juncea* (L.) Czerniak	Brassicaceae	H	Avalu, Sasavalu	Indian mustard, Brown mustard	+	+	+	SC-Seeds
51.	*Brassica nigra* (L.) Koch	Brassicaceae	H	Nallaavalu, Varnavalu	True mustard, Blackmustard	+	+	+	SC-Seeds
52.	*Brassica oleracea* L. var. *botrytis* L.	Brassicaceae	H	Gobipuvvu	Cauliflower	+	+	+	Veg-Flowers
53.	*Brassica oleracea* L. var. *capitata* L.	Brassicaceae	H	Cosu, Cabbage	Cabbage	+	+	+	Veg-Leaf
54.	*Lepidium sativum* L.	Brassicaceae	H	Adityalu, Chandrasoora	Garden cress	+	+		Med
55.	*Raphanus sativus* L.	Brassicaceae	H	Mullangi	Radish	+	+	+	Veg-Tubers
56.	*Sinapis alba* L.(=*Brassica alba* Rabenh.,non L.)	Brassicaceae	H	Tella avalu	White mustard	+	+	+	SC - Seeds
57.	*Ananas comosus* L.	Bromeliaceae	T	Anasa	Pineapple		+		Edible-Fruits
58.	*Cannabis sativa* L.	Cannabinaceae	H	Ganjai chettu, Bhangaku	Hemp, Soft hemp, True hemp	+		+	Narcotics
59.	*Carica papaya* L.	Caricaceae	T	Boppaya, Parangi	Papaya	+	+	+	Edible-Fruits
60.	*Casuarina equisetifolia* L.	Casuarinaceae		Sarugudu	Ironwood	+	+	+	Wood
61.	*Celastrus paniculatus* Willd.	Celastraceae	C	Jyothismathi, Teega-palleru, Malkangini	Staff Tree, Peng		VJN& WG		Med

Contd...

Table 10.1–Contd...

Sl.No.	Name of the Species	Family	H	Telugu Names	English Names	RS	CA	TN	Use Value
62.	*Beta vulgaris* L.	Chenopodiaceae	H	Beet root	Beet root, Garden beet	+		+	Veg-Tubers
63.	*Chenopodium album* L.	Chenopodiaceae	H	Pappukura, Chakravartikoora	Whitegoose	+		+	Veg-Leaf
64.	*Spinacia oleracea* L.	Chenopodiaceae	H	Palakoora, Palakalaku	Palak, Garden spinach	+	+	+	Veg-Leaf
65.	*Terminalia catappa* L.	Combretaceae	T	Badam chettu Desi badam	Indian almond Tropical almond	+		RR& NZB	Edible-Nuts
66.	*Terminalia chebula* Retz.	Combretaceae	T	Nallakaraka, Karakkai chettu	Yellow myrobalan Chebulic, Myrobalan			NZB	Med
67.	*Ipomoea batatas* (L.) Lam.	Convolvulaceae	H	Chilakada dumpa, Genusu gadda	Sweet potato	+	+	+	Veg-Tubers
68.	*Benincasa hispida* (Thunb.) Cogn.	Cucurbitaceae	C	Budida gummadi	Ash gourd, White gourd, Kushkanda	+	+	+	Veg-Fruit
69.	*Citrullus colocynthis* (L.) Scharder	Cucurbitaceae	C	Paparakaya, Indravaaruni	Colocynth	ATP			Med
70.	*Citrullus lanatus* (Thunb.) Matsum.& Nakai (=*C.vulgaris* (Schrad.) Schrad.)	Cucurbitaceae	H	Pucchakaya, Tarbuja	Watermelon, Indian squash	+	+	+	Edible-Fruits
71.	*Coccinia grandis* J. Voigt (=*C.indica* Wight & Arn.)	Cucurbitaceae	H	Donda, Kakidonda	Little gourd	+		+	Veg-Fruit
72.	*Cucumis melo* L.	Cucurbitaceae	H	Siddotam, Karbuja	Musk melon	+	+	+	Edible-Fruits
73.	*Cucumis sativus* L.	Cucurbitaceae	H	Dosa, Keeradosa	Cucumber	+	+	+	Veg-Fruit
74.	*Cucurbita maxima* Duchesne	Cucurbitaceae	H	Teeya gummadi	Winter squash, Giant pumpkin	+	+	+	Veg-Fruit
75.	*Cucurbita pepo* L.	Cucurbitaceae	H	Gummadi	Pumpkin, Vegetable marrow	+		+	Veg-Fruit

Contd...

Table 10.1–Contd...

Sl.No.	Name of the Species	Family	H	Telugu Names	English Names	RS	CA	TN	Use Value
76.	*Lagenaria siceraria* (Molina) Standley	Cucurbitaceae	C	Anapakaya, Sorakaya	Bottle gourd, Calabash cucumber	+	+	+	Veg-Fruit
77.	*Luffa acutangula* (L.) Roxb.	Cucurbitaceae	C	Beera	Ribbed gourd, Vegetable sponge	+	+	+	Veg-Fruit
78.	*Luffa aegyptiaca* Miller (=*L.cylindrica* (L.) Roemer & Schultes)	Cucurbitaceae	C	Nethi Beera	Sponge gourd, Smooth gourd	+	+	+	Veg-Fruit
79.	*Momordica charantia* L.	Cucurbitaceae	C	Kakara	Bitter gourd, Balsam pear	+	+	+	Veg-Fruit
80.	*Momordica charantia* L. var. *muricata*	Cucurbitaceae	C	Kakara	Bitter gourd, Balsam pear		NLR		Veg-Fruit
81.	*Momordica cymbalaria* Hook.f. (=*M. tuberosa* (Roxb.) Cogn.)	Cucurbitaceae	C	Adavikakara, Kasara	-	CTR			Veg-Fruit
82.	*Trichosanthes anguina* L.	Cucurbitaceae	C	Potlakaya	Snake gourd, Viper gourd	+	+	+	Veg-Fruit
83.	*Dioscorea alata* L.	Dioscoreaceae	C	Pendalam	Water yam, White yam	+	+		Veg-Tubers
84.	*Dioscorea esculenta* (Lour.) Burkill	Dioscoreaceae	C	Akatayi teega	Lesser yam	+		+	Veg-Tubers
85.	*Muntingia calabura* L.	Elaeocarpaceae	T	Cherry	Cherry tree, Capulin	CTR			Edible-Fruits
86.	*Drypetes roxburghii* (Wall.) Hurusawa (=*Putranjiva roxburghii* Wall.)	Euphorbiaceae	T	Putrajivika, Kudrajuvvi, Kaduru	Drypetes	+		+	Wood
87.	*Jatropha gossypifolia* L.	Euphorbiaceae	S	Yerranepalamu, Seemanepalamu	Bellyache bush, Bronze leaved physic nut		VSKP		FW

Contd...

Table 10.1–Contd...

Sl.No.	Name of the Species	Family	H	Telugu Names	English Names	RS	CA	TN	Use Value
88.	*Manihot esculenta* Crantz	Euphorbiaceae	S	Karrapendalam	Cassava, Tapioca, Manioc	+	+		Veg-Fruit
89.	*Phyllanthus amarus* Schumann & Thonn.	Euphorbiaceae	H	Nelausiri, Budatri, Kirinelli	Stone breaker		+	+	Med
90.	*Phyllanthus emblica* L. (=*Emblica officinalis* Gaertner)	Euphorbiaceae	T	Usiri, Nelli, Triphalamu, Datri	Indian gooseberry Amla		+	+	Med/ Edible fruits
91.	*Ricinus communis* L.	Euphorbiaceae	T	Amudamu	Castor oil	+	+	+	Oil
92.	*Senna alata* (L.) Roxb. (=*Cassia alata* L.)	Fabaceae-Caesalpinioideae	S	Sima avisa	Ring worm cassia	CTR			Med
93.	*Senna alexandrina* Gars. ex Miller (=*Cassia anguistifolia* Vahl)	Fabaceae-Caesalpinioideae	S	Senna, Sunamukhi, Swarnapathri	Indian senna	+	+	+	Med
94.	*Saraca asoca* (Roxb.) De Willd.	Fabaceae-Caesalpinioideae	T	Asoca, Vanjulam	Asoca		VSKP		Med
95.	*Tamarindus indica* L.	Fabaceae-Caesalpinioideae	T	Chinta	Tamarind	+	+		Edible-Fruits
96.	*Arachis hypogaea* L.	Fabaceae-Faboideae	H	Verusenaga	Ground nut, Peanut	+	+	+	Oil
97.	*Cajanus cajan* (L.) Millsp.	Fabaceae-Faboideae	S	Kandi, Togari	Pigeon pea, Red gram	+	+	+	Pulses
98.	*Canavalia gladiata* (Jacq.) DC.	Fabaceae-Faboideae	C	Tamba, Tamma, Sambapikka	Sword Jack bean	+	+	+	Veg-Fruit
99.	*Cicer arietinum* L.	Fabaceae-Faboideae	H	Sanaga	Chickpea, Bengal gram	+		+	Pulses
100.	*Crotalaria juncea* L.	Fabaceae-Faboideae	H	Janumu, Sanapusti	Sunnhemp, Bombay hemp	+	+		Fibre
101.	*Cyamopsis tetragonoloba* L.	Fabaceae-Faboideae	H	Goruchikkudu, Choudekayalu	Cluster bean, Guar	+		+	Veg-Fruit

Contd...

Table 10.1–Contd...

Sl.No.	Name of the Species	Family	H	Telugu Names	English Names	RS	CA	TN	Use Value
102.	*Gliricidia sepium* (Jacq.) Kunth ex Walp.	Fabaceae-Faboideae	T	Seema kanuga, Akumanu	Spotted gliricidia	+	+	+	GM
103.	*Glycine max* (L.) Merr.	Fabaceae-Faboideae	H	Soya chikkudu	Soya bean, Cinderella crop	+	+	+	Pulses
104.	*Glycyrrhiza glabra* L.	Fabaceae-Faboideae	T	Atimaduram, Yastimaducam	Liquorice, sweet root			+	Med
105.	*Indigofera tinctoria* L.	Fabaceae-Faboideae	S	Neelimandu	Indigo	KDP			Med-dye
106.	*Lablab purpureus* (L.) Sweet (=*Dolichos lablab* L.)	Fabaceae-Faboideae	C	Chikkudu, Anapa Pandiri chikkullu	Field bean, Indian bean	+	+	+	Veg-Fruit
107.	*Lens culinaris Medikus*	Fabaceae-Faboideae	H	Siri senagalu, Misoorupappu	Lentil			+	Pulses
108.	*Macrotyloma uniflorum* (Lam.) Verdc. (=*Dolichos uniflorus* Lam.)	Fabaceae-Faboideae	H	Ulavalu	Horse gram	+	+	+	Pulses
109.	*Mucuna pruriens* (L.) DC.	Fabaceae-Faboideae	C	Doolagondi, Doola-dummu, Kapikachu	Horse eye bean, Velvetbean	+	+	+	Med
110.	*Phaseolus vulgaris* L.	Fabaceae-Faboideae	H	Beenesu	Bangalore beans, Garden bean, French bean	+		+	Veg-Fruit
111.	*Pisum sativum* L.	Fabaceae-Faboideae	H	Batani	Common pea	+		+	Pulses
112.	*Millettia pinnata* (L.) Panigrahi (=*Pongamia pinnata* (L.) Pierre)	Fabaceae-Faboideae	T	Kanuga, Karanji	Pongam, Karanja		+	+	FW
113.	*Psoralea corylifolia* L.	Fabaceae-Faboideae	H	Bavanchalu, Sevri, Bakuchi	Psoralea	+	+	+	Med
114.	*Pterocarpus santalinus* L.f.	Fabaceae-Faboideae	T	Raktachandanam, Errachandanamu	Redsanders	+		+	Med-Wood

Contd...

Table 10.1–Contd...

Sl.No.	Name of the Species	Family	H	Telugu Names	English Names	RS	CA	TN	Use Value
115.	*Sesbania grandiflora* (L.) Poiret	Fabaceae-Faboideae	T	Avisa, Tellasumintha, Avushika	Sesban, Agati Swamp pea	+	+	+	Veg-Leaf
116.	*Sesbania sesban* (L.) Men.	Fabaceae-Faboideae	S	Jeelugu, Chittiavise	Smalla sesban	+	+	+	GM
117.	*Trigonella foenum-graceum* L.	Fabaceae-Faboideae	H	Menthulu	Fenugreek	+		+	Veg-Leaf
118.	*Vigna aconitifolia* (Jacq.) Marechal	Fabaceae-Faboideae	H	Kumkuma pesalu	Moth bean, Mat bean	+	+	+	Pulses
119.	*Vigna mungo* (L.) Hepper (=*Phaseolus mungo* L.)	Fabaceae-Faboideae	H	Minumulu, Uddibedalu	Blackgram	+	+	+	Pulses
120.	*Vigna radiata* (L.) Wilczek	Fabaceae-Faboideae	H	Pesalu,Uttalu	Mung bean, Green gram	+	+	+	Pulses
121.	*Vigna trilobata* (L.) Verdc.	Fabaceae-Faboideae	H	Pillipesara	African gram, Pillipesary		NLR		Pulses
122.	*Vigna unguiculata* (L.) Walp. (=*V.sinensis* (L.) Savi ex Hassk.)	Fabaceae-Faboideae	H	Alasandalu, Bebbarlu	Cowpea	+	+	+	Pulses
123.	*Acacia planifrons* Wight & Arn. (=*A. planiformis* Wight & Arn)	Fabaceae-Mimosoideae	T	Godugu thumma, Godugujali, Sale	Umbrella thorn tree	CTR			FW
124.	*Leucaena leucocephala* (Lam.) de wit	Fabaceae-Mimosoideae	T	Subabul, Kaniti	Horse tamarind	+	+	+	FO & GM
125.	*Prosopis juliflora* (Sw.) DC.	Fabaceae-Mimosoideae	T	Sarkari tumma, Jalichettu	Mesquite		+		FW
126.	*Samanea saman* (Jacq.) Merr.	Fabaceae-Mimosoideae	T	Nidraganneru, Bellapuchettu	Raintree, Monkeypod	+		+	FW
127.	*Pelargonium graveolens* L' Herit.	Geraniaceae	S	Jeranium	Rose geranium			HYD	AO

Contd...

Table 10.1–Contd...

Sl.No.	Name of the Species	Family	H	Telugu Names	English Names	RS	CA	TN	Use Value
128.	*Curculigo orchioides* Gaertner	Hypoxidaceae	H	Nelatadi, Musli, Adavitati	Pelargonium			WGL	Med
129.	*Mentha arvensis* L.	Lamiaceae	H	Pudina	Japanesemint, Mint	+	+	+	SC-Leaves
130.	*Mentha piperita* L.	Lamiaceae	H	Pudina	True peppermint, Mint	+		+	SC-Leaves
131.	*Ocimum basilicum* L.	Lamiaceae	H	Sabja, Bhutulasi, Surabhi	French basil, Romanbasil			HYD	Med
132.	*Ocimum tenuiflorum* L. (=*O. sanctum* L.)	Lamiaceae	H	Tulasi, Brunda, Lakshmi tulasi, Krishna tulasi	Holy basil, Sacred basil	+		+	Med
133.	*Origanum majorana* L. (=*Majorana hortensis* Moench)	Lamiaceae	H	Maruvam	Sweet majoram, Common majoram	+			SC-Leaves
134.	*Plectranthus amboinicus* (Lour.) Sprengel (=*Coleus amboinicus* Lour.)	Lamiaceae	H	Vamaku, Pashanabedi, Karpuravalli	Coleus	+	+	+	Med
135.	*Plectranthus barbatus* Andrews (=*Coleus forskohlii* (Willd.) Briq.)	Lamiaceae	H	Pasanabedi, Kolius	Coleus		PKSM	KMM	Med
136.	*Pogostemon cablin* (Blanco) Benth.	Lamiaceae	S	Patchouli	Patchouli			HYD	Med
137.	*Cinnamomum camphora* (L.) Nees & Eberm.	Lauraceae	T	Karpuram	Camphor tree		+	+	Med
138.	*Cinnamomum verum* C. Presl (=*C. zeylanicum* Bl.)	Lauraceae	T	Dalchina chekka, lavangapatta	Cinnamon	+	+		SC-Bark
139.	*Allium cepa* L.	Liliaceae	H	Vullipaya	Pyas, Onion	+	+	+	SC-Bulb
140.	*Allium sativum* L.	Liliaceae	H	Thellavaya	Garlic	+	+	+	SC-Bulb
141.	*Aloe vera* (L.) Burm.f. (=*A.barabadensis* Miller)	Liliaceae	H	Kalbanda, Kumari Kootikalabanda	Barbados aloe, Indian aloe, Ghrit-kumari	+	+	+	Med

Contd...

Table 10.1–Contd...

Sl.No.	Name of the Species	Family	H	Telugu Names	English Names	RS	CA	TN	Use Value
142.	*Asparagus racemosus* Willd.	Liliaceae	C	Pilliteegalu, Satavari, Sallaigaddalu	Asparagus			+	Med
143.	*Chlorophytum arundinaceum* Baker	Liliaceae	H	Safed musli, Musali, Kusala	Safed musli	+		+	Med
144.	*Chlorophytum borivilianum* Santapau & Fernandes	Liliaceae	H	Musali, Safed musli, Tella nelatadi	Safed Musli	+		+	Med
145.	*Gloriosa superba* L.	Liliaceae	C	Nabi, Adavinabi, Vankavajram	Gloriosa, Glorililly	+	+	+	Med
146.	*Linum usitatissimum* L.	Linaceae	H	Avisalu	Linseed oil, Flax	+	+	+	Oil
147.	*Lawsonia inermis* L.	Lythraceae	S	Gorinta, Maldaku	Henna, Camphire		GNT		Med
148.	*Abelmoschus esculentus* (L.) Moench	Malvaceae	H	Benda	Okra, Ladies finger	+	+	+	Veg-Fruit
149.	*Abelmoschus moschatus* Medikus	Malvaceae	H	Kasturibenda, Maskdana	Musk mallow, Ambretto	+	+	+	Med
150.	*Gossypium arboreum* L.	Malvaceae	S	Pratti	Tree cotton	+	+	+	Cotton
151.	*Gossypium barbadense* L.	Malvaceae	S	Pamida pratti, Paidipatti	Pima cotton	+	+	+	Cotton
152.	*Gossypium herbaceum* L.	Malvaceae	S	Patti	Levant cotton	+	+	+	Cotton
153.	*Gossypium hirsutum* L.	Malvaceae	S	Patti	Upland cotton	+		+	Cotton
154.	*Hibiscus cannabinus* L.	Malvaceae	H	Gongura, Pundi	Kenaf	+	+	+	Veg-Leaf
155.	*Hibiscus sabdariffa* L.	Malvaceae	S	Tellagogu, Pundikona	Jamaican sorrel, Roselle	CTR& KNL			Veg-Leaf
156.	*Melia dubia* Cav. (=*M. composita* Willd.)	Meliaceae	T	Pedda vepa, Munnutikaraka	White cedar			HYD	Wood
157.	*Swietenia mahogani* (L.) Jacq.	Meliaceae	T	Mahogani	Mahogany	CTR			Wood
158.	*Artocarpus heterophyllus* Lam.	Moraceae	T	Panasa	Jack fruit	+	+	+	Edible-Fruits

Contd...

Table 10.1–Contd...

Sl.No.	Name of the Species	Family	H	Telugu Names	English Names	RS	CA	TN	Use Value
159.	*Ficus carica* L.	Moraceae	S	Anjura, Mandimeda	Edible fig, Common fig	+	+	+	Edible-Fruits
160.	*Morus alba* L.	Moraceae	T	Pippili, Resmi	Mulbery, White mulbery	+	+		Edible-Fruits
161.	*Morus australis* Poiret.	Moraceae	T	Reshmi	Mulberry		NLR VSKP		Med
162.	*Moringa oleifera* Lam. (=*Moringa pterigosperma* Gaertn.)	Moringaceae	T	Munaga	Drumstick	+	+	+	Veg-Fruit
163.	*Musa paradisiaca* L.	Musaceae	A	Arati	Banana, Apple of paradise	+	+	+	Edible-Fruits
164.	*Eucalyptus camaldulensis* Dehn.	Myrtaceae	T	Eucalyptus	Red gum	CTR			AO
165.	*Eucalyptus citriodora* Hook.	Myrtaceae	T	Nimma thylam	Lemon scented eucalypt	+	+		AO
166.	*Eugenia bracteata* (Willd.) Roxb. ex DC.	Myrtaceae	T	Kondanedu, Arivita			+		Edible-Fruits
167.	*Psidium guajava* L.	Myrtaceae	T	Jama	Guava	+	+	+	Edible-Fruits
168.	*Syzygium cumini* L.	Myrtaceae	T	Neredu, Allaneredu	Java plum,Jambalona	+			Edible-Fruits
169.	*Jasminum brevilobum* DC.	Oleaceae	C	Jazi malli	Jasminum	+	+	+	Orn-Flowers/EO
170.	*Jasminum calophyllum* Wall. ex DC. (=*J. courtallense* Wt.)	Oleaceae	C	Sanna jaji	Jasminum	+	+		Orn-Flowers/EO
171.	*Jasminum fluminense* Vell. (=*J. azoricum* L.)	Oleaceae	C	Sanna jazi	Lemon scented jasmine	+		+	Orn-Flowers/EO
172.	*Jasminum multiflorum* (Burm.f.) Andr.	Oleaceae	C	Vira jaji, Kundamu, Adavimalli	Jasminum	+	+		Orn-Flowers/EO
173.	*Jasminum sambac* (L.) Aiton	Oleaceae	C	Gundu Malli, Boddumalli	Arabian jasmine	+	+	+	Orn-Flowers/EO

Contd...

Table 10.1–Contd...

Sl.No.	Name of the Species	Family	H	Telugu Names	English Names	RS	CA	TN	Use Value
174.	*Vanilla planifolia* Jacks. ex Andrews	Orchidaceae	C	Vanilla	Vanilla beans, Vanilla		+	+	SC-Fruits
175.	*Averrhoa bilimbi* L.	Oxalidaceae	T	Bimbli	Tree sorrel	+			Edible-Fruits
176.	*Averrhoa carambola* L.	Oxalidaceae	T	Tamarta, Carambola	Carambola, Country gooseberry	+	+		Edible-Fruits
177.	*Papaver somniferum* L.	Papaveraceae	H	Gasagasalu	Poppy seeds				SC-Seeds
178.	*Passiflora foetida* L.	Passifloraceae	C	Passiflora	Passion fruit		SKLM		Edible-Fruits
179.	*Sesamum indicum* L. (=*Sesamum orientale* L.)	Pedaliaceae	H	Nuvvalu	Sesame, Gingelly	+	+	+	Oil
180.	*Piper betle* L.	Piperaceae	C	Tamalapaku, Tadameku Nagavalli	Betle leaf plant Betel pepper, Paan	+	+	+	Narcotics
181.	*Piper longum* L.	Piperaceae	C	Pippallu	Long pepper		VSKP		SC-Seeds
182.	*Piper nigrum* L.	Piperaceae	C	Mirialu, Nallamirialu	Black pepper		EG		SC-Seeds
183.	*Plantago ovata* Forssk.	Plantaginaceae	H	Isabgol, Aswakarna, Aswagola	Isabgol, Psyllium		+	+	Med
184.	*Plumbago zeylanica* L.	Plumbaginaceae	S	Chitramoolam, Chitrica	Whiteleadwort	ATP			Med
185.	*Avena sativa* L.	Poaceae	H	Yawalu	Oats	+	+	+	Cereal
186.	*Brachiaria mutica* (Forssk.) Stapf (=*Panicum muticum* Forssk.)	Poaceae	H	Para gaddi	Para grass	+	+	+	FO
187.	*Cenchrus ciliaris* L. (=*Pennisetum cenchroides* Rich)	Poaceae	H	-	Buffel grass, Anjan grass	+	+	+	FO
188.	*Cymbopogon citratus* (DC.) Stapf (=*Andropogon citratus* DC.)	Poaceae	H	Nimma gaddi, Vasanagaddi	Lemon grass	+	+	+	AO

Contd...

Table 10.1–Contd...

Sl.No.	Name of the Species	Family	H	Telugu Names	English Names	RS	CA	TN	Use Value
189.	*Cymbopogon martinii* (Roxb.) Watson	Poaceae	H	Rushagaddi, Palm rose	Russagrass, Roshagrass		WGL	HYD	AO
190.	*Cymbopogon nardus* (L.) Rendle var. *confertiflorus* (steud.) Stapf ex Bor	Poaceae	H	-	Citronella grass		VSKP		AO
191.	*Cymbopogon nardus* (L.) Rendle var. *nardus* (=*Andropogon nardus* L.)	Poaceae	H	Ganjini	Citronella grass	+	+		AO
192.	*Cymbopogon winterianus* Jowitt	Poaceae	H	Citronella	Java citronella		+	+	AO
193.	*Dendrocalamus strictus* (Roxb.) Nees	Poaceae	T	veduru	Bamboo	KNL			Wood
194.	*Echinochloa colona* (L.) Link (=*Panicum colonum* L.)	Poaceae	H	Jerripothu gaddi, Vudaragaddi	Shama Grass		+		Millet
195.	*Echinochloa frumentacea* (Roxb.) Link	Poaceae	H	Oodarlu, Oodalu	Billion dollar grass, Japanese barnyard millet	+	+	+	Millet
196.	*Eleusine coracana* Gaertner (=*Cynosurus coracanus* L.)	Poaceae	H	Thidalu, Ragulu, Chollu	Fingermillet, Ragimillet, African millet	+	+	+	Millet
197.	*Oryza sativa* L.	Poaceae	H	Vari, Vadlu, Chauval	Paddy, Rice	+	+	+	Cereal
198.	*Panicum miliaceum* L.	Poaceae	H	Warigalu, Barigalu	Prosomillet, Bread millet	+	+	+	Millet
199.	*Panicum sumatrense* Roemer & Schultes (=*P. miliare* Lam.)	Poaceae	H	Gangasamulu	Small millet, Little millet	+	+	+	Millet
200.	*Paspalum scrobiculatum* L.	Poaceae	H	Argulu, Arikelu, Nitarigaddi	Kodomillet, Ditch millet	+	+	+	Millet
201.	*Pennisetum clandestinum* Hochst.	Poaceae	H	-	Kikiyu grass		+		FO

Contd...

Table 10.1–Contd...

Sl.No.	Name of the Species	Family	H	Telugu Names	English Names	RS	CA	TN	Use Value
202.	*Pennisetum glaucum* (L.) R.Br. (=*P.americanum* (L.) Schum.)	Poaceae	H	Sajjalu, Ganti	Bajra, Pearlmillet	+	+	+	Millet
203.	*Pennisetum purpureum* Schum.	Poaceae	H	Napier gaddi	Napier grass, Elephant grass			+	FO
204.	*Saccharum officinarum* L.	Poaceae	H	Cheruku, Karumbu	Sugarcane, Noblecane	+	+	+	SU-P
205.	*Setaria italica* (L.) P. Beauv.	Poaceae	H	Korralu	Italian millet, Foxtailmillet	+	+	+	Millet
206.	*Setaria verticillata* (L.) P. Beauv.	Poaceae	H	Chiklenta	Burmillet, Bristle grass, Hooked grass		+		Millet
207.	*Sorghum bicolor* (L.) Moench (=*S.vulgare* Pers.)	Poaceae	H	Jonna, Joargaddi	Jowar	+	+	+	Millet
208.	*Triticum aestivum* L. (=*T.vulgare* Vill.)	Poaceae	H	Godhumalu	Common wheat, Rice wheat	+	+	+	Cereal
209.	*Triticum dioccum* Schuebler	Poaceae	H	Godhumalu	Emmer wheat, Rice wheat	+		+	Cereal
210.	*Vetiveria zizanioides* (L.) Nash	Poaceae	H	Vattiveru, Ittagaddi	Khus-Khus grass, Cus-cus grass, Vetiver		+	+	AO
211.	*Zea mays* L.	Poaceae	H	Mokkajonna, Bailumo-kkalu, Dabbajonna	Maize, Corn	+	+	+	Cereal
212.	*Rumex vesicarius* L.	Polygonaceae	H	Chukkakura	Bladder dock	+	+	+	Veg-Leaf
213.	*Punica granatum* L.	Punicaceae	T	Danimma	Pomegranate	+	+	+	Edible-Fruits
214.	*Ziziphus mauritiana* Lam. (=*Z. jujuba* (L.) Gaertner, non Miller)	Rhamnaceae	T	Regu, Reni, Badari	Indian jujube	+	+		Edible-Fruits
215.	*Rosa centifolia* L.	Rosaceae	H	Gulabi, Taruni	Cabbage rose	+		+	Orn-Flowers
216.	*Coffea arabica* L.	Rubiaceae	H	Coffee	Arabian coffee		EG, SKLM		Bevarage

Contd...

Table 10.1–Contd...

Sl.No.	Name of the Species	Family	H	Telugu Names	English Names	RS	CA	TN	Use Value
217.	*Guettarda speciosa* L.	Rubiaceae	T	Panneruchettu, Pannichettu	Scarlet bush		+	+	AO
218.	*Ixora acuminata* Roxb. (=*I. alba* L.)	Rubiaceae	S	Tella ramabhanamu	White flowered ixora	CTR			Med
219.	*Morinda citrifolia* L.	Rubiaceae	S	Noni	Noni	+	+	+	Med
220.	*Mussaenda tomentosa* Wight	Rubiaceae	S	Mussaenda	Mussaenda	CTR			Med
221.	*Citrus aurantifolia* (Christm.) Swingle	Rutaceae	T	Nimma	Lime, Acid lime	+	+	+	Edible-Fruits
222.	*Citrus aurantium* L.	Rutaceae	T	Narinja	Sour orange, Siveile orange	+	+	+	Edible-Fruits
223.	*Citrus grandis* (L.) Osbeck (= *C.maxima* Merr.)	Rutaceae	T	Pampalamasam, Pamparapanasa	Shaddock, Pomelo, Forbidden fruit			MDK, NZB	Edible-Fruits
224.	*Citrus limon* (L.) Burm.f.	Rutaceae	T	Dabba, Peddanimma, Gajanimma	Nimbu, Lemon	+	+	+	Edible-Fruits
225.	*Citrus limonia* Osbeck.	Rutaceae	T	Rangapur nimma	Rangapur lime, Lemandarin	ATP			Edible-Fruits
226.	*Citrus medica* L.	Rutaceae	T	Madiphalamu, Naradabba	True Citron	+	+	+	Edible-Fruits
227.	*Citrus reticulata* Blanco	Rutaceae	T	Kamalapandu, Santralu	Mandarin orange, Loose skinned orange	+	+		Edible-Fruits
228.	*Citrus sinensis* (L.) Osbeck	Rutaceae	T	Bathai, Chini, Mosambi	Sweet orange	+	+	+	Edible-Fruits
229.	*Murraya koenigii* (L.) Sprengel	Rutaceae	T	Karivepaku, Surabinimba	Curry leaf	+	+	+	SC-Leaves
230.	*Santalum album* L.	Santalaceae	T	Chandanam, Srigandham	White sandal wood	+			Wood
231.	*Sapindus trifoliata* L. (=*Sapindus emarginata* Vahl)	Sapindaceae	T	Kunkudu, Rita	Soapnut, Soap berry	+	+		SP

Contd...

Table 10.1–Contd...

Sl.No.	Name of the Species	Family	H	Telugu Names	English Names	RS	CA	TN	Use Value
232.	*Manilkara zapota* (L.) P. Royen (=*Achras zapota* L.)	Sapotaceae	T	Sapota	Sapodilla	+	+	+	Edible-Fruits
233.	*Bacopa monnieri* (L.) Pennell	Scrophulariaceae	H	Jalabrahmi, Neelisambrani	Bacopa			HYD	Med
234.	*Capsicum annuum* L.	Solanaceae	H	Mirapa, Mirchi,	Capsicum, Pepper Simlamirchi	+	+	+	SC-Fruits
235.	*Capsicum frutescens* L.	Solanaceae	H	Mirapa, Mirchi.	Hot chilly, Pepper	+	+	+	SC-Fruits
236.	*Lycopersicon esculentum* Miller	Solanaceae	H	Tomato, Rama mulakkaya	Tomato	+	+	+	Veg-Fruit
237.	*Nicotiana tabacum* L.	Solanaceae	H	Pogaku	Tobacco	+	+	+	Narcotics
238.	*Solanum melongena* L.	Solanaceae	H	Vankaya	Brinjal, Egg plant	+	+	+	Veg-Fruit
239.	*Solanum tuberosum* L.	Solanaceae	H	Bangaladumpa, Urlagaddalu, Alu	Potato	+	+	+	Veg-Tubers
240.	*Solanum viarum* Dunal	Solanaceae	S	Kasivanga	Tropical soda apple		+	+	Med
241.	*Withania somnifera* (L.) Dunal	Solanaceae	S	Aswagandha, Dommadolu, Bommadolu	Withania, Indian gingseng	+	+	+	Med
242.	*Guazuma ulmifolia* Lam.	Sterculiaceae	T	Bhadraksha, Rudraksha	Bastard cedar	+	+	+	Orn-Flowers
243.	*Sterculia urens* Roxb.	Sterculiaceae	T	Errapoliki, Tapasi, Kondatamara	Gum karaya	CTR			Med-Gum
244.	*Tamarix aphylla* (L.) Karsten	Tamaricaceae	T	Erravsaru, Sarru			+	+	Wood
245.	*Berrya cordifolia* (Willd.) Burr.	Tiliaceae	T	Kalateku, Konda kanchaku	Trincomali wood tree	CTR			Wood
246.	*Corchorus capsularis* L.	Tiliaceae	H	Gonunara	Bengal jute, White jute, Jute	+	+	+	Fibre
247.	*Tectona grandis* L.f.	Verbenaceae	T	Teku	Teak	+	+	+	Wood
248.	*Citheroxylum subserratum* Sw.	Vitaceae	S	Vathangi			EG		Orn-Flowers

Contd...

Table 10.1–Contd...

Sl.No.	Name of the Species	Family	H	Telugu Names	English Names	RS	CA	TN	Use Value
249.	*Vitis vinifera* L.	Vitaceae	C	Draksha	Grape, Grapevine	+	+	+	Edible-Fruits
250.	*Curcuma amada* Roxb.	Zingiberaceae	H	Mamidiallam	Mangoginger	+	+		SC-Rhizome
251.	*Curcuma aromatica* Salisb.	Zingiberaceae	H	Kachoram	Wild turmeric, Zeodari	CTR			SC-Rhizome
252.	*Curcuma longa* L. (=*C. domestica* Valet.)	Zingiberaceae	H	Pasupu	Turmeric	+	+	+	SC-Rhizome
253.	*Curcuma zeodora* (Chrism.) Roscoe	Zingiberaceae	H	Kuchuru	Zedoary		+		SC-Rhizome
254.	*Elettaria cardamomum* (L.) Maton	Zingiberaceae	H	Yalakulu	Queen of spices, True cardamom	CTR			SC-Fruits
255.	*Kaempferia galanga* L.	Zingiberaceae	H	Kacchuralu, Ekangi, Chandramula	Spike ginger lilly, Aromatic ginger		+	+	Med
256.	*Zingiber officinale* Roscoe	Zingiberaceae	H	Allam, Sonti	Ginger, Dried ginger	+		+	SC-Rhizome

Regions: RS: Rayalaseema; CA: Coastal andhra; TN: Telangana. Districts: ADB: Adilabad; ATP: Anantapur; CTR: Chittoor; EG: East Godavari; GNT: Guntur; HYD: Hyderabad; KDP: Kadapa; KMM: Khammam; KNL: Kurnool; KNR: Karimnagar; KSN: Krishna; MBNR: Mahaboobnagar; MDK: Medak; NLG: Nalgonda; NLR: Nellore; NZB: Nizamabad; PKSM: Prakasam; RR: Ranga reddy; SKLM: Srikakulam; VJN: Vijayanagaram; VSKP: Visakhapatnam; WGL: Warangal; WG: West Godavari. Habit: A- Arborescent; C: Climber; H: Herb; S: Shrub; T: Tree. Use Value: AO: Aromatic oil; EO-Essential oil; FO: Fodder; FW: Fuel wood; GM: Green manure; Med-Medicinal; Orn: Ornamental; SC: Spices and condiments; SP: Soap preparation; SUP- Sugar production; Veg: Vegetable.

(Rajeswaramma, 2001; Suryanarayana and Rao, 2007), Kurnool (Raju and Pullaiah, 1995), Anantapur (Pullaiah and Yesoda, 1989), Chittoor (Madhavachetty *et al.*, 2008), Nizamabad (Pullaiah and Rao, 1995), Medak (Pullaiah *et al.*, 1998) and Ranga Reddy (Pullaiah and Silar Mohammed, 2000). Other references include: Narapareddy (2001) and Anonymous (2002) and Agrodiversity of Ananatapur district (Praveena, 2006).

Results

A total of 256 cultivated plant species are currently known in cultivation in the state of Andhra Pradesh. Abstract of crop diversity is presented in Table 10.2. The list of species along with family, Telugu names, English names, use value, and region of cultivation is provided in the Table 10.1. Of the 256 species, 69 are trees, 27 shrubs, 129 herbs, and 31 are climbers. Medicinal crops are represented by 51 species followed by vegetable crops, 44. Crops yielding sugar, bevarages and narcotics are at minimum number. Faboideae- Fabaceae dominates the crop plants and are represented maximum with 35 species. Grasses comprise 27 species.

Table 10.2: Crop Diversity in Andhra Pradesh

Crop Group	*No. of Species*
Cereal and millets	15
Pulses	11
Spices-condiments	27
Edible fruits and nuts	38
Vegetables	44
Sugar yielding plants	01
Oils	08
Aromatic oil	11
Beverages-Narcotics	05
Fibres	08
Timber/Fuel wood	14
Medicinal	51
Flowers-ornamental	15
Fodder and Green manure	7
Others	1

References

Anonymous. 2002. *Agro-Biodiversity in Andhra Pradesh*. Deccan development society, Hyderabad.

Anonymous. 2007. State-specific interventions for higher agricultural growth. ICAR publication. www.icar.org.in

Bhairavamurthy, P.V. and Krishna Mohan, R. 1991. A contribution to the Flora of Prakasam district, Andhra Pradesh, *J.Econ. Tax. Bot.* 15: 49-63.

Cromwell Elizabeth. 1999. *Agriculture, Biodiversity and Livelihoods: Issues and Entry Points Final Report.* Overseas Development Institute. London (UK)

Don Ellison. 1999. *Cultivated Plants of the World - Trees, Shrubs & Climbers.* New Holland, London.

Lakshmaiah, A. 2006. Agrestals of Rayalaseema region, Andhra Pradesh. Ph.D thesis submitted to Sri Krishnadevaraya University, Anantapur.

Lakshminarayana, K., P. Venkanna and T.Pullaiah .1997. *Flora of Krishna District (Andhra Pradesh).* M D Publications PVT LTD, New Delhi.

Madhava Chetty, K., K.Sivaji and K. Tulasi Rao .2008. *Flowering Plants of Chittoor (Andhra Pradesh).* Students Offset Printers, Tirupati.

Narapareddy, D. and B. Sitharam. 2001. *Medicinal Plant Cultivation and Marketing Opportunities.* Sri Karthik Publications, Tirupati.

Praveena, K. 2006. *Agro bio-diversity of Anantapur district, Andhra Pradesh.* M. Phil. dissertation. Sri Krishna Devaraya University, Anantapur.

Pullaiah, T. 1997. *Flora of Andhra Pradesh, India.* Vol. 3. Scientific Publishers, Jodhpur.

Pullaiah, T. and B. Ravi Prasad Rao .1995. *Flora of Nizamabad (Andhra Pradesh).* Bishen Singh Mahendra Pal Singh, Dehradun (India).

Pullaiah, T. and E. Chennaiah .1997. *Flora of Andhra Pradesh, India.* Vol.1. Scientific Publishers, Jodhpur.

Pullaiah, T. and M. Silar Mohammed. 2000. *Flora of Ranga Reddy District (Andhra Pradesh).* Regency Publications, New Delhi.

Pullaiah, T. and S. Sandhya Rani. 1999. *Trees of Andhra Pradesh, India.* Regency Publications, New Delhi.

Pullaiah, T., C. Prabhakar and B. Ravi Prasad Rao .1998. *Flora of Medak (Andhra Pradesh).* Daya Publishing House, Delhi.

Pullaiah, T., Ramakrishnaiah, V., Sandhya Rani, S. and Rao, P.N. 2000. *Flora of Guntur district, Andhra Pradesh.* Regency Publications, New Delhi.

Pullaiah,T. and D. Ali Moulali. 1997. *Flora of Andhra Pradesh, India.* Vol. 2. Scientific Publishers, Jodhpur.

Pullaiah,T. and N.Yesoda.1989. *Flora of Anantapur (Andhra Pradesh).* Bishen Singh Mahendra Pal Singh, Dehradun.

Raju, R. R. V and T. Pullaiah.1995. *Flora of Kurnool (Andhra Pradesh).* Bishen Singh Mahendra Pal Singh, Dehradun.

Rana, R.S. 1993. India takes a lead in conservation of plant genetic resources. *Indian farming.* 43(7): 37- 44.

Rao, Ravi Prasad, B. 2010. *The Plant Directory.* Anantha Biodiversity Consortium-Dharani, Anantapur.

Rajeswaramma, P. 2001. Agrestals of Nellore district, Andhra Pradesh-India. Ph.D thesis submitted to Sri Krishnadevaraya University, Anantapur.

Rao, R.S., S. Sudhakar and P. Venkanna .1999. *Flora of East Godavari (Andhra Pradesh).* INTACH.

Rao, R.S, and S. Hara Sreeramulu. 1986. *Flora of Srikakulam (Andhra Pradesh).* Indian Botanical Society, Meerut.

Rao, R.S., P.Venkanna and T. Appi Reddy. 1986. *Flora of West Godavari (Andhra Pradesh).* Indian Botanical Society, Meerut.

Subba Rao, G.V. and G.R. Kumari. 2008. *Flora of Visakhapatnam (Andhra Pradesh) Vol.2.* Botanical Survey of India, Kolkata.

Sudhakar Reddy,C, Girish S. Pujar, S. Sudhakar, Shilpababar, K.Sudha, B.Shivam trivedi, B.Gharai and M.S.R. Murthy. 2008. Mapping the vegetation types of Andhra Pradesh. *Proc. A. P. Akademi of Sciences* 12 (1&2): 14-23.

Suryanarayana, B. and A. Sreenivasa Rao. 2007. *Flora of Nellore (Andhra Pradesh).* Gurudev Prakashan, Maharashtra.

Vavilov N.I. 1951. The origin variation and breeding of cultivated plants (Translated from Russian by K. Starrchester) *Cronica Botanica.* 13: 1-364.

Venkaiah, M. 2004. *Studies on the Vegetation and Flora of Vizianagaram district, Andhra Pradesh.* Andhra university press and Publications, Visakhapatnam.

Venkateswarlu, J., P.V. Bhiravamurty and Piratla Narasimha Rao. 1972. *Flora of Visakhapatnam (Andhra Pradesh).* Vol.1. A. P. Akademi of Sciences, Hyderabad.

Index to Scientific Names and Subject Index

D

H

N

Q

R

S

Z